Exploring Digital and Information Technology

Portfolio Manager: *Pat Keeney*

Product Developer: *Rebecca Hoogerhyde*

Marketing Manager: *Victor Hugo*

Content Specialist: *Jeff Pitcher*

Content Contributors: *Ted Tedmon, Casey Wilhelm*

Designer: *Olivia Barbour*

Content Licensing Specialist: *Caitlin Smith*

Cover Image: *PopTika/Shutterstock*

Compositor: *SPi Global*

All credits appearing on page are considered to be an extension of the copyright page.

Cover Credits: PopTika/Shutterstock

mheducation.com/prek-12

Send all inquiries to:
McGraw Hill
120 S. Riverside Plaza, Suite 1200
Chicago, IL 60606

ISBN: 978-1-26-631861-0
MHID: 1-26-631861-5

Printed in the United States of America

1 2 3 4 5 6 7 8 9 LWI 29 28 27 26 25 24

Table of Contents

1 Computer Hardware

What To Expect

After completing this chapter, you will be able to:

- describe types of computers and computer functions;
- differentiate between different types of personal computers;
- describe large computer systems;
- explain computer processors and the relevant parameters;
- identify the memory and storage systems in computers;
- identify the various computer ports;
- describe computer protocols;
- define the terms *video card* and *pixel* and explain the parameters of computer graphics;
- explain how a computer works.

Chapter Topics

- What Is a Computer?
- Computer Components
- Computer Storage
- Connections
- Wireless Connections
- Graphics
- How a Computer Works
- Quantum Computing

Let's Explore Computers

Take a step back in time to when people didn't have computers. In those days, different jobs required a bunch of different equipment. For writing, there were typewriters. If you wanted to record moving pictures, you needed a video camera, and for still pictures, you would use a regular camera. Drawing pads were there for making illustrations, and for games, you might have a deck of cards. And, of course, there were big old telephones for making calls. We might still use some of these tools today, but computers have combined many of them.

Imagine having to lug around a video camera, a deck of cards, a drawing pad, and a typewriter all the time! That would be pretty tough. Most of the time, people would keep these items stored away, pulling them out only when they needed them. Each had its own special use and couldn't do the job of the others. If you were doing one thing and wanted to switch to another, you had to put one piece of equipment away and get out a different one. Not very handy!

Then the computer came along. It was a game-changer! It combined all these different tools into one neat, efficient, and incredibly useful device. And the cool part is, the computer didn't lose any effectiveness by bringing all these things together.

A computer is way more than just a typewriter. It's a video camera, a drawing pad, a telephone, a game console, and a music player—all in one! It lets you go from writing to video calling to drawing without a hitch. No need to change equipment.

Because it can do so many things in one, the computer is a pretty amazing piece of tech. It has changed the way we learn, talk, work, and have fun. It's like having a magic box that instantly gives you any tool you need. The computer lets us do things that were impossible with old individual tools.

As we dig deeper into computers, we're going to learn how this all-in-one device works and how it can do so many different things without losing any power. We'll see how computers have changed our world, letting us do many things at once and making life a lot more interesting. Keep in mind this picture of the computer as a handy, all-around tool as we go on to learn more about the wonders of computing.

What Is a Computer?

Define the Term *Computer*

Computers are digital devices that accept input, process and store the input, and provide output. A computer is made up of multiple components that create user functionality. **Computers** consist of hardware and software that create data and process information. Hardware is the physical elements of the device, including the computer's processor, memory, storage, ports, and peripheral devices. Software is the virtual elements, including the OS and a variety of software applications.

The Four Basic Computing Functions

A computer is a digital device that completes four basic functions. First, computers accept input. They accept data from many sources, including keyboards, cameras, microphones, and even other computers. Next, computers compile and transform data into useful information; this is called the process function. **Computers** also provide output, which means the ability to display information. This output comes in many forms: images, video, and audio. Finally, computers store data for future use.

Different Types of Computers

Computers are digital devices that come in many shapes, sizes, and designs.

Examples of computers:

- laptops
- desktops
- all-in-ones
- tablets
- smartphones
- mainframes
- servers
- gaming consoles
- embedded devices

Computers are found within countless other items. For example, if a car has a modern ignition system, the computer within the ignition system completes a systems check, unlocks the ignition, signals the fuel system (run by another computer), signals the starter system, and unlocks the transmission.

What Is Meant by the Term *Computer Hardware*?

Computer hardware refers to the physical parts of a computer system. Examples of hardware include the system unit (also known as the tower or chassis) a printer or monitor, the keyboard, and a mouse. Communication devices such as routers and modems are also considered computer hardware.

The System Unit of a Computer

The system unit is the part of the computer that houses the motherboard, which is the circuit board that holds the computer's main microprocessor. This microprocessor is called the central processing unit. The location of the system unit depends on the type of computer. For example, in a desktop computer, the **system unit** is the chassis or tower unit. With a laptop computer, the system unit is housed beneath the keyboard. In a smartphone, tablet, or all-in-one computer, the system unit is housed beneath the monitor or screen.

Laptop Computers and Their Uses

Laptops are mobile hinged and folding computers with an integrated keyboard. **Laptops** come with a full operating system (OS) and remain a top-selling personal computer. Laptop computers are great for people who need a portable digital device that has more computing power than a tablet or phone. Most laptops come with the capability to add more memory. For example, if a new version of Microsoft Office requires additional random access memory (RAM), most laptops can be upgraded. Because they can be carried around the house, to class, to the library, or to a coffee shop, laptops are the most popular computer choice among students.

Tablet Computers and Their Uses

Tablet computers are highly mobile, which makes them great for travel. **Tablet computers** use touchscreen technology as the main input method. Unlike laptops, tablets cannot be upgraded. Some tablet computers have a smaller operating system (OS) than laptop computers, making them less powerful.

Because they are light yet powerful, tablets are popular for e-mail, gaming, and travel. Tablets are often used by doctors making their rounds or by servers taking orders in restaurants. By adding a separate keyboard, a tablet gains much of a laptop's functionality. But the smaller OS in a tablet may limit what programs can be run on it.

Desktop Computers and Their Uses

Desktop computers are stationary devices that consist of a separate case called the system unit or chassis. Peripheral devices, such as a monitor, keyboard, and mouse, are connected to the system unit. The system unit houses the main components of the computer. Because of its size, a desktop's system unit allows for relatively easy hardware upgrades. Desktops also have a relatively large footprint. A footprint refers to the amount of space a device takes up on a desk, table, or other area. **Desktop computers** are great for families with small children or for businesses because these computers don't get carried around and dropped. **Desktop computers** are sold by many different companies, including Apple, HP, and Dell. Gamers and computer experts like desktops because they can be easily upgraded. Global shipments of desktop PCs have declined recently. Last year there were 268.2 million shipments of desktop PCs compared to 300 million desktop PCs in 2021.

All-in-One Computers and Their Uses

All-in-one computers provide the power of a desktop but with elegance. Because the motherboard is housed within the monitor, there's no bulky chassis or tower system unit. **All-in-one computers** are designed to be stationary without taking up much desk space. Many all-in-one computers have touchscreen interfaces, and all come with full (not just mobile) operating systems. These computers are used by those who like the power of a desktop computer but prefer the neat, clean look and smaller footprint. **All-in-one computers** are not as popular with computer experts or serious gamers because they're not as easy to upgrade as a traditional desktop computer with a separate system unit (chassis).

Embedded Computers

An **embedded computer** is a digital device that accepts input, processes data into information, provides output, and can store data—but it is part of a larger device or system. Examples of embedded computers can be found in digital watches, programmable thermostats, and even programmable coffeepots! Many embedded computers are found in medical devices, like blood pressure monitors. Embedded computers have relatively small operating systems called real-time operating systems (RTOS).

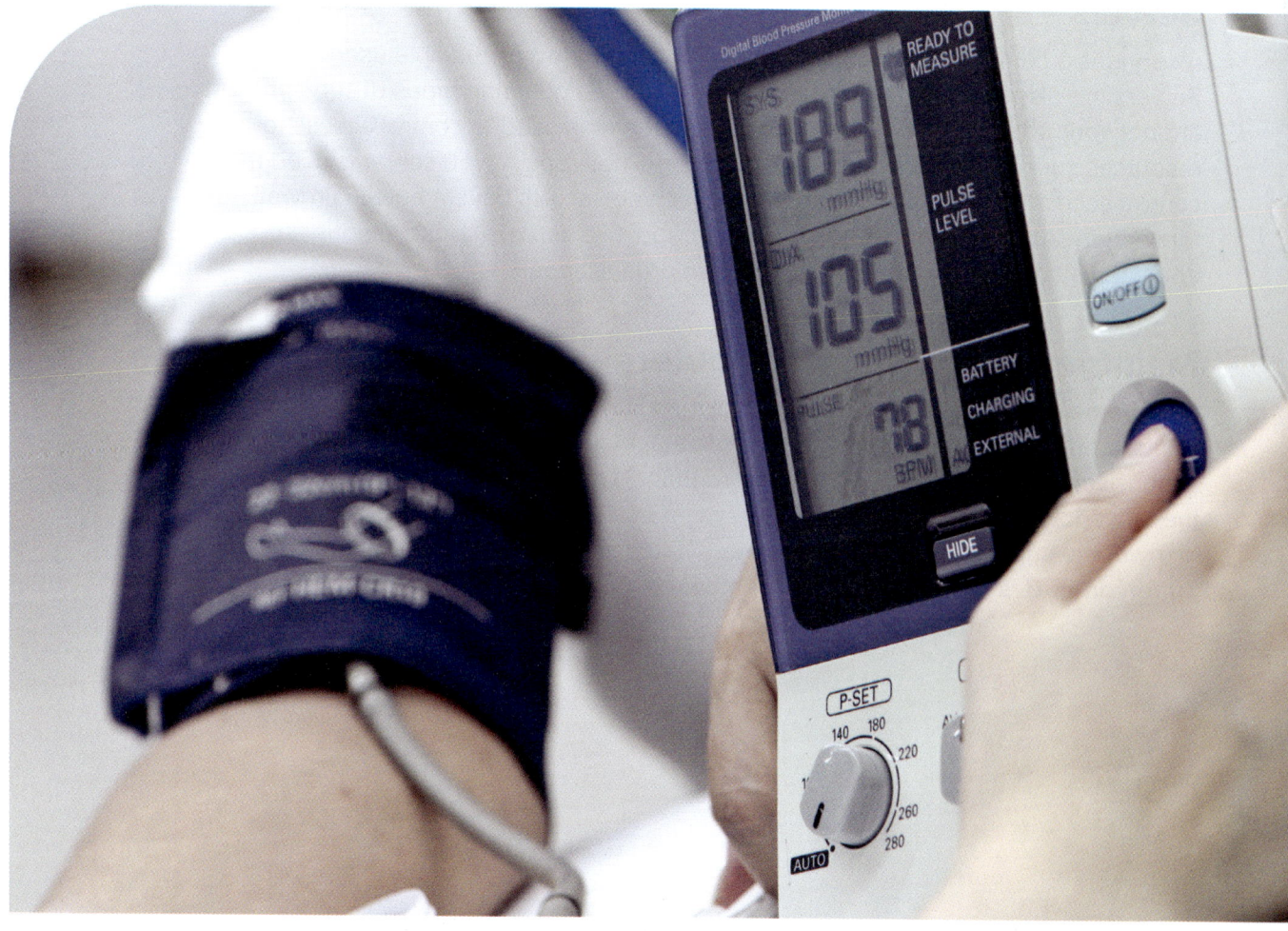

2-in-1 Computers

2-in-1 computers are touchscreen laptops with keyboards that are either detachable or that can be flipped out of the way. This gives them the mobility of tablets while still using full operating systems. Like tablets, **2-in-1 computers** have solid-state hard drives, making them very fast and providing long battery life, but this also means they typically have less storage capacity than laptops.

Smartphones

Smartphones are handheld computers with cellular networking capability.

Common features of smartphones include:

- Internet capability
- Touchscreen
- Front- and rear-facing cameras
- Speaker

The main limitations of smartphones are their relatively small screen size and limited computing power. To maximize limited computing power, smartphones use mobile versions of applications (apps). Smartphones use durable, solid-state storage technology to maximize battery life.

Servers

A **server** is a computer system in a network that is shared by multiple users. Servers are also referred to as host computers. Three common types of servers are network, web, and file servers. A network server is used to control access to resources such as printers and applications on a network. A web server is used to control access to resources on the web. Most web servers have a built-in firewall for security. A file server is used for data storage. File servers house various documents and information resources for an organization.

Mainframes and Supercomputers

Mainframe computers are more powerful than servers. Mainframes are used by organizations to process large amounts of data. **Mainframe computers** are very expensive, costing from $100,000 to more than $1 million. **Supercomputers** are the most powerful type of computer. They can evaluate complex data very quickly. Many of the supercomputers in the United States are owned by the government. **Supercomputers** can cost $1 million or more.

Computer Components

What Is a Digital Device Processor?

A processor or **central processing unit (CPU)** is the brain of the computer where most calculations take place. The processor is the most important component of a digital device and is integrated into a chip called a microprocessor. Without a processor, the digital device would not function. The processor has two main components: the arithmetic logic unit (ALU) and the control unit. The ALU performs the following logic and mathematical tasks: arithmetic (addition, subtraction, multiplication, and division), comparison (equal to, less than, the same as, and more), and logic (*and*, *or*, *not*). The control unit locates, analyzes, and executes each program instruction residing in memory. Essentially, it tells the computer's memory, ALU, and input and output devices how to respond to a program's instructions.

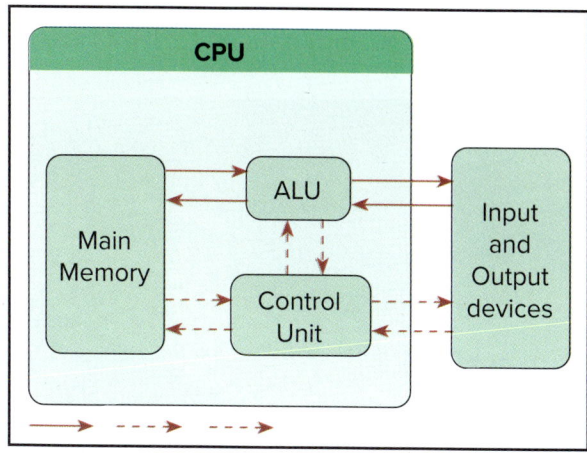

What Is the Machine Cycle?

Every time a processor executes an instruction from a program, it goes through the same four steps. This process is called the machine cycle. The steps in the **machine cycle** are:

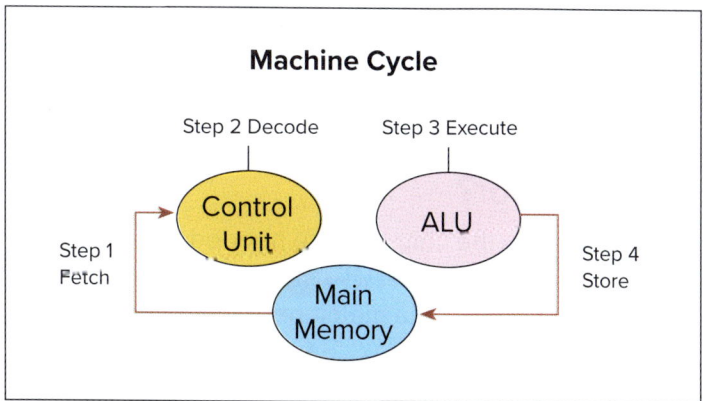

1. Fetch—The processor fetches the instruction or data from RAM.

2. Decode—The processor decodes the instruction or data into a readable form the computer can understand.

3. Execute—The processor executes the instruction.

4. Store—The results of the processing are stored in RAM. Once the results are stored, the processor fetches the next instruction or piece of data.

What Is Clock Speed?

Clock speed, also referred to as clock rate, is the speed at which the processor performs the operations required to run a digital device and instructions executed by the digital device. **Clock speed** is measured in the number of machine cycles the processor can run per second. Cycles per second are referred to as hertz (Hz). Modern processors operate at clock speeds of billions of machine cycles per second, or gigahertz (GHz).

What Is Overclocking?

Overclocking refers to running the processor faster than recommended by the manufacturer. This can increase performance and speed, but it can also void the manufacturer warranty. Some serious gamers and computer experts overclock. It requires additional CPU cooling.

What Is a Single-Core Processor?

A *core* on a CPU refers to the components on the chip that are needed for processing. A **single-core processor** has only one core. Usually, this means one ALU and one CU. Note that the number of cores isn't a measure of the computer's speed. It is just a description of the processor.

What Is a Multicore Processor?

A **multicore processor** has two or more cores that are responsible for processing. A multicore processor can execute two or more sets of instructions at the same time. This is called **hyperthreading**. **Hyperthreading** affords faster processing by allowing a new set of instructions to start before the prior set of instructions has finished being processed. Multicore processors allow for faster multitasking and overall processing of instructions.

Processor Manufacturers

Intel Corporation—Founded in 1968, Intel is the world's largest manufacturer of processors.

AMD—Founded in 1969, AMD's technology is featured inside gaming consoles and home entertainment systems including Microsoft's Xbox One and Sony's PS4™.

NVIDIA—Founded in 1993, NVIDIA focuses on graphical processing unit (GPU) cores.

Memory

Computer memory is found in microchips on the motherboard. *Memory* can refer to a variety of memory types including caches and registers. Memory composed of solid-state electronics is fast and energy efficient because it does not involve mechanical moving parts. Memory is sometimes referred to as volatile because its data are lost upon computer shutdown. Storage retains data upon computer shutdown and is sometimes called nonvolatile.

What Is RAM?

Random access memory (RAM) is a type of memory found on the motherboard of a digital device. RAM is electronic and has no moving parts. RAM is also called *main memory*, which means the memory is available to programs to execute tasks. RAM is a temporary storage area and is cleared when a device is powered off.

History of RAM

Professor Robert Dennard of IBM developed dynamic random access memory (DRAM) in 1968, and it was released commercially by Intel in 1970. DRAM allowed processors to work at much higher speeds. Static random access memory (SRAM) is faster than DRAM and generally used in cache memory. Professor Dennard was awarded the Kyoto Prize in 2013.

How Much RAM You Will Need

It is important to consider the amount of **RAM** that comes with a digital device. For example, most smartphones come with at least 2GB of **RAM** installed. Most laptops come with at least 4GB of **RAM** installed. The amount of **RAM** directly impacts processing speed.

Different computers use different types of **RAM**. Either bring your laptop to the computer store or visit a site such as Crucial.com to ensure you are buying the correct variety.

Cache

Cache is high-speed storage usually located directly on the CPU. **Cache** is smaller but faster than **RAM**. It is used to quickly access repeated instructions. The word *cache* is pronounced "cash."

Memory Cache and Disk Cache

Cache refers to types of memory: memory cache and disk cache. **Memory cache** is high-speed memory used by the CPU to store frequently accessed data and instructions. **Memory cache** is referred to by levels. L1 cache is small and integrated into the CPU. L3 cache is larger and next to the CPU. Computer advertisements often display the amount of L3 cache included. **Disk cache** is high-speed memory used by the hard drive to store frequently accessed data.

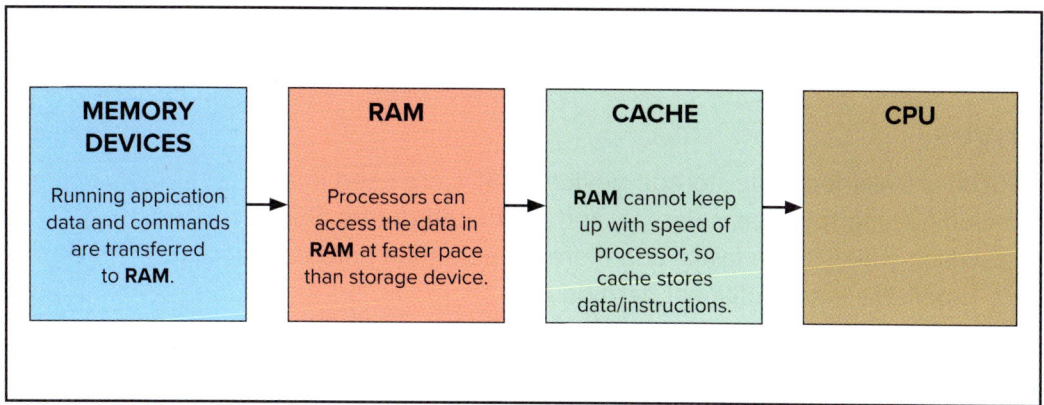

ROM

Read-only memory (ROM) is a storage area in a digital device that is installed by the digital device manufacturer. Most **ROM** cannot be altered or removed, which is why it is called *nonvolatile memory*. Information that is commonly stored in **ROM** includes a digital device's boot instructions.

Computer Storage

Digital devices need to store some information indefinitely. Storage devices allow data and information to be retrieved for future use. Data remain intact when the computer is turned off.

Differentiate Between Internal and External Storage

Internal storage is storage that is integral to the computer. This is called the computer's hard drive. **External storage** devices reside outside the computer. Examples of external storage devices include external hard drives and USB or thumb drives. Optical drives (such as a laptop's DVD drive) and memory cards (such as a microSD card for a phone) are usually considered external storage. Though they're housed within the computer, they aren't part of the computer itself.

Hard Drive Capacity

Hard drive capacity is the amount of storage available to save data and information. Capacity is usually measured in gigabytes (billions of bytes) and terabytes (trillions of bytes). Computer applications, such as Microsoft Office, and multimedia files take up a lot of storage capacity. One megabyte of storage on the first hard drive in IBM's RAMAC in 1957 would have cost $200,000 in today's dollars. That same megabyte today costs as little as four one-thousandths of a cent.

A Traditional Hard Drive

A **hard drive** is the computer's primary storage device. A hard drive uses fixed disk platters to store data and information. The terms *hard drive* and *hard disk* are used synonymously. Hard drives are the primary storage device in a variety of digital devices, including laptops and servers.

Solid-State Drives

Solid-state drives (SSDs) are all-electronic storage devices. SSDs are used in a variety of products, including smartphones, cameras, and tablet computers. SSDs have no moving parts, which makes them faster and more durable than hard drives. SSDs tend to be more expensive than traditional hard drives.

SSD Capacity

SSD capacity is the amount of storage you have available to save data and information. Capacity is usually measured in gigabytes (billions of bytes). The more **SSD capacity** you need, the more money it will cost. **SSD capacity** is one of the most determinant factors of price when buying a tablet or smartphone.

The Hard Drive

Identify the Capacity of a Hard Drive: Windows 10

The hard drive in your computer stores a large amount of data. **Hard drive capacity** is usually measured in gigabytes (GB). It is a good idea to check the capacity of your hard drive.

Here is how:

1. Click the Search box button.

2. Type "This PC" in the [Ask Me Anything box] and then click on the This PC app.

3. Look in the Devices and Drives section at Local Disk (C:) to find the hard drive capacity and available space on the hard drive.

Defragment and Optimize a Hard Drive: Windows 10 and 11

It is important to keep your hard drive running at peak efficiency. One thing you can do to maintain the efficiency of your hard drive is to use a disk utility that comes with Windows: the Defragmenter.

Here are the steps to defragment a hard drive:

1. Open Disk Defragmenter by clicking the Search window. Type "Disk Defragmenter," then select Disk Defragmenter.

2. Under Current Status, select the disk you want to defragment.

3. To determine whether the disk needs to be defragmented or not, click Analyze Disk.

4. Click Defragment Disk.

Use Disk Cleanup: Windows 10 and 11

Another Windows disk utility that can help maintain the efficiency of your hard drive is **Disk Cleanup**.

1. In the Search Box (on the taskbar—usually the bottom left of your window), type "**Disk Cleanup**" and select the **Disk Cleanup** app.

2. If necessary, select the drive you wish to clean up.

3. Under Files to Delete, select the files that you wish to remove (such as Temporary Internet Files).

4. Click OK, then Delete Files.

Connections

Ports

A **port** is a slot or hole that matches the cord or expansion card being connected to the port. Input and output devices plug into the ports located on your digital device. A connector plugs into a port. Common types of ports include USB, Thunderbolt, and **HDMI**. For example, a thumb-drive connector plugs into a USB port.

Connectors

A **connector** is the specialized end of a cord, plug, or expansion card that connects into a port. On a cable, the connector is the end that connects to a port.

USB

Universal Serial Bus (USB) is a type of interface that enables communication between digital devices. USB allows for data transfer between devices and for devices to be electrically charged. USB drives are sometimes referred to as thumb drives, flash drives, or jump drives. USB drives use solid-state flash memory to store information on an internal memory chip. USB drives allow fast data transfer, are inexpensive, and are durable.

HDMI Connectors

HDMI is a standard interface for audio-video connectivity. **HDMI** allows for the transmission of high-definition audio and video signals.

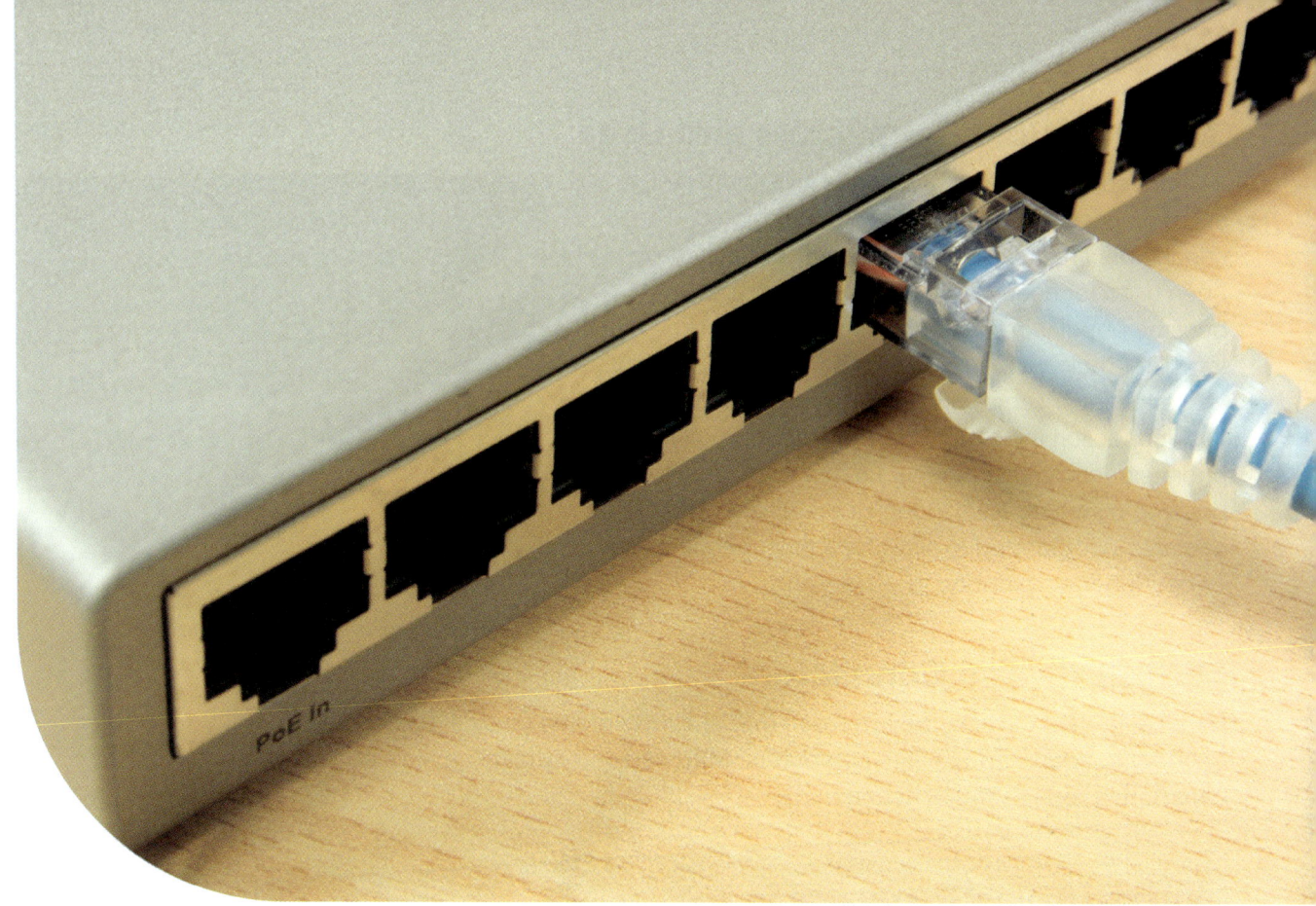

Ethernet

Ethernet ports and **Ethernet** cables are used for local area networks or for connecting devices in home networks. A common use in the home is to connect a modem to a router using **Ethernet** cable to access the Internet. **Ethernet** cable comes in different categories. CAT5 cable can transfer data at up to 100 megabits per second. CAT5e can transfer data at up to 1 gigabit per second, while CAT6 cable can transfer data at up to 10 gigabits per second. CAT8 can transfer data at up to 40 gigabits per second.

Lightning Connectors

Lightning ports and connectors were developed by Apple for use in iPods, iPads, iPhones, and Apple Watch docks. The **lightning connector** is an eight-pin connector, and, with adapters, it can connect to USB, **HDMI**, or VGA connectors. Most iPhone users like the fact that a lightning connector can be inserted in a lightning port in either direction (unlike the USB port). Apple plans to replace the lightning connector and port with the USB-C connector and port in October 2024.

VGA Connectors

Video Graphic Array connectors, often called VGA or HD-15 connectors, are analog connectors that were once very popular for connecting monitors and televisions to computers. VGA connectors were largely replaced by Digital Visual Interface (DVI) connections and **HDMI** connections in the early 2000s. Although they are analog only, VGA connectors are capable of high resolution and frame rates.

3.5-mm Audio Connectors and Uses

3.5-mm connectors are among the most popular for audio devices such as earbuds or headphones. These connectors are small and convenient, and they have been used for nearly 70 years. These connectors became popular in the 1950s with the introduction of the transistor radio. In the 1980s, the Sony Walkman portable cassette player came with a 3.5-mm audio port, making this connection the standard for all headsets. When cell phones became popular, the 3.5-mm audio connection became commonplace, though it evolved and had more contacts incorporated to allow for a microphone as well as headphone capability.

MIDI Ports and Connectors

Musical Instrument Digital Interface (MIDI) allows digital musical devices to connect to and interface with computers. MIDI can carry digital music signals and is common in today's music scene.

Wireless Connections

Protocol

A **protocol** is a set of rules for communication between devices that determines how data is transmitted and received. Many protocols are used to help streamline the communication of all digital devices. For example, **Wi-Fi** uses the 802.11 protocol. This protocol ensures that devices enabled with **Wi-Fi** can connect to any **Wi-Fi** network.

Wireless Ports

Wireless ports allow for the transmission of data between both fixed and mobile devices using short-range radio waves or light waves. Most **Bluetooth** transmissions remain strong up to 35 feet away and then start to reduce in signal strength. Infrared (IR) requires line-of-sight transmission and is commonly used for TV controllers.

Wi-Fi

Wi-Fi is a wireless local area network. The word **Wi-Fi** is a play on the old audio term *hi-fi*. A **Wi-Fi** transmission site is called a **hotspot**. Most computers, tablets, smartphones, and gaming platforms are **Wi-Fi** capable. **Wi-Fi** uses the 802.11 protocol with 2.4 GHz and 5.0 GHz radio waves.

Bluetooth

Bluetooth-enabled devices are found in numerous items, including basketballs, key fobs, and pet collars. **Bluetooth** is a short-distance wireless communication technology that uses short-wavelength radio waves to transfer data. Because they use short-wavelength radio waves, **Bluetooth** signals easily pass through walls, furniture, and so on.

IrDA

IrDA stands for the Infrared Data Association, which establishes the protocols for infrared communication transfer. **IrDA** is a wireless connection that uses infrared (relatively long) radio waves to transmit data. Longer-wavelength waves don't easily pass through walls or furniture (line-of-sight only). **IrDA** devices are often used with a wireless mouse, keyboard, remote control, and so on.

Graphics

Graphics Output

Graphics are the depiction of image data on a display or output device. **Graphics** are usually created by a separate processor within the computer known as the graphics processing unit (GPU).

What Is a Video Card?

A **video card** is a piece of hardware inside your computer that helps display images on your screen. Think of it like a mini-computer dedicated just to graphics! It has its own "brain" called a GPU, which handles the heavy lifting of creating visuals. The video card also comes with its own special memory to store graphics data and ports, like **HDMI** or DisplayPort, to connect to monitors. Plus, it usually has a fan or cooling system to keep it from getting too hot. All these parts work together to make your games, videos, and other graphics look smooth and vibrant on your screen. Gaming computers usually have upgraded graphics cards.

What Is Screen Size?

Screen size is the actual viewable area of a display device.

Screen size is measured diagonally from one corner of the screen's viewable area to the other.

Screen size can be a major factor when determining which digital device to buy.

What Is a Pixel?

A **pixel** is the smallest element in an electronic image. It is simply a dot on the screen of a display device. The more pixels in an image, the better the image quality. Different displays can have differently shaped pixels.

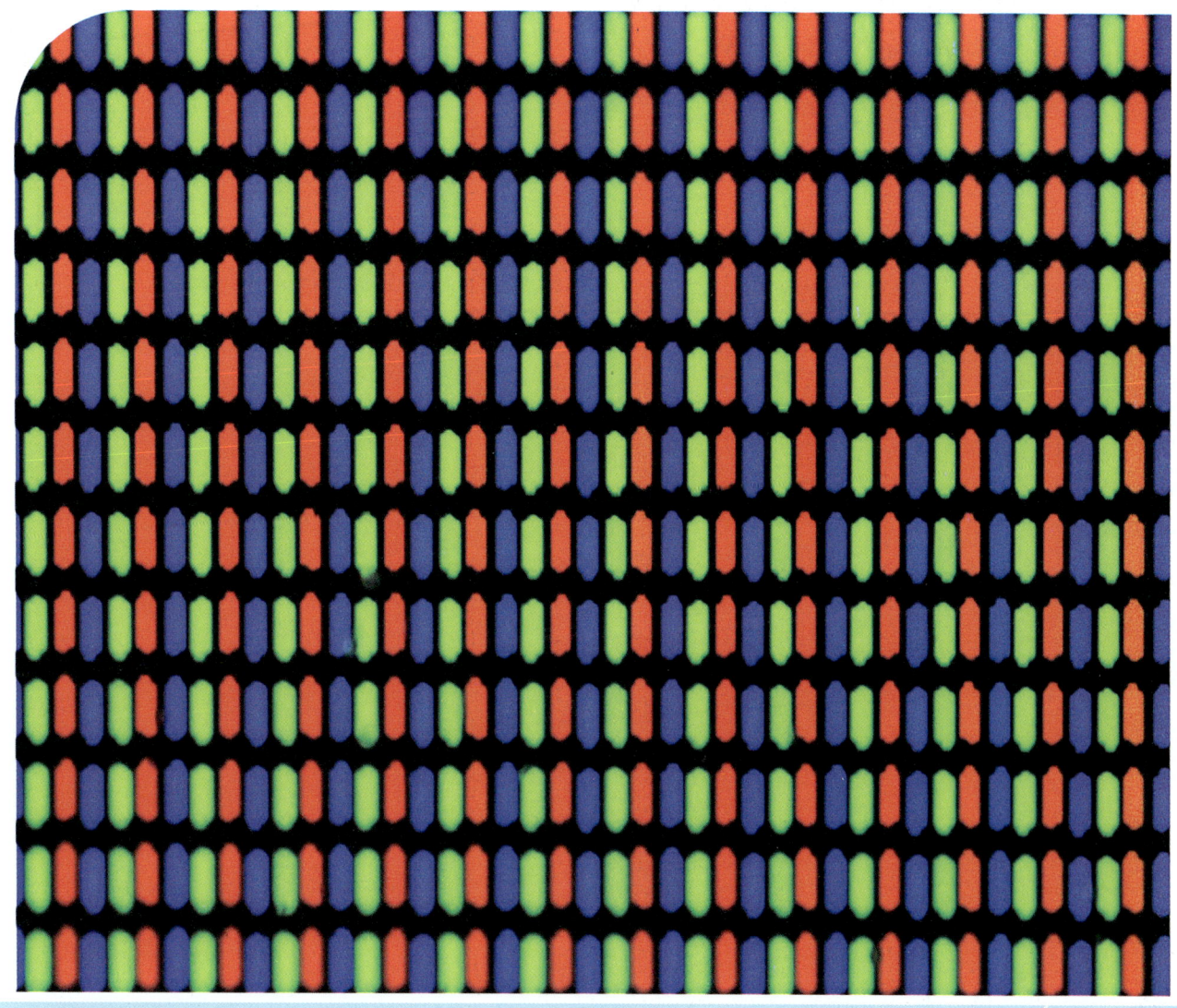

What Is Resolution?

Resolution is the clarity of an image. The resolution of monitors and other display devices is measured in pixels. The more pixels in a display device, the higher the resolution. Larger displays need more pixels to achieve the same resolution.

What Is Native Resolution?

Native resolution is the maximum resolution of a display device. A laptop with 1920 (horizontal pixels) × 1080 (vertical pixels) has 2,073,600 total pixels. The first number is horizontal resolution, and the second is vertical resolution. A display is said to be pixel perfect when the incoming video signal has the same number of pixels as the native resolution of the display.

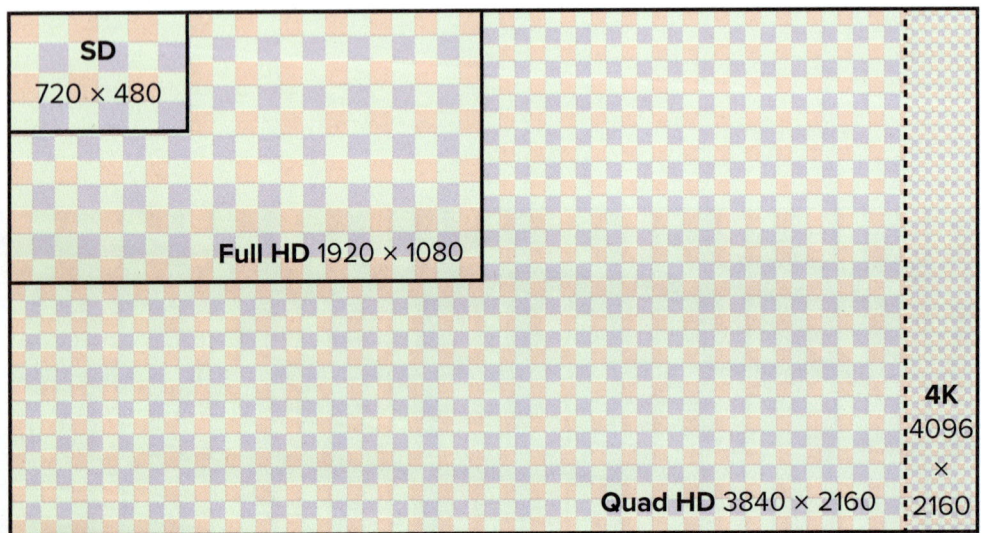

How a Computer Works

Computer System Design

In the diagram below, the red areas are the computer's two processors: the central processing unit, known as the CPU or the processor, and the graphics processing unit, known as the GPU. Memory is depicted in light green. The system cache provides memory for data that the CPU is using repeatedly. The random access memory (**RAM**) holds data for the CPU's use. Note that the GPU has its own memory. Storage is shown in dark orange. This is the computer's hard drive that keeps data even after the computer is powered down. Knowing a basic computer layout makes understanding how they work much simpler.

Computer System Design Graphic

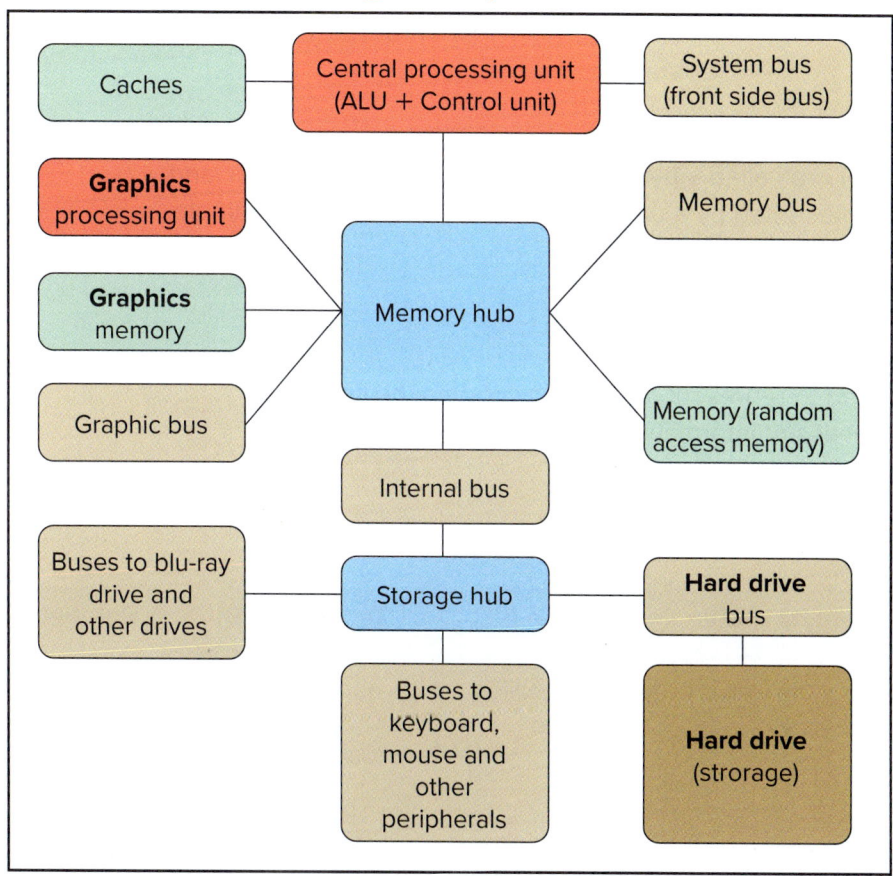

Bits

Computers communicate using their own language. This language is called *binary*. Binary language consists of two digits: 0 and 1. Each 0 or 1 is called a *binary digit*, or a *bit* (b). **Bits** are the smallest unit of data a computer can process.

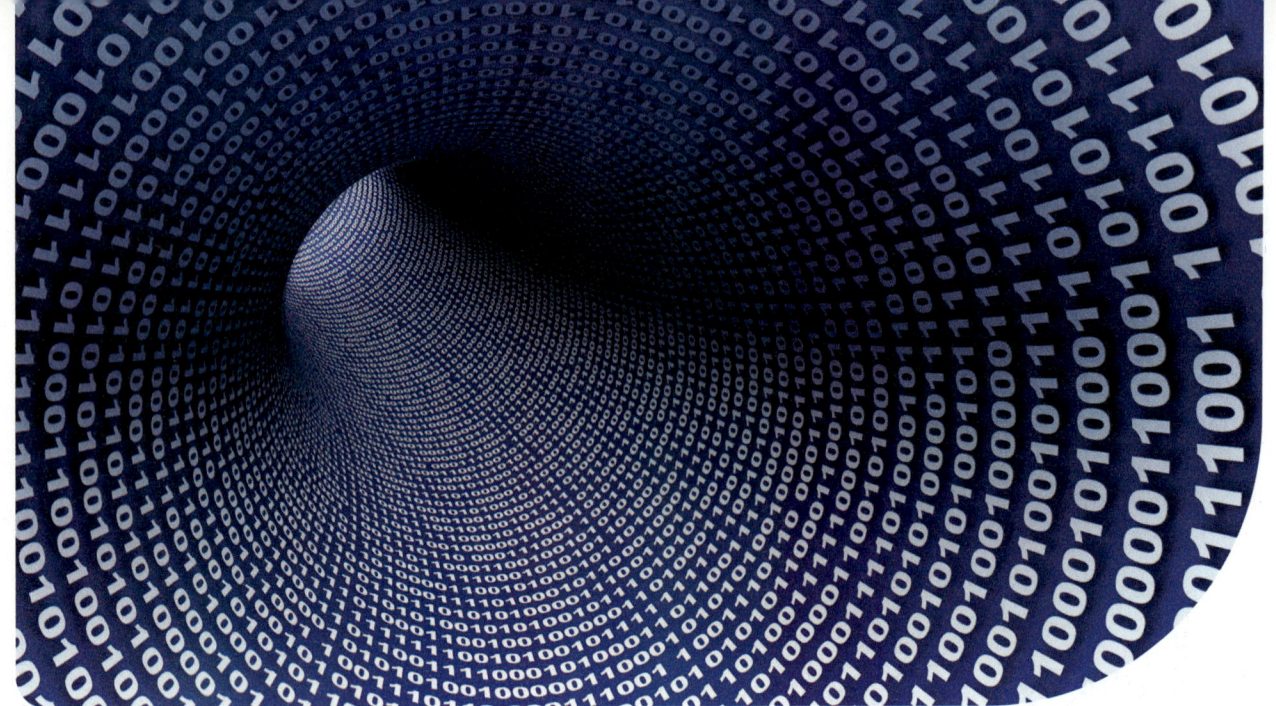

Bytes

Eight bits grouped together are called a **byte**. Each number, special character, and letter of the alphabet is represented by a unique combination of bits. For example, an ampersand (&) is represented as 00100110.

Storage Capacity

Bits and bytes are also used to represent the quantity of data the computer processes. Common quantities of information include:

Kilobyte (KB)	Megabyte (MB)	Gigabyte (KB)	Terabyte (KB)	Petabyte (KB)
—about 1 thousand bytes	—about 1 million bytes	—about 1 billion bytes	—about 1 trillion bytes	—equals 1,000 terabytes
—1 KB is equal to about one page of text.	—1 MB is equal to about 875 pages of text.	—1 GB is equal to 341 digital images (3 MB average file size).	—1 TB is equal to about 349,000 digital images (3 MB average file size) or 40 single-sided Blu-ray discs.	—1 PB is equal to about 358,000,000 digital images (3 MB average file size) or about 42,000 single-sided Blu-ray discs.

Word Size

Word size refers to the amount of data a processor can hold and process at one time. Today's processors generally have word sizes of 32 or 64 bits. A 64-bit processor can process information faster than a 32-bit processor can.

The Boot Process

The **boot process** is the loading of an OS into the main memory of a device. The boot process consists of four steps:

1. The device is powered on, and the CPU activates the basic input/output system (BIOS).

2. The Power-On self-test is executed.

3. The OS is loaded into **RAM**.

4. The OS checks the registry settings and loads saved configuration settings.

Computer Speed

The speed of a digital device's processor is determined by several factors. Three of the most important determinants of a **computer's speed** are clock speed, word size, and bus width.

Hertz

Heinrich **Hertz** proved the existance of electromagnetic waves. **Hertz** is a measurement for how frequently something happens. It's shortened to "Hz." When something happens once every second, it's at a frequency of 1 Hz.

What Are Megahertz and Gigahertz?

Megahertz is abbreviated MHz and is equal to one million cycles (or instructions) per second. 1 MHz is one million instructions per second. Gigahertz is abbreviated GHz and is equal to one billion cycles (or instructions) per second. A 3.8 GHz processor can execute 3.8 billion instructions per second.

Bus

A **bus** is a channel over which information flows. Think of a bus as a highway on which data travels in a computer. A bus has two parts: the address bus and the data bus. The address bus transfers information and instructions about where the data should go inside the digital device. The data bus transmits actual data. The most important bus is the one connecting the central processing unit to the memory. Some call this the *system bus* or the *front-side bus*.

Bus Width

Bus width is the amount of data that can be transmitted at a given time. The wider the bus, the more information that can travel along it, which creates faster transmission speeds. **Bus width** is measured in bits. A 64-bit bus can transmit 64 bits of data.

ASCII

ASCII, the American Standard Code for Information Interchange (pronounced AS-key), is a common encoding standard. **ASCII** code can represent the 26 uppercase and lowercase letters in the English language, as well as numbers and many different symbols. **ASCII** can represent a total of 256 characters.

What Is Unicode?

Unicode is a common encoding standard that can represent all world languages. **Unicode** uses at least 16 bits per character. **Unicode** has virtually limitless characters, allowing for the large number of characters in certain foreign languages.

Quantum Computing

Quantum computing uses the power of atoms and molecules for memory and processing tasks. One of the keys of quantum computing is the quantum bit or qubit. Qubits are the basic units of information that are similar to the 0s and 1s (bits) that represent transistors in today's computers. Qubits are more powerful because a qubit can represent both a 0 and 1 at the same time and can affect other qubits using quantum entanglement. This allows quantum computers to take shortcuts to the correct answers for certain types of calculations. Quantum computers could be much faster at running artificial intelligence (AI) programs and could also create more secure data encryption.

Chapter Review

1. Describe the four main computing functions.

2. Explain what is meant by *computer hardware*. Identify computer hardware that is commonly used today.

3. Describe the system unit of a computer. What major computer components are commonly housed in the system unit?

4. Differentiate between a laptop and desktop computer.

5. What differentiates a smartphone and a personal media player (PMP)? How are they different? How are they similar?

6. What is a server computer? How is one used? Discuss when you might use a server computer.

7. Describe a computer processor. Why are computer processors so important? Who are the main manufacturers of computer processors? Do all computers need a computer processor to run correctly?

8. Explain the four steps of the machine cycle. How does a machine cycle apply to computer processors?

9. What is the difference between a single-core and multi-core processor?

10. Explain computer memory. Why do computers need memory?

11. What is **RAM**? What are the various sizes of **RAM** used in computers? Do a smartphone and tablet use **RAM**? How much **RAM** do most laptops, tablets, and smartphones come with?

12. Describe cache. How is cache used in a computer?

13. Explain **ROM**. What makes **ROM** different from **RAM**?

14. Compare and contrast USB and Firewire/Thunderbolt. What types of devices use each type of port/connector?

15. What are **Ethernet** ports/connectors/cables? What are **Ethernet** cables/ports used for?

16. Discuss computer storage capacity. What is the unit of measure for computer storage? List some of the various computer storage capacities. What storage capacity do most laptops, tablets, and smartphones come with?

17. When discussing display devices, how are pixel and resolution related? Is resolution an important consideration when choosing which computer to buy?

18. What is the difference between a bit and a byte? How large is each? What are bit and bytes used for in a computer?

Ethics Discussion

Digital devices are produced using multiple production inputs, including natural resources and labor. Responsible production practices are one consideration when purchasing a new digital device. Conduct research on the Internet about the ethical production of digital devices/computers. What are some of the issues that surround the ethical production of digital devices/computers? What ethical factors would you consider if you were purchasing a new digital device/computer?

2 Computer Software

What To Expect

After completing this chapter, you will be able to:

- describe the interaction between hardware and software;
- differentiate between system and application software;
- recognize functions of various computer components;
- classify different operating systems and their uses;
- assess software and hardware compatibility and performance;
- evaluate personal needs in selecting computer hardware and software.

Chapter Topics

- System Software
- Application Software
- Software Licenses
- Preparing a Computer
- Buying a Computer
- Specifications for Purchasing PC Hardware and Software
- Software Considerations
- Self-Assess and Identify Information Needs

Copyright © McGraw Hill   ra2 studio/Shutterstock

Let's Explore Software

Let's think about cooking in the kitchen as an analogy for software, and the right hardware you need to run the software. In the kitchen, the pots and pans and utensils are the hardware: they are physical objects that you can touch. But what determines the type of dish you'll cook? The recipe! In the world of computers, the recipe is the software that gives the computer (our kitchen tools) the rules or instructions on what to do. You use software to do different things with the same hardware. Software turns the computer into a multi-functional tool capable of performing many tasks. One moment it can be a writing pad; the next, it can help you solve complex math problems; and in another, it becomes a movie screen! There are two major types of system software: system software and application software.

- **System software** provides the basic instructions to run the computer. Think of system software as the fundamental rules that define pots and pans as objects you can use in cooking, how cooking works, and how cooks interact with kitchen utensils. The system software ensures that the computer "cooks" correctly and safely and understands its basic directions.

- **Application software** allows you to perform specific tasks, such as creating documents or playing games. It's like the recipe you decide to cook.

You can cook a whole bunch of dishes with one type of pot. But you might be thinking not every type of pot works for every kind of dish. You want to get pots and pans that are best for the kinds of dishes you want to cook. The relationship between the pots and a recipe is much like the relationship between hardware and software. Choosing the right pots and pans and the right recipes allows you to enjoy your favorite dishes; choosing the right computer and software lets you perform your desired tasks on your computer. Now, let's talk about the different roles in our kitchen and what to consider when buying a computer.

- CPU (Chef): Role and Consideration: This is the main cook, executing the recipes. If you want to whip up some really demanding dishes like a soufflé (think games or video editing), you'll need a fast and skilled chef. We are looking for a CPU with higher GHz and more cores!

- RAM (Countertop): This is where our chef keeps the ingredients for the current dish and does their work. If you love preparing multiple dishes at once (multitasking), you'll need a spacious countertop. So, more RAM is essential—16GB or even more!

- Hard Drive (Refrigerator): It stores all your ingredients, files, and programs. If you have a lot to store, you are going to need a bigger fridge. SSDs are the fancy fridges—faster and more reliable but pricier!

- Graphics Card (Oven): It finalizes or bakes the dish, rendering those beautiful graphics. For gourmet dishes or high-quality games and graphic design, you'll need a high-quality oven, a top-notch graphics card!

Remember, it's all about balancing your needs with your budget and choosing the right components to suit the dishes (tasks) you want to prepare most often!

System Software

Operating System, Your Computer Platform

This is the main software that tells your computer how to work. Another term for operating system (OS) is *computer platform* because, like a real-world physical platform, it's the software that supports all the other software.

Popular Platforms

There are many operating systems that run computers. We're going to focus on the major operating systems: **Windows**, MacOS, **iOS**, and **Android**.

Windows

You'll mostly find this one on regular computers. Its logo looks like four squares put together. A lot of businesses and schools use **Windows** because it works with many programs and isn't as expensive as other operating systems. Microsoft chose the name "**Windows**" due to the computing boxes, or "windows" that represent one of the main elements of the OS.

MacOS

MacOS stands for Macintosh Operating System and is only available on Apple computers. Its logo is an apple with a bite taken out. People like using MacOS because it's easy to use and has an attractive design.

Apple and Macintosh operating systems are designed and distributed by Apple, Inc., and have a reputation of being very easy to use. Apple OS are found in a variety of digital devices, including iPhones, iPads, and laptops. Macintosh computers are also the industry standard for photo editing, film and movie development, and graphic design. Therefore, graphic designers and media producers tend to use Apple products.

Advantages—Mac OS

- hard to get viruses
- built for art and design
- looks great

Disadvantages—Mac OS

- used by fewer people
- costs more than other computers
- only works with Apple-approved software
- hard to customize

iOS

This one stands for iPhone Operating System, and—no surprise—it's only on iPhones, which are popular in the United States. The iPhone is part of Apple Computer, so, its logo is also an apple with a bite out of it.

The **iOS** includes many Apple-specific apps, including FaceTime, Safari, and Siri.

Apps that that aren't made by Apple cannot run on **iOS** without being reviewed and approved by Apple. This approval process ensures a smooth interface with the **iOS**.

Android

Android is a Google product that has a green robot for its logo. Almost all phones and tablets, except for Apple's, use **Android**. In countries outside the United States., more people use **Android** than any other system. In the United States, it's close to a tie between **Android** and **iOS**.

Advantages—Android

- works great with Gmail and Google
- lots of free apps
- easy to make your own apps

Disadvantages—Android

- can be harder to use than iPhones in some instances
- free apps have ads

Linux Operating System

The **Linux operating system** is a free, open-source software that anyone can use and modify. Most servers, mainframes, and supercomputers use the Linux OS. It also runs most video game consoles and embedded computers, and is the underlying software for the **Android** OS. Despite its popularity among computer manufacturers, only a small percentage of personal computers use the Linux OS.

Advantages—Linux

- totally free
- hard to get viruses
- easier to customize

Disavantages—Linux

- harder to learn
- doesn't work with all programs

Network Operating System

Unlike regular systems such as **Windows**, which are for just one computer, a network operating system (NOS) helps many computers work together on a network. Businesses use this kind of system to run their networks and share software, printers, and databases. **Utility programs** perform a specific maintenance-type task on a computer and are included with both **Windows** and macOS. Maintenance-type tasks performed by utility programs include disk cleanup, file maintenance, file search, and system backup and restore.

Applets

Applets support a larger application program. They are commonly used to execute tasks in the **Control Panel** of Microsoft **Windows**, making them a kind of utility program. Think of applets as the small, specific rules or guidelines that help manage and run the bigger ball game. They're like the quick directions a referee gives during a game to ensure everything plays out fairly and smoothly.

Application Software

Computer Applications (Apps)

Applications, or *apps* for short, run on phones, computers, and the internet. They are programs that perform like tools such as drawing pads, calculators, word consoles, or movie screens. In our game analogy, you can think of apps as the different kinds of games you can play with a ball.

Productivity Software

Productivity software helps people get their work done more easily. Apps designed for individuals and for businesses help with tasks, such as writing documents, organizing numbers, and making presentations.

Word Processing Programs

People use word processing programs to write many kinds of documents. These can also be used to design brochures or even basic websites. The most common programs are Microsoft Word, Apple Pages, and Google Docs.

Spreadsheet Programs

Spreadsheets are used to organize numbers and text in boxes called cells. They can do math for you. Some common terms are

- *worksheet*: one page in a file;
- *workbook*: several worksheets combined;
- *cell reference*: the location of a cell;
- *formula*: math used to do calculations;
- *function*: a formula the program already knows.

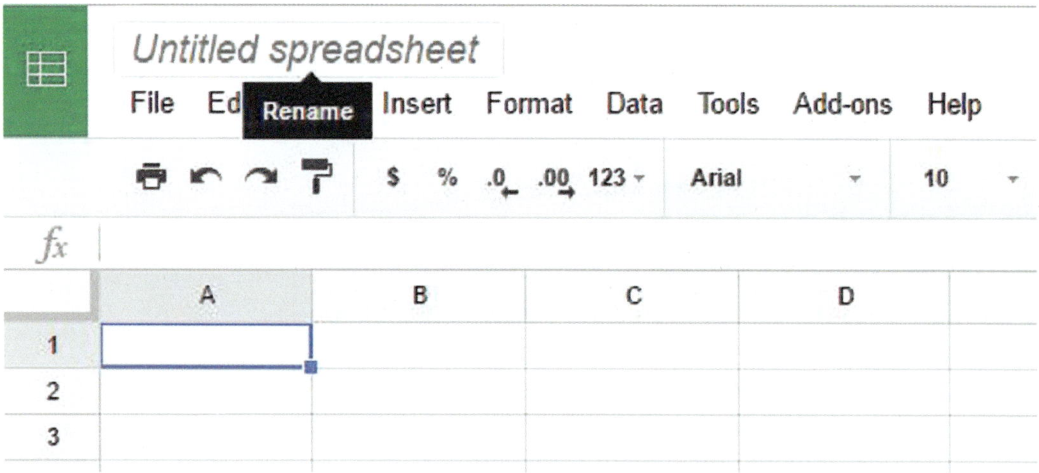

Presentation Programs

These programs help you make slides that you can display on a screen. The text should be big and simple. Some terms you should know are

- *slide*: one page of your presentation;
- *slide master*: a repository that holds the design for all slides
- *slide show*: all the slides in order
- *speaker notes*: extra notes to help you talk during your presentation

Database Programs

Databases store lots of information in a structured way. They're important for websites and other web services. Some common terms are

- *table*: a section in a database;
- *record*: one row in a table;
- *field*: one column in a table;
- *primary key*: a unique identifier for each record;
- *query*: criterium for pulling out certain data;
- *reports*: organized data for printing or looking at;
- *forms*: ways to put new information into the database.

Software Licenses

A **software license** is a legal document that governs the use or redistribution of software. In the United States under copyright law, all registered software is copyright protected, in both source code and object code forms (except open-source public domain software). Typical software licenses grant the licensee (or end user) the right to use one or more copies of the software.

Common software licenses

- **End-user license agreement (EULA):** A legal contract between the software manufacturer and the user of an application. The EULA explains how the software can and cannot be used and any restrictions on use imposed by the application developer.

- **Single-user license:** A single-user license restricts the use of the software to one user at a time. If you purchase software for personal use, it is probably a single-user license.

- **Network license:** Network licenses give anyone on a network the right to use the software. Network software is not installed on each individual device. Network licenses are common in colleges and businesses.

- **Site license:** Site licenses are different from network licenses because the software is installed on the device of qualified users who request the software inside an organization. Site licenses typically have time limits and must be renewed on a regular basis.

Preparing a Computer

Installing Software

When you get a new phone, computer, or tablet, it usually comes with some programs already on it. These are called preinstalled programs. Most people add, or install, more programs or games that they like.

Installing means you're setting up your device to use new features: software, like new games or programs, or hardware, like a printer or a webcam.

Why do you want to install new stuff? Well, it helps you get the most out of your device. For example, you might need a certain program for schoolwork or a game you really want to play. By installing them, you make your device more useful for you.

Generally, there are two primary ways to install software: downloading it from the internet or using a physical device, like a compact disc (CD), digital video disc (DVD), or universal serial bus (USB) drive.

Uninstalling Software

When you get a new device like a computer, tablet, or phone, it often comes with programs you didn't choose. These might be things you don't want or need. Getting rid of these programs is called **uninstalling**.

Uninstalling is the opposite of installing. It means you're taking software off your device. This can be helpful for a few reasons. For one, it clears up space on your device's storage, like its hard drive. This helps your device run better and leaves room for things you want.

Steps for Installing and Uninstalling Software

Operating System	How to Install Software	How to Uninstall Software
Windows	1. Download from internet and run installer. 2. Insert a CD/DVD and follow instructions. 3. Use a USB stick to create a bootable drive and install.	1. Open "Control Panel" 2. Go to "Programs and Features" 3. Select the program 4. Click "Uninstall."
Mac	1. Download from internet and drag to Applications folder. 2. Insert a CD/DVD and follow instructions.	1. Go to "Launchpad," hold the program icon, click the "X." 2. Drag program to the trash from the Finder and empty trash.
Android	1. Open "Play Store" 2. Search for the app 3. Click "Install."	1. Open "Settings" 2. Go to "Apps" 3. Select the app 4. Click "Uninstall."
iPad (iOS)	1. Open "App Store" 2. Search for the app 3. Click "Get" 4. Click "Install."	1. Press and hold the app icon 2. Click the "X" that appears 3. Then confirm you want to delete it.

The Control Panel and System Preferences

The **Control Panel** is the system management tool in the **Windows** operating system. **System Preferences** is the system management tool in the Macintosh OS. System management tools allow you to change computer settings, manage tasks, install and uninstall programs, and many other options.

Identifying Key System Programs

Prior to uninstalling software, it's important to identify those programs that the computer needs to run correctly. These programs include the computer's operating system software and the drivers that run the computer's hardware.

Use caution when uninstalling software that has been published by your computer's manufacturer or software that has been published by the manufacturer of the computer's OS.

Buying a Computer

What Should You Think about When Buying a Computer?

When you're trying to pick out a computer, think of it like picking out your favorite game console or toy. You want to make sure it can do all the things you like!

1. **Type of Device**

   - Do you want something you can carry around easily like a tablet or a laptop, or do you want a desktop that stays in one place?

2. **Primary Use**

   - What are you going to use it for the most?
     - Schoolwork?
     - Chatting with friends?
     - Gaming?
     - Or a little bit of everything?

3. **Battery Life**

   - If you're going to be on the go, you'll want something with a long battery life so you don't have to charge it all the time.

4. **Budget**

   - How much do you have to spend?
   - You want to find something cool but also something that doesn't break the bank.

5. **Storage**

   - Do you want a solid-state drive (SSD)? They are like the fast, super-cool storage spaces that make everything run smoothly, but they can be a bit more expensive.

Device Connections

And don't forget to check out the different spots where you can plug things in!

- **USB ports:** They let you connect other gadgets, like a mouse or keyboard.
- **HDMI ports:** These are very helpful if you want to show your friends a video or a presentation on a bigger screen, like a TV!
- **SD card slot:** If you like taking pictures with a digital camera, you'll want a slot to put your memory card in so you can see your pictures on your computer!

Think about It

- Does the device you are looking at match what you need and what you can spend?
- Is it going to make your life easier and be fun to use, or is it going to be a hassle?

How Much RAM Will You Need?

It is important to determine how much random-access memory (RAM) you need to install in your new computer. The more you multitask (for example, using Instagram, working in Microsoft Excel, streaming music, and running different programs all at the same time), the more RAM you will need.

Most computers today range from 4GB to 16GB of RAM. The more RAM a computer has, the higher the price of the computer.

One way to save money is to purchase a computer that has expandable memory and then to add RAM as needed.

How Much Storage Does Your Computer Need?

A computer's **storage capacity** refers to the size of the computer's hard drive. Most laptop computers have between 500GB and 1TB hard drives, whereas most tablet computers have hard drive capacities of between 16GB and 256GB.

To keep this in perspective, a standard-definition movie takes about 1GB to 2GB of storage, whereas a high-definition movie takes about 3GB to 5GB.

Solid-state hard drives tend to be smaller and more expensive. They are also much faster and require less energy, providing longer battery life.

Don't overspend on your computer's storage capacity because you can always store files on the Cloud or on flash drives.

What Screen Size Do You Need?

It is important to determine what screen size best suits your needs.

When choosing the screen size for your needs, it may be helpful to visit a local computer retailer and use your eyes to help you decide.

If you plan on doing a lot of multitasking or watching videos, a larger display may be your best bet.

Touch Screens

A **touchscreen** is a type of display screen that is sensitive to the touch of a finger or stylus. Touchscreens are used for a variety of applications, such as automated teller machines (ATMs), global positioning systems (GPSs), and point-of-sale terminals. Touchscreens are included in a variety of devices, including phones, tablets, and laptops.

Touchscreens come in a few different categories. One of the most popular categories of touchscreens for laptops, tablets, and phones is the projective capacitive (Pro-Cap) or touch-sensitive screen. Pro-Cap screens use a grid sandwiched between two glass panels. Voltage is applied to the corners of the screen, and when a finger touches the screen, it draws a tiny amount of current. The controller computes the X–Y location from the change in capacitance at the touch points.

Screen Resolution

Screen resolution tells you how many pixels your screen can display horizontally and vertically. This determines the sharpness and clarity of an image.

A **pixel** is the smallest element in an electronic image. The more pixels, the higher the resolution, which results in a sharper and clearer image.

Specifications for Purchasing PC Hardware and Software

Hardware and Software Specs

When purchasing a new laptop computer, it is essential to consider several different variables to ensure you are getting a device that fits your needs and performs optimally under various use conditions.

Remember, you will likely use this computer for many years, so you'll want to spend some time researching different laptop computers, comparing specifications and prices, and determining present and future needs.

Size

Laptop computers come in various sizes, starting at 11.6 inches and going up to 17.3 inches. The device's size corresponds to the screen size (or size of the viewing area of the laptop). The larger the size/screen size of the laptop, the larger the footprint (or area the device takes up on a table). Unlike RAM and software, you cannot upgrade or change the size of a laptop, so make sure you are choosing a laptop with the most optimal size.

Form Factor

Form factor refers to the shape, size, and hardware specifications of a laptop. Form factor describes the physical aspects of a computer, or in this case, laptop computers. Form factor covers four categories of computers, including ultrabooks, notebooks, convertibles/2 in 1, and laptops. The following table demonstrates the similarities and differences of these devices.

	Ultrabook	Notebook	2 in 1 Laptop
Screen Size	10–20 in.	9–14 in.	11–14 in.
Weight	2–5 lb	1–3 lb	1.5–3 lb
Thickness	0.7 in. for displays 14 in. or less	0.9 in. for displays more than 14 in.	0.5–1 in.
RAM	4–16 GB	2–4 GB	4–16 GB
Storage	256 GB–1 TB	256 GB–512 GB	128 GB–1 TB
Cost	$400–over $1,000	$150–$500	$350–over $3,000 $300–over $3,000

Software Considerations

Purchasing new software may seem like a straightforward process. However, there are some considerations to ponder before purchasing new software.

Compatibility

New software may require a specific operating system and specific computer capabilities. Therefore, you should compare the software requirement specifications to the current specifications of your operating system.

School and Student Options

Many software programs are available for free or at a reduced price. Make sure to check with your school or the software provider to see if there are any options available.

How Much Training?

Some software programs have a steep learning curve, so before purchasing new software, check to see if there is training required and what training options are available.

Is the Software Up to Date?

Software updates ensure you are using the most up-to-date version of the software. Conduct research to see how often the software is updated.

Self-Assess and Identify Information Needs

Self-assessing your information needs is evaluating what you will need to accomplish tasks using computer hardware and software. Self-assessment is essential because individual needs will vary based on different factors, including your college major/area of emphasis, career plans, cost, and software needs.

Area of Study

Technology requirements vary across different areas of study. It is important to research the specific needs for your area of study before selecting a device. For example, many business courses use PCs and software designed to run optimally on them, including Microsoft Office. However, device-specific software has been minimized due to the proliferation of cloud-based applications such as Microsoft Office 365. On the other hand, graphic design and animation courses often require software that runs optimally on **Mac** computers.

Career Plans

Your chosen career field will often dictate your information needs. An important consideration in your self-assessment should include research on industry-specific needs regarding technology and information. Some industries and careers rely heavily on **Windows** operating system—based PCs while others depend on **Mac** computers that run the MacOS. While operating systems fundamentally execute the same set of tasks and functions, the interface and commands are quite different. You should become comfortable using the operating system that is most widely used in your chosen career field.

Cost

A crucial step in your information needs assessment is the analysis of cost. There often is a difference between the cost of digital devices and software packages. A basic cost–benefit analysis can aid in your assessment. Essentially, you should research the cost of various digital devices and then list each device's benefits. For example, **Mac** computers are often more expensive than PCs. However, **Mac** computers have some benefits when compared to PCs. Additionally, the type and version of software should be investigated. Comparing cost to benefits will help you identify which software and device will best fit your needs.

Software Needs

When conducting a self-assessment, you should determine what types of software are needed and required for your program of study and chosen career field. For example, many business schools require the use of Microsoft Excel and Access. Additionally, they may also cover analytics and data modeling program work. Research on the types of software needed in your industry and your program of study can set you up for success.

Chapter Review

1. Describe what system software is used for in a computer.

2. List the different roles in a computer compared to a kitchen, as mentioned in the text.

3. Name the major operating systems mentioned in the text.

4. How does having the right software affect the versatility of a computer?

5. Why is it important to consider your budget when selecting computer components?

6. Based on the text, how do system software and application software collaboratively contribute to the functioning of a computer?

7. Why might someone need to consider the size of a device when purchasing a computer?

8. Explain the importance of software updates.

9. Describe the importance of software compatibility with hardware.

10. Why is considering training requirements essential when selecting software?

11. List types of software that are essential for a typical student.

12. Why is it necessary to uninstall unwanted software from your device?

13. How does the choice of operating system affect the performance and user experience of a computer?

14. What considerations should be made regarding screen size and resolution when purchasing a computer?

15. How does application software enable a computer to be versatile?

16. How does having adequate RAM affect the functioning of a computer?

17. How do the roles of a CPU and a graphics card differ in a computer system?

18. What is the significance of considering the form factor when purchasing a laptop?

19. What is the importance of having a large hard drive in a computer system?

20. How does the choice of a network operating system affect resource sharing in a network?

Ethics Discussion

Two issues threaten the privacy of our information. The gathering of people's personal information—often without their knowledge—and the increased use of that information for many purposes. Conduct research on the internet about information privacy. What are the ethical issues that surround information privacy? What are the information privacy laws in the United States? Who enforces these laws? What information is public knowledge? How can you safeguard your personal information?

3 Computer Input

What To Expect

After completing this chapter, you will be able to:

- define an input device and provide examples of some input devices;
- discuss adaptive input devices and provide examples;
- describe combination input devices.

Chapter Topics

- Input Devices
- Keyboards
- Pointing Devices
- Audio Input
- Image Input Devices
- How Cameras Work
- Radio Frequency Identification
- Adaptive Input Devices
- Combination Input Devices

Let's Explore Input Devices

Input devices are the controls or the "drivers" of your computer. They help you navigate and interact in the digital world. Imagine a bumper car arena. The bumper cars are buzzing with energy, ready to zoom around when the ride begins, but they can't move without someone in the seat working the controls. Imagine your computer as one of those bumper cars. It's powered up and ready, but it needs someone—you—to drive it! You're in the driver's seat, and input devices are your controls. You use the accelerator to make it move, the brake to make it stop, and the steering wheel to decide its direction. In much the same way, you use input devices like keyboards, mice, and touchscreens to "drive" your computer, giving it directions on what to do next.

Examples of Input Devices:

- Keyboard: Like the accelerator, it helps you input commands and text to make the computer "move."
- **Mouse** or Touchpad: Acting as your steering wheel, it navigates the pointer on the screen, helping you click, select, and move things around.
- Touchscreen: This one is a bit like having a steering wheel and accelerator in one.You can navigate and give commands with just a tap or a swipe.
- Camera & **Microphone**: These let your computer "see" and "hear," allowing you to input visual and audio data, like talking to a friend or taking a photo.

Adaptive Input Devices: If the standard controls are not comfortable for everyone then adaptive input devices are used. These are specialized devices designed for people who find the standard ones challenging to use. It's like having custom controls on a bumper car that make driving easier for everyone!

- Voice Recognition Systems: This acts like a voice-controlled steering wheel, letting them navigate and command with their voice.
- Eye-Tracking Devices: These let people control the computer with their eye movements.
- Sip-and-Puff Systems: These allow people to control the computer by sipping or puffing on a tube.

Combination Input Devices:

Sometimes, you might need more advanced controls, like a steering wheel with built-in buttons. These are combination input devices that can do multiple things. Smartphones and tablets have touchscreens, cameras, and microphones all in one! Gaming controllers have buttons, joysticks, and triggers, allowing you to control games in various ways.

Remember, just like every driver might have a unique way of controlling their bumper car, every computer user might use different input devices to interact with their computer. And thanks to adaptive and combination devices, everyone can enjoy the ride, navigating the digital world in their own way!

Input Devices

Input devices are like the controls for your computer. Without them you can't do anything. It's like a bumper car that's powered up and ready to run when the ride starts, but it won't do anything unless you give it input. You push the accelerator to make the car go faster, you push the brake to slow down, and you move the steering wheel to turn the car. Just like someone can use these controls to drive a bumper car safely and have fun, input devices like keyboards, mice, and touchscreens help you navigate and interact with your computer. You're in the driver's seat, steering your computer's actions with the help of these tools.

Examples of input devices:

- keyboard
- mouse
- touchpad
- touchscreen
- camera
- microphone
- game controller.

Keyboards

Keyboard Options

The type of keyboard that comes with your computer depends on the computer that you purchase. Nearly all phones and tablets use virtual keyboards on a touchscreen.

Laptop computer keyboards are limited by the size of the monitor's screen. Thus, the larger the monitor, the more room there is for additional function keys on the laptop's keyboard. Desktop and all-in-one computers provide the most flexibility because the keyboard is already a separate component. No matter what kind of computer you're using, you can always add an external keyboard if you desire.

QWERTY Keyboards

The **QWERTY keyboard** is the most common text-entry input device for computers. It's named for the first six top-row letters from the left.

It was developed to keep nineteenth-century typewriters from jamming if keys were pressed too quickly.

The standard **QWERTY keyboard** is one of the world's great examples of the value of open-source standards and protocols, as this keyboard layout is used around the world.

Alternative Keyboards

With improving technology, your keyboard options keep improving. Many desktop users like using ergonomic keyboards to help reduce repetitive stress injuries such as carpal tunnel syndrome. Projection laser keyboards allow any flat surface to be a keyboard. There are even waterproof roll-up keyboards that you can easily stuff into your backpack!

Windows Keyboard Shortcuts

There are numerous keyboard shortcuts that can speed your work. Here are some of the most common:

Common Windows Shortcuts:

- left button double click = select word
- left button triple click = select paragraph
- CTRL + A = select all
- CTRL + C = copy
- CTRL + V = paste
- CTRL + S = save
- CTRL + X = cut
- CTRL + P = print
- CTRL + Z = undo
- CTRL + B = bold
- CTRL + U = underline
- CTRL + I = italics
- CTRL + ESC = Start menu

Note: Pressing CTRL and ALT and DEL simultaneously accesses the Task Manager; doing it twice reboots your computer.

Mac Shortcuts

There are numerous handy keyboard shortcuts for Macs. A few of the more commonly used shortcuts are listed here.

Common Mac Shortcuts

- Command + A = select all
- Command + C = copy
- Command + V = paste
- Command + S = save
- Command + X = cut
- Command + Z = undo
- Command + F = find in most apps
- Command + Shift + 3 = snipping tool
- Command + Shift + 4 = screenshot

Pointing Devices

Mouse

The mouse was developed so that objects such as images or icons could be selected or dragged across a bit-mapped screen. Apple Macintosh computers popularized the GUI and the mouse to computer buyers. Now nearly every desktop or all-in-one computer comes with a mouse.

Mouse Options

There are numerous options available besides the mouse that came with your desktop computer or the trackpad on your laptop. Most of these options attach to the computer, either wired or wirelessly through your computer's USB port. Wireless connections include an IrDA (infrared) or **Bluetooth** dongle. A trackball has a rolling ball housed within a stationary device. Some computer users prefer an ergonomically shaped mouse that fits their hand more comfortably.

Touchpads

Touchpads take up very little space. Because of this, most laptop computers come with touchpads. Most touchpads use a capacitance system to track finger movement across the touchpad. Fingers conduct electrons, and the touchpad senses this electron movement. Therefore, a pencil eraser would not work on a touchpad. Gloves or even extremely dry fingers can prove problematic with touchpads.

Touchpad Options

There are many different touchpad options to give users significant control over their computer's pointers.

Artists and graphic designers often use specialized stylus-equipped touchpads that allow them to create brush strokes, calligraphy, and many other artistic options.

Game Controllers

Game controllers provide an interface that optimizes a user's ability to interact with a specific gaming platform.

The PS4 and Xbox One appeal to similar markets and therefore have similar controllers.

Nintendo's Switch has a hybrid console that can be used as both a home console and a portable device. Its Joy-Con controllers can be used in several ways: attached to the main unit for handheld mode, joined together as a single controller, or separated into individual controllers for two players. When separated, they can also be used as independent controllers in each hand.

Game Controller Options

There are nearly as many game controllers as there are electronic games. **Game controllers** attempt to optimize the user experience. Examples include

- steering wheels and pedals for racing and flying games;
- musical instruments;
- nearly every sort of weapon;
- dance pads.

Audio Input

Different devices provide the ability to input sound or audio. They include microphones and MIDI.

Microphones

Microphones change sound waves into electrical signals. Most microphones work by using a sound-sensitive diaphragm over an electrically charged plate (capacitance plate). The diaphragm's movement causes the electrons to move, creating an electrical current. In cell phones and hearing aids, this diaphragm/capacitance plate combination is housed in a microelectrical-mechanical system (MEMS) chip. In cell phones, the MEMS chip includes an analog-digital converter.

Microphone Options

There are many microphone options available on the market today. The right microphone choice depends on the input requirements and your budget.

Most microphones connect using a USB port or a 3.5-mm (earbud) port. Nearly all laptops have built-in microphones, usually near the webcam. Many gamers use headsets that include headphones and a microphone.

MIDI

The **Musical Instrument Digital Interface (MIDI)** allows you to turn your computer into a sound studio.

MIDI includes the protocols for translating music into digital signals. It includes 16-channel inputs, allowing for 16 different instruments or voice inputs. This greatly reduces the cost of quality music creation.

Biometrics

Biometrics, which means "life measure," uses human traits for identification. To be useful, everyone should have a unique trait that does not change over time.

Traits that have proven useful include fingerprints, eye retinas and irises, facial bone structure, palm prints, stride, typing patterns, and speech recognition.

Fingerprint Scanners

There are three main types of fingerprint scanners. The most common is the capacitance sensor, which works like a laptop's touchpad or a smartphone's touchscreen. It uses electron movement to sense fingerprint patterns. The second type of sensor is an optical sensor that literally takes a photo of your fingerprint. This type has the disadvantage of not working well when the sensor or fingers are dirty. The third type is an **ultrasonic sensor** that uses the reflection of sound waves to detect the patterns on your fingerprint.

Retinal and Iris Scanners

There are two main eye recognition systems. With iris recognition, your iris is photographed. The patterns of colors that make up your iris are compared to a database of irises until your match is verified.

The back of your eye (the retina) has many blood vessels. The pattern made by these blood vessels is unique to your eye. In retinal scanning, a low-power infrared beam shone in the eye is reflected. Because blood vessels absorb more light than the rest of the retina, different blood vessel patterns can be detected.

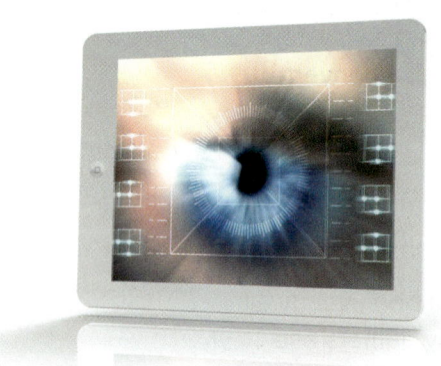 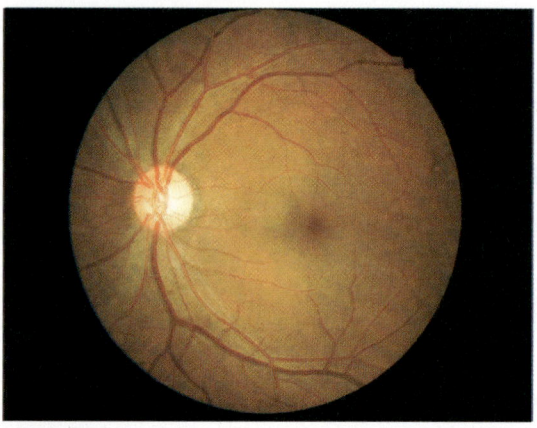

Facial Recognition

Facial recognition software attempts to identify individuals by specific facial features. As this technology develops, several methods are emerging, including 3D mapping, which maps the depth of eye sockets, forehead dome, and so on; relative feature charting, which measures the distance between eyes, mouth, nose, and so on; skin texture; or combinations of these.

Signature Recognition

There are two main signature technology methods. In the static method, signatures are written on paper and then scanned. The software then maps the signature.

The **dynamic method** uses a stylus and touchpad. This method maps the signature and measures pressure, pen angle, pen up/down, speed, and so on.

Signature Recognition Graphic

Genuine Signature   Forgery

Voice Recognition

Speech recognition technologies record, map, and analyze voice patterns. Voice frequency, pitch, speech rhythms, and other characteristics are charted and compared to existing records.

Speech recognition is increasingly used in telephone banking and for other financial transactions. It is also used by governmental agencies for security purposes.

Stride or Gait Recognition

Individuals can be identified by their walking stride using technology called **biometric gait analysis**.

Gait analysis looks at step length, step width, walking speed, and the angles formed by the trunk, hip, knee, ankle, and foot, which are measured and compared with stored identification data.

This form of identification is valuable because it can be used at night or through walls with infrared (IR) cameras or even with satellite imagery.

Image Input Devices

A wide range of items are used to create images and input them into the computer. Options include cell phone cameras, point-and-shoot digital cameras, and DSLR cameras.

Cell Phone Cameras

Smartphone cameras work by focusing incoming light onto an active pixel sensor array that captures images.

Smartphones usually come with both front- and rear-facing cameras as well as a light-emitting diode (LED) flash for nighttime photos.

Although these cameras often have many features, such as stabilization, slow motion, and panoramic views, most have the disadvantage of not being able to zoom without losing resolution. This is because it's hard to extend the camera's focal length while keeping the smartphone slim.

Point-and-Shoot Cameras

Point-and-shoot digital cameras work in the same way as smartphone cameras. Incoming light is focused onto an active pixel sensor array that captures images.

The advantage of the point-and-shoot camera is that it has more room for a larger sensor screen, allowing for higher-resolution photos. Because they can be thicker, most point-and-shoot cameras also come with optical zoom.

Digital Single-Lens Reflex Cameras

Digital single-lens reflex (DSLR or SLR) cameras combine the photographic quality of a single-lens reflex camera with the flexibility of digital technology.

In an SLR camera, mirrors allow the user to look directly through a high-quality lens at the object to be photographed. When the photo is taken, the mirror pops up to allow the light to be focused on a sensor array (this is what makes that clicking sound).

Many DSLRs have full (35 mm) sensor arrays, enabling photographers to take extremely high-resolution photos. DSLRs also allow users to swap out lenses for different settings or light conditions.

How Cameras Work

Although it's not entirely necessary to use a camera, understanding how a camera works may help you take better pictures.

APS CMOS

Special computer chips called active pixel sensor complementary metal-oxide semiconductor (APS CMOS) chips make inexpensive digital cameras possible. With APS technology, light is continually sensed as it hits the sensor array. This allows for high-speed photography and video, all with the same chip. For this reason, most digital cameras and all smartphone cameras and webcams use APS CMOS chips.

Resolution

Many factors come into play when discussing digital camera photo quality. One of the most important factors is resolution, or how many pixels are used to create the photo image. Broadly speaking, the more pixels used, the higher the photo quality. For enlargement or cropping purposes, higher resolution has significant advantages, but for online sharing and storing purposes, lower resolution may be preferable.

Zooming

In photography, **zooming** means to make a distant object appear closer in an image. This can be done two different ways: with optical zooming or with digital zooming. In optical zooming, the distance between the camera's lens and the camera's sensor, known as the focal length, is increased. This means a smaller or more distant image is photographed with the same number of pixels.

In digital zooming, a smaller or more distant portion of the photo is enlarged. This reduces the photo's resolution.

Choosing a Camera

Choosing a digital camera is a difficult process because there are hundreds of different options. First, decide which type of camera is right for you.

Do you want one that easily fits in a pocket, or are you interested in high-quality photography? Ideally, you should try your various options before purchasing. For ease of use, the point-and-shoot cannot be beaten.

If you like more zoom, but still want the ease of a point-and-shoot, a superzoom might be a good option. For high-quality photos, a DSLR or a mirrorless camera with a large sensor will provide the best shots. Remember not to overspend because tomorrow's electronics will be better and less expensive than today's!

Webcams

A webcam is a camera that takes photos or streams video into a computer or over the internet. Nearly all laptops and phones come equipped with webcams.

Webcams allow users to video-teleconference and use apps such as Skype.

They have gained a certain notoriety because they can be hacked, and users can unknowingly be monitored and photographed.

Webcam Options

There is a huge variety of webcams to choose from. Usually, webcams are purchased by desktop users because most laptops come with webcams incorporated in their monitors. The choice and price of the webcams are determined by image quality, microphones, and style. Remember not to overspend on webcams with high image quality if used primarily for video teleconferencing; most likely you won't be using enough bandwidth for a high-definition image, anyway.

Video Teleconferencing

Video teleconferencing (VTC) uses cameras, microphones, displays, speakers, and software to allow users to both see and hear each other. Many families, professionals, and even gamers use VTC to allow more detailed and intimate conversations than simple voice conferencing. Software such as Microsoft's Skype, Google Meet, Zoom, and Apple FaceTime provide essentially free VTC capability to anyone with a smartphone, tablet, or laptop and a broadband connection.

Extreme Action Video Webcams

Extreme action video cameras, such as GoPro's Hero 4, the iON Air Pro, and the SONY Action Cams, are relatively rugged, water-resistant video cameras that allow users to film high-definition video under harsh conditions, such as on ski mountains, on or in the ocean, or at high altitudes. These cameras are usually mounted on helmets, and most incorporate video stabilization, background noise dampening microphones, and even Wi-Fi capability. Millions of YouTube videos have been filmed with these types of cameras.

Scanners

Image scanners turn hard copy, such as a paper document, into a digital format that can be viewed or manipulated on a computer. These include flatbed scanners, usually incorporated with printers, and handheld or wand scanners. Other types of scanners include barcode scanners, OMR scanners, and MICR readers.

Barcode Scanners

There are several different barcode readers, but the type most used for checkout in stores measures the reflection of a laser beam. The black bars of the Universal Product Code (UPC) absorb more of the laser's light than the surrounding area, and this difference is detected by receptors, called photodiodes, in the scanner.

This code is entered into a database that identifies the item's name, cost, and any other information. Another common barcode reader uses camera-type charged coupled device (CCD) technology that can detect colors.

Optical Mark Recognition

Optical mark recognition (OMR) scanning is most found on standardized tests. The test form has oval bubbles that a student fills with a No. 2 pencil.

The scanner uses a laser that is reflected by the mark made by the No. 2 pencil, and this reflection is read by the scanner.

Because OMR technology makes test grading faster and more accurate, most standardized tests now have multiple choice or true/false questions.

Optical Character Recognition

Optical character recognition (OCR) is like document scanning technology on steroids. OCR allows any text, even handwritten scrawls, to be digitized.

OCR software incorporates algorithms that estimate the likelihood of the word that is being written based on character shape, prior entries, and frequency of usage. The more often users enter data into OCR software, the more likely the software will correctly recognize the intended entry.

Magnetic Ink Character Recognition

Magnetic ink character recognition (MICR) is used on checks so that they can be processed quickly by cashiers and banks.

The characters at the bottom of the check are written with ink containing iron oxide that can be detected by MICR readers. This information typically includes the bank's routing number, checking account identification, and the check number itself.

Radio Frequency Identification

Radio frequency identification (RFID) technology uses radio waves to transmit data to a scanner. There are two categories of RFID tags. Passive tags absorb energy while being scanned and then transmit data back to the scanner. Active tags have batteries and transmit data continuously.

RFID tags, particularly passive tags, can be extremely small and inexpensive. They have a variety of uses, including inventory, shipment tracking, and even pet identification.

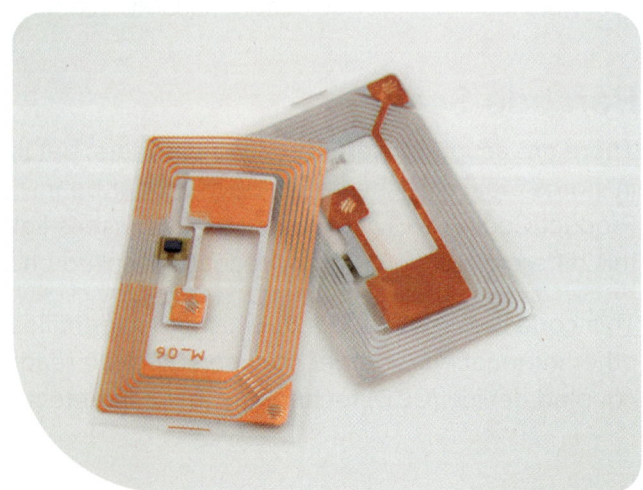

Copyright © McGraw Hill Albert Lozano/Shutterstock

Near Field Communications

Near field communication (NFC) is a wireless technology with a short transmission range that is included in new smartphones and other devices, including tablets and home appliances.

NFC is used for smartcard identification, digital wallet payments, transportation tickets, and other instances where short-range data transfer is needed. It can be used to transfer data between devices as well as initiate data transfer between networks.

Bluetooth

Bluetooth provides short-distance wireless communication between electronic devices. **Bluetooth** protocols allow devices to pair (link) easily. It uses extremely low-power radio signals (2.45 GHz frequency range), so it rarely interferes with other devices. However, certain baby monitors, garage door openers, and older microwave ovens may interfere with **Bluetooth** signals. Because of its low power, **Bluetooth** signals have a maximum range of about 30 feet.

Bluetooth Devices

In its infancy, **Bluetooth** was primarily used for making wireless connections for keyboards, mice, and printers. With each passing month, it seems like more and more devices use **Bluetooth** technology, primarily to connect to smartphones.

Speakers and headphones now often use **Bluetooth** technology to connect to smartphones and MP3 players. Many athletes and athletic equipment now connect to phones with apps specifically designed for different activities. A lot of cars have **Bluetooth** connections, as do many digital cameras.

Importantly, several personal medical devices, such as blood sugar monitors and heart-rate monitors, use **Bluetooth** technology.

Adaptive Input Devices

Adaptive input technology, also known as assistive input technology, provides alternative methods of inputting data into a computer. The most common provide ways to enter text aside from using a traditional keyboard.

Speech recognition and predictive text software has become increasingly sophisticated. Devices include oversized keyboards and mice.

Many software and hardware innovations that were originally intended to be assistive have become mainstream software and hardware and are used by nearly everyone.

The most common adaptive (assistive) input technologies aid in entering text, examples include oversized keyboards and speech recognition software.

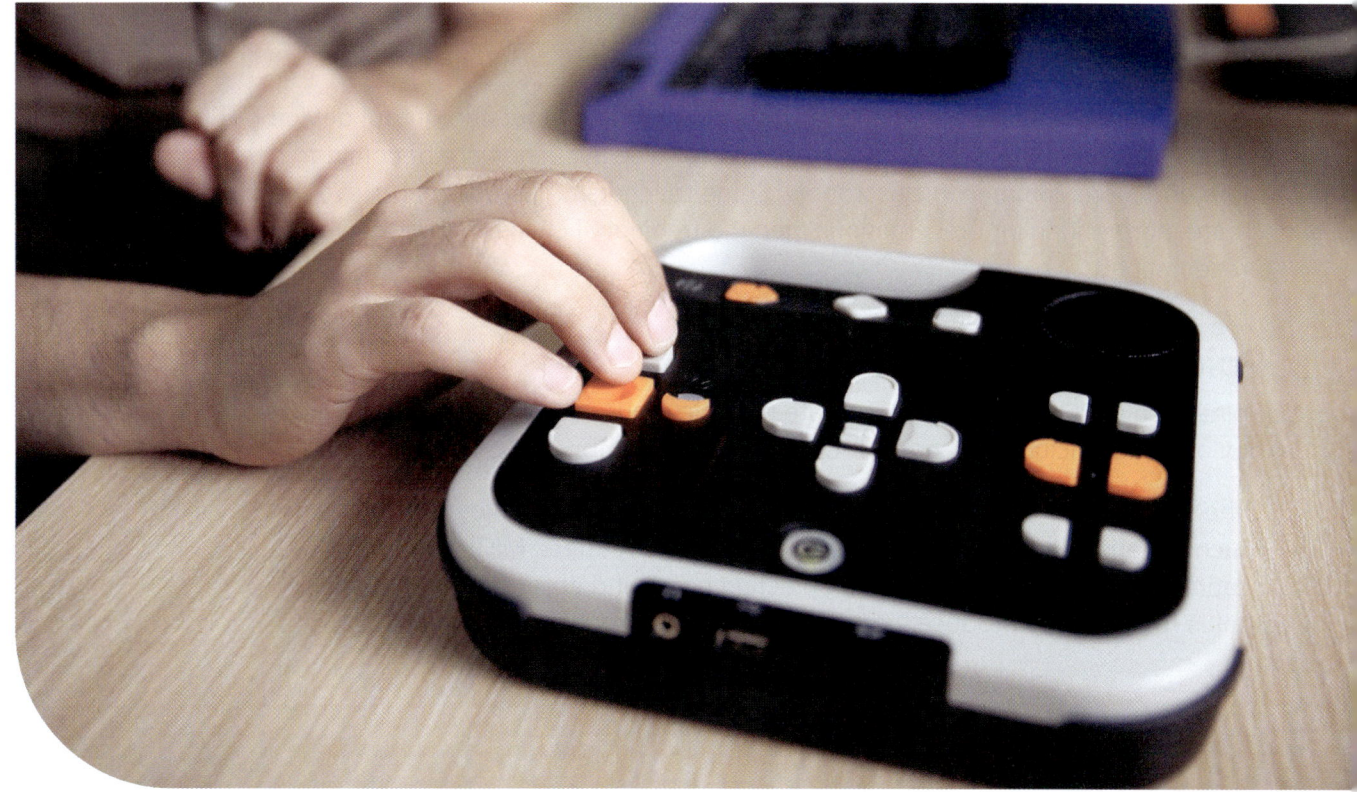

Adaptive Input Examples

Adaptive input devices include common-word keyboards that enable people with amyotrophic lateral sclerosis (ALS) and others with limited mobility to enter text into a computer.

Technologies include electronic pointing devices that are used by eye movements, nerve signals, and even brain waves. Other input devices are used by sipping or puffing air.

Joysticks can be manipulated by feet, and wands and sticks can be worn on the head and manipulated by the mouth or chin.

Some input devices are relatively low-tech, such as a mini touchpad on the roof of the mouth activated by a metal button placed on the tongue.

Combination Input Devices

Smartphone Input Devices

Even a device as common as a smartphone includes many input devices. Along with a microphone and touchpad, many smartphones have motion sensors, buttons, and switches on the sides, front and back cameras, and micro-USB and other ports for entering input or for attaching peripheral input devices.

Smartboards

A smartboard or interactive whiteboard is a combination input and output device that allows users to interact with large computer displays. They are becoming more and more common in classrooms. Some teachers and students believe they create more engaging presentations than the traditional chalkboard or PowerPoint presentation.

Point-of-Sale Devices

Point-of-sale (POS) devices incorporate multiple input devices. Some of the most obvious ones include barcode (UPC) readers and number keypad entries.

Others include UPC scanning wands, credit card readers with keypads or touchscreens, a cash and change reader, QR code readers, MICR readers for scanning checks, touchscreens for selecting items and entering item amounts, scales for weighing produce, and even cameras for recording transactions.

Chapter Review

1. What is meant by the term *input device*? List five input devices that are integrated in a smartphone.

2. Why are the most common keyboards called QWERTY keyboards? Discuss the history of the **QWERTY keyboard**.

3. What do the following keyboard shortcuts accomplish in most Windows-based programs?

   a. Ctrl+V

   b. Ctrl+X

   c. Ctrl+S

   d. Ctrl+P

   e. Ctrl+A

4. What do the following keyboard shortcuts accomplish in most Mac-based programs?

   a. Command+V

   b. Command+X

   c. Command+S

   d. Command+C

   e. Command+A

5. What is meant by the term *biometrics*? Identify three different biometric technologies currently used.

6. Give three examples of image input devices.

7. List at least four input devices that are found at a brick-and-mortar retail store point-of-sale (POS) or checkout station.

8. What is meant by NFC when referring to making payments? Does NFC make transaction processing safer?

9. What is **Bluetooth** technology? List three examples of **Bluetooth** devices.

10. Describe an example of when an adaptive input technology device came into use by nearly all users.

Ethics Discussion

Biometrics uses individual physical or behavioral characteristics to identify and verify a living person. This makes tracking, locating, and identifying people possible. Conduct research on the internet about biometric technology. How are biometrics used by the US government? How are biometrics used by law enforcement? What is your opinion of using biometrics to track and identify people? Identify three ethical considerations surrounding the use of biometrics.

4 Computer Output

What To Expect

After completing this chapter, you will be able to:

- describe output and define output devices and peripherals, as well as the ways to connect to them;
- define computer monitors, display parameters, and types of televisions;
- describe printers and the types of printing;
- discuss audio output, sampling rate, and speaker options;
- describe projectors and discuss output devices;
- discuss ergonomics in output devices.

Chapter Topics

- Output Devices
- Televisions
- Projectors
- Printers
- Audio Output
- Visual Output

Copyright © McGraw-Hill/Scanrail/Shutterstock

Let's Explore Output

When you use a computer, you input information and are asking it to produce something for you – the "thing" it produces is the output. If you input "2 + 3" into a calculator, the answer "5" will display on the screen. That answer, 5, is the output. It's not just answers, any information that comes out of the computer is output. In the early days of computers, the way they "talked" to us was really simple. It was often just a bunch of blinking lights or beeping sounds to let us know they finished a task or to give us a coded message. This was way different from the many ways we interact with computers today, as we live in a world where we rely on digital interactions more and more.

At first, getting something back from a computer meant getting printouts. This was a big step—it let users have a physical piece of the information the computer processed. The real game-changer was when computer monitors came into the picture. This allowed us to see and interact with computers directly. And wow, have monitors changed! They got better and better at showing clear, detailed, and colorful images. Pixels, the tiny dots that make up the image on the screen, began making pictures clearer and showing finer details. We also got to experience images in full High Definition (HD), ultra High Definition (UHD), and even full ultra High Definition (FUHD), making our viewing experiences seem more real. As computers got better at showing us things, the ways they could connect to other devices also improved. High-Definition Multimedia Interface (HDMI) and Universal **Serial Bus** (USB) ports are everywhere now, allowing lots of different technologies to connect and share information easily.

Computer output isn't just about what we see; it's also about what we hear. Computers became multimedia devices, able to play sounds and music that are more and more like real life. The term 'sampling rate' is important here—it's about how often audio is captured and recreated, with higher rates giving us sound that is closer to real life.

And let's not forget about printers! **Printers** have become extremely advanced as well! We now have inkjet, laser, and thermal printers, each using different ways to put images on paper. And there's more! We also have **3D printing**, letting us create real threedimensional objects, opening up all sorts of new possibilities.

Understanding how computer output has changed is really important. From simple beeps and blinks to 3D prints and high-definition displays, we've come a long way. This journey shows how clever humans are, always finding new ways to see clearer, hear better, and experience more in the digital world.

Output Devices

Output devices are like the TV screens in a video game arcade. They show you the results! When you play a game or watch a video on a computer, you're seeing those things on an output device such as a monitor or a screen. Those devices help you see and hear what your computer has produced from the data you gave it.

Peripherals

All the devices you plug into a computer are called *peripherals*. Some peripherals are used to put information into a computer, like a keyboard or a mouse. Others, like speakers or a printer, get information from the computer. You connect peripherals with cords or even wirelessly, in the same way some of your toys might connect!

Ports and Cables

Ports are the different kinds of plugs or sockets you see on an electronic device. Computers have various ports for different types of devices, and each port goes with a different type of cable. For instance, you might plug your headphone cable into one kind of port, while your phone charger goes into a different type.

Over time, companies realized that it's easier for everyone if these ports are somewhat similar or "standardized." This means that instead of every brand having a completely unique port, manufacturers try to use common ports that many devices can share. This is why you might notice the same USB port on different brands of computers or phones. But while there's been a push to standardize, there are still many different types of ports, because technology keeps evolving and improving.

HDMI Port

HDMI ports allow for the transfer of high-quality video and audio signals. **HDMI ports** can be found on nearly all new televisions, computer monitors, DVRs, Blu-ray and DVD players, and gaming devices.

HDMI Cable

HDMI cords or HDMI cables provide a way to transfer high-quality video and audio signals to and from an output device. These cords are divided into two major categories: standard HDMI for screen resolutions of up to 720 horizontal pixels and high-speed HDMI for 1080 horizontal pixels and more.

USB Ports and Cables

In the 1990s there were tons of different connectors for various computer devices—and things were starting to get messy! A group of companies, including large ones like Microsoft and Intel, worked together to make a universal port. They made the USB port, and it's become an industry standard. USB stands for Universal **Serial Bus**. Some people probably think that sounds like a vehicle that travels around the world and delivers breakfast to children. Right? No? Okay. Let's discuss what that means.

- **Universal:** "Universal" isn't meant to be taken literally. It doesn't work with EVERYTHING. But the term speaks to the goal of making a cable that has the potential to work with any kind of computer.
- **Serial:** This refers to how the data is sent. **Serial** means the data is sent in sequence—one bit at a time.
- **Bus:** In computer lingo, a "**bus**" is a system that transfers data between digital devices.

USB ports have been an industry standard connection for transferring data and power to and from computers. USB connections have replaced older serial bus ports and parallel ports. USBs are used on nearly all computers, smartphones, and game consoles. There are several varieties of USB ports including USB 2.0 for transfer speeds up to 480 megabits per second, USB 3.0 for transfer speeds up to 5 gigabits per second, microUSB ports for Android smartphones, and USB to lightning port connectors for iPhones.

USB-C Port

USB-C is a small, oval-shaped connector that's reversible, meaning you can plug it in either way (there's no wrong side up). It's become popular for its compact size and the ability to handle multiple tasks like charging, transferring data, and even sending video to monitors.

Computer Displays - Monitors

A computer **monitor** is an electronic, fixed-screen device that visually displays computer output. Examples of monitors include the touchscreen monitors found on tablet computers, smartphones, and some laptop computers. Many users prefer liquid crystal display (LCD) monitors, whereas some like the crispness of a light-emitting diode (LED) display.

Pixels

A **pixel**, or picture element, is the smallest controllable element of an image. Most computer monitors use pixels made up of three colors: red, green, and blue. Some monitor manufacturers have made pixels with four colors, adding yellow.

Resolution

The **resolution** of an output device, such as a monitor, is simply the number of pixels in the display. The higher the resolution, the higher the image quality.

Aspect Ratio

Aspect ratio compares the height of a screen or monitor by its width. **Aspect ratio** is written as two numbers separated by a colon and is spoken as the ratio of the height to the width. For example, a traditional TV or computer monitor's aspect ratio is "4:3", whereas a high-definition TV (HDTV) has an aspect ratio of "16:9". **Resolution** is displayed in width by height. For example, a standard 1080p monitor is 1,920 pixels wide and 1,080 pixels in height. This equals a ratio of "16:9" and 2,073,600 total pixels.

Screen Size

Monitors and televisions seem to get bigger and bigger every year. **Screen size** is measured from the diagonal corners of the screen display. **Screen size** is among the most important factors in television selection.

Contrast Ratio

Contrast ratio refers to the difference in the brightness of an image created on a display. It compares the brightness of the white pixels to the darkness of the black pixels. The higher the contrast ratio, the more accurate an image, or the better the image quality.

Copyright © McGraw Hill   nito/Shutterstock

Color depth

Color depth, also known as bit depth or color bit depth, indicates the accuracy of the color of each pixel. **Color depth** describes the number of bits supporting each pixel. Devices noted for high color accuracy typically have a color depth of at least 24 bits.

Dot Pitch

Dot pitch is a description of the quality of an image produced by a monitor or television screen. **Dot pitch** is the distance between the center of two pixels, or the distance between two subpixels of the same color.

Frame Rate

Frame rate refers to how frequently a frame in a video is advanced. **Frame rate** is measured in frames per second, or FPS. The standard FPS for movies, known as the Cinema Standard frame rate, is 24 FPS, whereas the standard for television, known as the National Television System Committee Standard frame rate, is 30 frames per second. Some movies, such as Peter Jackson's *The Hobbit*, have been filmed at 48 FPS, but this hasn't become an industry standard.

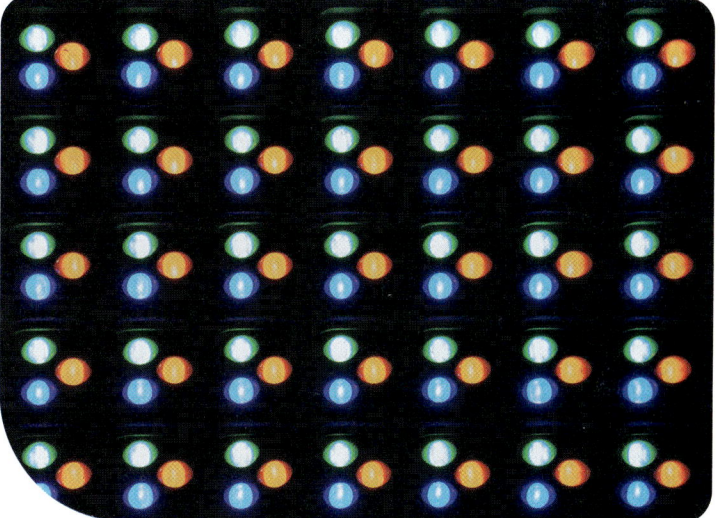

Refresh Rate

The **refresh rate** is the speed at which each pixel is updated. In general, the higher the refresh rate, the higher the video quality. **Refresh rate** is measured in cycles (refreshes) per second, or hertz (Hz). Videos replay best when the refresh rate is an even multiple of the frame rate. Because the two frame rate standards are 25 (for cinema) and 30 (for television), most modern televisions and monitors have refresh rates that are multiples of 120 Hz.

Liquid Crystal Displays

LCD stands for **Liquid Crystal Display**. LCDs are the most common type of display for computer monitors and televisions. An LCD is made of several layers of filters, plus a screen of transparent electrodes and a layer of liquid crystal molecules. When powered, the electrode "twists" or "bends" the liquid crystal molecule, allowing light to either pass through or be blocked from a pixel filter. This determines the color and intensity of each pixel.

LED Displays

LED stands for **Light Emitting Diode**. LED displays are actually LCDs with a screen of white LEDs to backlight the display. LEDs convert electricity into light. The greater the amount of electricity passing through the diode, the brighter the light. Because LEDs are microscopic, LED displays can be thin enough to hang on walls like pictures.

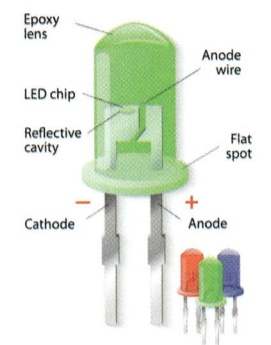

LIGHT-EMITTING DIODE

Epoxy lens
Anode wire
LED chip
Reflective cavity
Flat spot
Cathode
Anode

OLED Displays

Organic LED displays, or OLEDs, use red, green, and blue LEDs to create each pixel. These displays currently cost more than LCD displays, but mass production will bring down costs. Because LEDs are microscopic, OLED displays can be curved and paper-thin.

High-Definition Television Monitors

High-definition televisions or computer monitors have 1,080 horizontal lines of pixels. This is also known as full HD or FHD. An HDTV has an aspect ratio of 16:9, giving it a resolution of 1,920 × 1,080 pixels, also known as 2.1 megapixels.

Ultra-High-Definition Television

Ultra-high-definition television, known as UltraHD, UHD, 4K UHD, or 8K UHD, provides the latest in high-resolution displays. UHD TVs have an aspect ratio of 16:9 (the same as HD televisions) but have at least four times more pixels. This higher resolution provides lifelike displays, even when sitting extremely close to the screen. The market trend in televisions has led to larger screens with greater resolution. There is no reason to think this trend will end soon.

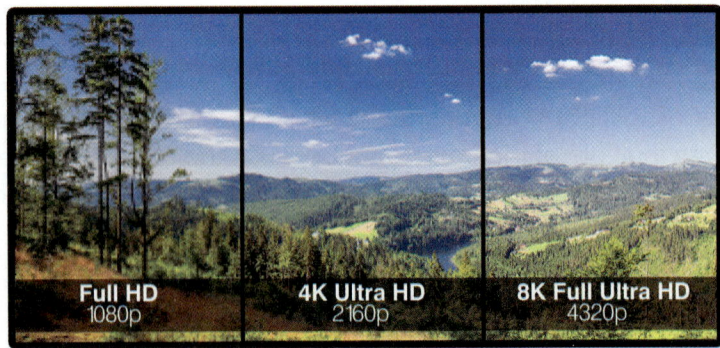

Full HD
1080p

4K Ultra HD
2160p

8K Full Ultra HD
4320p

Projectors

Projectors use light to display output on a surface. They allow large audiences to view the video output of a digital device. Because light is being projected, projectors require a darkened environment. **Projectors** are popular in classrooms, theaters, and outdoor venues, where lighting can be controlled and where video and images need to be viewed by larger audiences. Large-screen TVs have decreased home projector use.

Printers

Printers recreate a computer image on paper or another substance. Printer resolution is described as dots per inch (dpi). Printer speed is described as pages per minute. The most common printers (inkjet printers) use ink. Other common printers include laser printers that use toner and thermal printers, which use heat-sensitive paper.

Inkjet Printers

Inkjet printers are the world's most popular type of printer. They print by spraying droplets of ink onto paper. Manufacturers often sell printers below production costs and then mark up the price of the ink.

Thermal Printers

Thermal printing is most commonly used for creating receipts. Thermal paper turns black when heated, making it much less expensive and cleaner than using ink or toner. Some thermal paper incorporates different colors, usually red and black, and heats the paper at different temperatures.

Laser Printers

Laser printers are the most popular printers for use in business offices because they use relatively inexpensive toner rather than liquid ink. **Laser printers** are also faster than inkjet printers when printing multiple copies. **Laser printers** use toner, spinning cylindrical drums, and a laser to transfer images to paper. Toner is a powdery, dry, plastic ink with a positive electrical charge. The cylindrical drum is negatively charged with electrons. The laser removes the unnecessary electrons, and paper is fed between the toner and the drum. The positively charged toner is attracted to the remaining electrons on the drum. The toner is then melted onto the paper to create the image.

Plotters

Plotters are used to print large graphic images such as designs of buildings, machinery, and ships. **Plotters** recreate images by mechanically moving a pen over a large sheet of paper.

3D Printing

3D printing allows designers to create three-dimensional objects from a computer image. Cross-sections of the object are first created, and then successive layers of material, usually plastic, are sprayed through a nozzle to recreate the object. Some 3D printer nozzles can even spray chocolate—imagine the possibilities! Input data, output dessert!

Audio Output

Audio output, or sound, is an essential component of most computer systems. Smartphones, tablets, and laptops include integrated speakers. To create sound, a computer uses digital-to-analog converters. Speakers then change the analog electrical impulses to sound by physically pushing air. To improve sound quality, particularly in the lower (bass) ranges, many users add external speakers to their computers.

Sampling Rate

Sampling rate refers to the accuracy of a digital recording. **Sound** is an analog signal. **Sound** has wavelength and frequency, or volume and pitch. Because computers can only process digital data, sound must be digitized. To digitize an analog signal, each incoming sound wave is measured, or "sampled." The more samples taken, the more accurate the sound recording. Your cell phone samples an analog signal about 44,000 times per second!

Various Speaker Options

The speakers that are incorporated with smartphones and laptops are necessarily tiny, resulting in low sound quality, particularly in the lower ranges. To improve the sound output of digital devices, many users add external speakers. There are countless speaker options available. To determine which option to choose, first determine your needs. Does your speaker option need to be portable? How important is sound quality? In general, lower range tones require a physically larger speaker. Do you need speakers that are Bluetooth-capable, or will you be using wires? Once you've determined your needs, price will be a significant factor. Bluetooth-capable speakers give smartphones added capability.

Output Ergonomics

Output device ergonomics refers to the art of designing products that present information in the most useful way. Ergonomic output can help improve mental concentration and physical capability by providing only the information that is needed at the time. Ergonomic designs have been incorporated into modern cars, aircraft, ships, and many other devices. Ergonomic output devices include heads-up displays (HUDs), integrated helmets, and numerous other applications.

Chapter Review

1. Describe what is meant by the term *computer output* and provide three examples of computer output that you might experience on a typical day.

2. Define the following as they pertain to computer monitors:
   a. pixel
   b. resolution
   c. aspect ratio
   d. screen size
   e. contrast ratio
   f. color depth
   g. dot pitch
   h. frame rate
   i. refresh rate

3. Define and compare full HD resolution (FHD), ultra-HD (UHD/4K), and full ultra-HD (FUH/8K).

4. What is HDMI? What devices use HDMI inputs?

5. What is a USB port? List three devices that may use USB ports.

6. What does *LCD* stand for? How does an LCD TV create an image?

7. What does *LED* stand for? How does an LED TV create an image?

8. How does an LCD/LED display differ from an OLED display?

9. What is meant by *UHD*? How does it differ from an HD display?

10. Briefly describe how the following create an image on paper:
    a. an inkjet printer
    b. a laser printer
    c. a thermal printer

11. What is **3D printing**? Name three uses of 3D printers.

12. List three advantages of using large monitors versus using a projector.

Ethics Discussion

Conduct research on the Internet about gray market and black market goods. What is a gray market good? What is a black market good? What are some of the ethical considerations that surround gray market goods? What are some of the ethical considerations surrounding black market goods? What steps have US companies taken to reduce the amount of gray market and black market goods?

5 Computer Storage

What To Expect

After completing this chapter, you will be able to:

- differentiate between memory and storage, define storage, and discuss the parameters of storage devices;
- describe the different types of storage devices;
- discuss file compression for different types of files;
- consider the emerging storage challenges.

Chapter Topics

- Bits Bytes and Beyond
- Memory and Storage
- Storage Media
- Storage Devices
- Optical Storage
- Cloud Storage
- File Compression
- Emerging Storage Technologies

Copyright © McGraw Hill · Glow Images

Let's Explore Computer Storage

Think of your computer like a big digital backpack. Just like you store books, notes, and pencils in a backpack, your computer saves pictures, songs, and documents in its storage space. But how much can it hold? This is what we call 'storage capacity.' Often, we measure this capacity in terms like gigabytes (billions of bytes) and terabytes (trillions of bytes).

Now, not all items take up the same space. Just as a thick textbook occupies more space than a single sheet of paper in your backpack, large computer programs, like Microsoft Office, or a movie, use more of this digital space than a single photo. Different computers or devices have different sized 'backpacks' or storage capacities. Let's explore how computers manage and measure this space.

Bits Bytes and Beyond

You might wonder, why are there 8 bits in a byte? With this system, we can make 256 different combinations.

Here's the math behind that:

1 bit can be in 2 positions (either 0 or 1).

2 bits can be in 2 × 2 = 4 positions (00, 01, 10, or 11).

3 bits can be in 2 × 2 × 2 = 8 positions (like 000, 001, 010... up to 111).

...and so on. By the time we reach 8 bits, we get to 2^8:

2 × 2 × 2 × 2 × 2 × 2 × 2 × 2 = 256 different combinations.

That means, with 8 bits (which is 1 byte), we can come up with 256 unique combinations, from 00000000 to 11111111.

A system called ASCII, pronounced "askee," uses this method to encode English letters, numbers and common symbols. ASCII, stands for the American Standard Code for Information Interchange. In ASCII, the letter "A" is represented as 01000001.

Bits

A "**bit**" stands for binary digit. It's the smallest piece of digital data and can be either a 0 or a 1.

Check out this punch card from the 1970s. These cards had spots that could either have a hole or be solid. Just like bits, each spot was either a 0 (no hole) or a 1 (with a hole). When a machine called a punch card reader checked the card, it used pins. If a pin went through a hole, it read a "1". If the pin hit a solid spot, it read a "0".

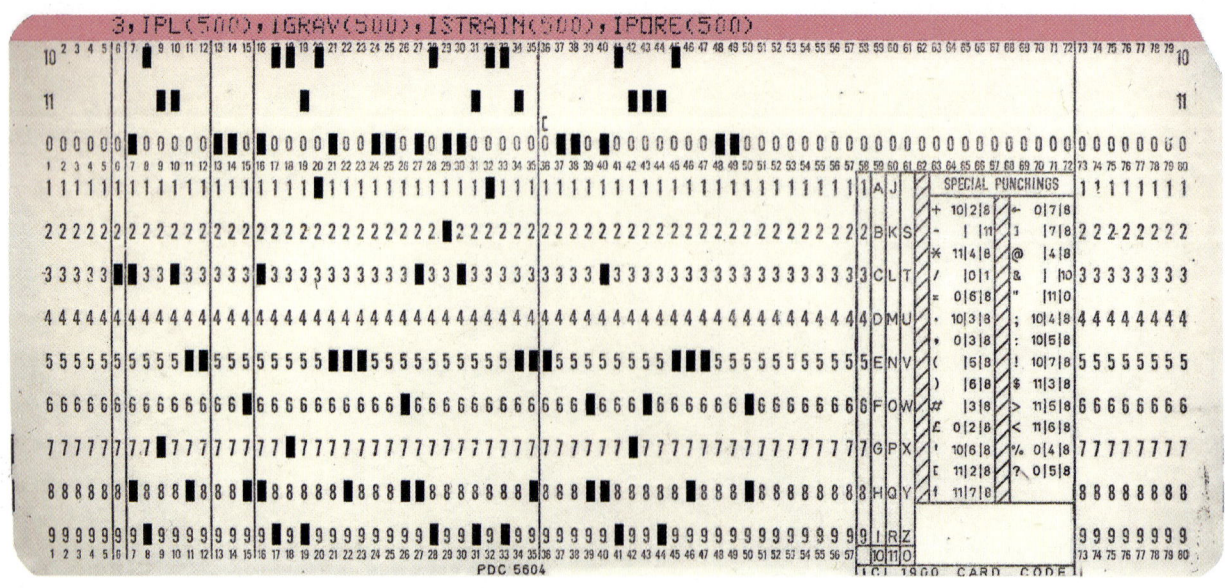

Now, imagine a door that can be open or closed. In a computer, we have something called a transistor that works in a similar way. If it's open, electrons can flow, like a "1". If it's closed, they can't, which is like a "0".

These basic open/close or on/off actions are how bits store information. And when we write about bits, we use a lowercase 'b'.

When we talk about storage sizes, it's key to know that each size is much, much bigger than the one before—it grows a lot each step, not just a little. To get a sense of it, let's think about how long it would take to download different sizes of data with high-speed internet.

Storage Unit	Short Form	What it is	Where it's Stored	Download Time
Bit	b	Tiniest piece of data, 0 or 1	In electronic switches like transistors, being ON or OFF	Almost no time (0.000008 seconds)
Byte	B	Made of 8 bits	On paper or in tiny memory chips	Almost no time (0.000008 seconds)
Kilobyte	KB	About 1,000 bytes	Small files like text or tiny pictures, on floppy disks or small memory cards	Really quick (0.008 seconds)
Megabyte	MB	About 1 million bytes	Music file or a good-quality photo, on thumb drives or CDs	8 seconds
Gigabyte	GB	About 1 billion bytes	A movie or computer program, on hard drives, DVDs, or big USB drives	About 2.25 hours
Terabyte	TB	About 1 trillion bytes	Huge amounts of data, like big movie collections, on modern hard drives or cloud servers	92 days
Petabyte	PB	About 1 quadrillion bytes	Enormous amounts of data, like what a big company might have, stored on lots of big servers or data centers	253 years

Bytes

A group of 8 bits is called a **byte**. The word "bite" was intentionally spelled as "byte" to avoid confusion with "bit". When we mention bytes, we use a capital 'B'. So, always remember: small 'b' for bits, big 'B' for bytes! A byte is the smallest unit of memory that most computer systems recognize.

Kilobyte

A **kilobyte** is a unit of storage that consists of approximately 1,000 bytes. It is represented by KB.

Megabytes

A **megabyte** (abbreviated MB) is equal to 1,000 kilobytes and comes before the gigabyte unit of measurement for storage. The prefix mega is synonymous with million. A megabyte is technically 1,000,000 bytes. Megabytes are used to denote the size of files stored on a digital device. For example, a JPEG image file might range in size from one to five megabytes.

Gigabyte

A **gigabyte** is a unit of digital information that consists of one billion bytes. It is represented by GB. The gigabyte is the standard unit of measurement for memory and storage in computers. Solid-sate drives (SSD) and traditional hard drives (HDD) storage capacities are often measured in gigabytes. A gigabyte is enough data to store approximately 250 MP3 music files.

Terabyte

A **terabyte** is a unit of digital data that consists of approximately one trillion bytes. It is the standard unit of measure for many of today's hard drives. It is represented by TB. A terabyte of data is enough to store approximately 230 standard 90-minute movies.

Petabyte

A **petabyte** is a unit of digital information that consists of 1,000 trillion bytes. It is represented by PB. A petabyte is a measure for significant data storage such as server banks. A petabyte of data is approximately enough to store 40,000 standard 90-minute movies.

Memory and Storage

Storage, sometimes referred to as secondary storage, allows computers to retain data when the device is turned off. Storage allows computers to keep programs, photos, videos, and other data files for future use. Computers use a variety of storage devices to store data and information. Traditional hard drives are still popular but are being replaced by solid-state devices (SSDs).

Storage Media

Storage Media for Data

Storage media refers to the technology used to retain data. It comes in various forms, each with its unique characteristics. Here are some common types:

- **Magnetic tape: Magnetic tape** is a type of storage media that uses a magnetic coating to record and store data. You can find it on the back of credit or debit cards, or on the platters of hard disks inside a computer.
    - Example: The magnetic stripe on the back contains important information about your account.
- **Optical storage: Optical storage** involves the use of lasers to read and write data on special discs. Some common optical storage formats include compact discs (CDs), digital versatile discs (DVDs), and Blu-ray discs.
    - Example: When you watch a movie on a **DVD** player, the **DVD** disc itself is the storage media that holds the movie's data.
- **Electronic storage: Electronic storage** relies on electronic components to store data. It includes devices like MicroSD (secure digital) cards, universal serial bus (USB) drives (often called thumb drives or flash drives), and solidstate drives (SSDs).
    - Example: Your smartphone may use a MicroSD card to store photos and videos, while a USB drive can hold documents and files you can transfer between computers.

These different types of storage media play essential roles in preserving and accessing data in various electronic devices, from your computer's hard drive to the tiny memory card in your phone.

Hard Drive Access Time

Access time refers to the time it takes data to reach a computer's processor. Computer access time is commonly measured in nanoseconds or milliseconds. The lower the access time, the faster data and information can be retrieved for processing. With a traditional magnetic hard drive, access time is considered the time it takes for the read/write head to locate the correct sector on the disc. Solidstate hard drives and flash drives provide data in significantly shorter access time because there are no spinning discs involved.

Storage Devices

A **storage device** is any piece of computer hardware that is used for storing data and information. Storage devices can hold and store information either inside or outside of a digital device. Examples of storage devices include external hard drives, optical drives, solid-state drives, and flash or thumb drives.

Hard Drives

A hard drive is the main **storage device** in many computers. The two main types of hard drives are hard disk drives (HDDs) and solid state drives (SSDs).

Hard Disk Drive

HDDs have been around for decades and rely on spinning disks, called platters, to read and write data. These platters are coated with a magnetic material and spin rapidly, while a read/write head moves over them to access the data.

The read/write head either transmits data to the computer's processor or changes (rewrites) the data. Because the disk must spin mechanically so that the read/write head can read the disk, hard disk drives have longer access times and require more electricity than solid-state hard drives.

- Advantages: HDDs are generally less expensive per gigabyte than SSDs, making them a cost-effective choice for large storage capacities.
- Drawbacks: They are slower than SSDs, can be noisier because of their moving parts, and are more vulnerable to physical shocks.

Solid-State Drive

Instead of spinning disks, SSDs use flash memory to store data. Think of them like super-sized and faster versions of USB thumb drives. With no moving parts, SSDs access and save data using electronic circuits.

- Advantages: SSDs are faster, quieter, and more durable in the face of physical shocks or drops. They also tend to have quicker boot times and can launch applications more rapidly.
- Drawbacks: Currently, they are more expensive per gigabyte compared to HDDs.

Redundant Array of Independent Disks

Redundant array of independent disks (RAID) allows you to store data redundantly (in multiple, various places) in a structured way to improve overall system performance. RAID disk drives are used frequently on servers but are usually not necessary for laptop or desktop computers. In a RAID, multiple layers of hard disk drives provide redundancy of data storage. Data are stored on multiple disks, and if one disk crashes, the data are recovered from a backup disk.

Optical Storage

An **optical drive** is a **storage device** that writes and reads data using lasers. Optical drives work by rotating a disc with a reflective coating. The coating is then read with a laser beam within the optical drive's read/write head. The speed of an optical drive is measured in revolutions per minute (RPM). The type of disc that is inserted into the optical drive (**DVD**, Blu-ray, or CD) is referred to as the optical media.

Common Optical Storage Terminology

Optical storage refers to storage that uses a laser to read or write to a disc (usually plastic) with a reflective coating. Burning a disc means to write data onto the disc using a laser. Ripping a disc means to copy an existing disc.

The three major categories of optical discs are compact discs (CDs), digital versatile discs (DVDs), and Blu-ray discs (BDs). The discs are sold in several varieties: read only memory (ROM), recordable (R), rewritable (RW), and recordable erasable (RE). Read-only memory (ROM) discs have the data stamped on the plastic and cannot be erased or rewritten. Recordable (R) discs can be burned once, at which point they become ROM discs. Rewritable (RW) discs can be burned multiple times. Recordable erasable discs refer to rewritable Blu-ray discs.

Compact Discs

Compact discs (CDs) store data on a plastic disc with a reflective coating. This coating has pits, which absorb laser beam light, and lands, which reflect laser beam light. The laser beam's reflection is detected by the CD player. Co-developed by Philips and Sony and released in 1982, CDs were originally developed only for sound recordings, but the technology was later adapted to store many types of data. CDs have a typical storage capacity of 700 MB. More than 500 billion music CDs have been sold worldwide!

DVDs

A **DVD** (digital versatile disc) stores information with a reflective coating on a plastic disc. The reflective coating has pits, which absorb laser beam light, and lands, which reflect laser beam light. The **DVD** player detects and reads the reflected laser light. A DVD's pits and lands are far more tightly packed than those of a CD.

The **DVD** was developed in 1995. The storage medium can store many types of data and is used for software, computer files, and video recordings watched by using **DVD** players. DVDs have a higher storage capacity than CDs but have the same physical dimensions.

How a Blu-ray Disc Stores Data

A Blu-ray player uses a shorter wavelength "blue" laser so that it can store more data. The **Blu-ray disc** stores information with a reflective coating on a plastic disc. The reflective coating has pits, which absorb laser beam light, and lands, which reflect laser beam light. The Blu-ray player detects and reads the reflected laser light. A Blu-ray's pits and lands are far more tightly packed than those of a **DVD**. For this reason, a Blu-ray drive can read a **DVD**, but a **DVD** drive might not be able to read a **Blu-ray disc**. Blu-ray was developed by the Blu-ray Disc Association and was released in 2006.

USB

USB drives (also called thumb drives or flash drives), like SSDs, are all-electronic storage devices. These drives use integrated circuits to store data. Unlike traditional transistors, flash drives use transistors that remain in the correct position when powered off. Flash based—same technology that is used in solid-state drives (SSDs) and microSD cards. If you find a thumb drive, don't use it. It may have been planted to install spyware! Besides, USB drives are very inexpensive.

Cloud Storage

Cloud storage refers to saving files to servers located on the internet. **Cloud storage** is a type of computer data storage where data are stored in logical pools. Physical storage of data spans multiple servers (and sometimes in multiple locations), and a data hosting company typically owns and manages the servers. **Cloud storage** providers are responsible for keeping the data available, accessible, and secure. There are many ways to use cloud storage. The most common way is for users to simply attach files to email and then send it to themselves. Another way to use cloud storage, more popular with businesses, is to purchase secure server space from an online supplier. Amazon is currently the largest supplier of cloud storage.

How Cloud Storage Works

Servers act like giant hard drives. Users upload files to the server. Files are backed up (stored on multiple servers). Users can download files to any device on the internet.

Advantages of Cloud Storage

- Automatic saves- cloud storage is that servers nearly always automatically make copies of data files and store these copies on other servers. This way, if a server crashes, the data are immediately recovered without any effort on the part of the user. Typically, the user will not even be aware that there was ever an issue.

- No physical copies- data does not need to be physically carried. Unlike a thumb drive or a **DVD**, a document stored on the cloud can't be left at home when it's needed at school.

Disadvantages of Cloud Storage

- Must be online—Unlike files stored on DVDs or thumb drives, files on the cloud are only available when you are connected to the internet.
- Files can be accessed by others—Like storing your valuables in a bank's safe deposit box, there is always a chance your files could be accessed inappropriately when they are on someone's server.
- Lack of control—Users lack some control over the security, processes, and procedures of the cloud storage site.

How to Choose the Amount of Storage Your Computer Needs

A computer's storage capacity refers to the size of the computer's hard drive. Most laptop computers have between 500 gigabytes to more than 1 terabyte hard drives, while most tablet computers have hard drive capacities of between 16 gigabytes to 256 gigabytes. To keep this in perspective, a standard definition movie takes about 1 to 2 gigabytes of storage, while a high- definition movie takes about 3 to 5 gigabytes. Do not overspend on your computer's storage capacity because you can always store files on the cloud or on flash drives.

File Compression

File compression reduces the size of data files. This saves storage space as well as transmission capacity and can speed up data transmission.

Image File Compression

There are two types of file compression: lossy and lossless. **Lossy compression** removes fewer valuable data, whereas **lossless compression** eliminates redundancy but retains all the original data.

Because a digital photo can be a large file, many users compress images before storing or sending them. Popular image compression formats include .jpg, .png, and .gif. The most popular lossy image compression format is .jpg. The most popular lossless image compression format is .png. Some still use .gif, though it has largely been replaced by .png. To make this even more confusing, GIFs are short video clips based on the same algorithms behind the .gif format.

Audio File Compression

Audio file compression allows extremely small devices to store large numbers of music files. The most popular compression format is a lossy compression format called M4A, a variation of the MP4 compression format. The M4A format has largely replaced the MP3 compression format. The most popular lossless compression format is called WAV.

Video File Compression

Video file compression allows you to store many movies on tablets and other digital devices. The most popular video compression format is MPEG-2, a lossy format that is used on commercial DVDs. MPEG-4 is a popular lossy format used by many individuals and can compress video files by factors of 20 to 200. Microsoft offers a popular video compression format known as AVI, and Apple has its QuickTime format called MOV

Emerging Storage Technologies

Emerging Storage Challenges

Emerging storage technologies are necessary to meet the challenges created by the vastly increasing demand for storage. For example, 700 hours of YouTube videos are uploaded every minute. That is the equivalent of all the video ever aired by the three major television networks being uploaded every three weeks! It takes nearly 3GB of space to store a single human genome. Doctors use genomic information to diagnose and treat patients, but will there be enough storage space to map the genomes of individual patients in the future?

Helium Hard Drives

A **helium hard drive** uses helium instead of air inside a hard drive—helium is less dense than air, so platters spin more smoothly. Platters can spin faster and stay cooler. Helium hard drives provide 20 to 30 percent more storage in the same space along with faster access time than air-filled hard drives.

Holographic Storage

Holographic data storage is a special way of saving information using a fancy laser beam. Instead of just one beam, it uses a split laser beam, which means it's like having two lasers in one. These lasers are used to put data onto something called "photoreceptive substrates." Think of photoreceptive substrates like a special kind of paper that can remember information when the laser beam hits it. What's cool about this laser is that it can hit the paper from different angles. Because of this, we can store lots of different pieces of information in the same spot on the paper. It's like writing on the same piece of paper from different sides, and all that information stays in one place! Theoretically, holographic storage should be capable of storing gigabytes of data in a single cubic millimeter of space.

Memristor Storage

Memristor storage uses an advanced circuit element called a memristor. Memristors have variable resistance based on the history of current that has passed through the element. A single memristor could replace many transistors on a chip. This means the chip could be far smaller and far faster. Flash technology, using transistors, is capable of access times of less than 25 nanoseconds. That is less than 25 billionths of a second! **Memristor storage** is theoretically capable of access times approaching 1 nanosecond!

Chapter Review

1. Describe the three types of memory used by a computer processor.

2. Computer memory and storage are often confused. Describe the difference between the two.

3. List three types of storage:

4. How much storage capacity does your computer have? Research how much is needed to store an average-sized HD movie.

5. If more computer storage space is needed, where can it be found?

6. What is file compression? When would it be necessary to compress a file?

7. Provide examples of file compression formats for the following file types:

   a. Image files:

   b. Audio files:

   c. Video files:

8. Complete the following list of storage capacities by size starting with bit and ending with petabyte.

   A. **Bit**: 1 character of data (a letter, number, or symbol)

   B.

   C.

   D.

   E.

   F.

   G. **Petabyte**: Approximately 1,000 terabytes

9. What is a traditional hard drive? How do traditional hard drives store information?

10. What is cloud storage? Name three advantages of cloud storage over other storage options. Name three providers of cloud storage.

11. What major problem or challenge are the emerging technologies described in the module trying to solve? Research the emerging technologies described and determine which is most likely to be available in the near-term. Describe your findings.

Ethics Discussion

An important issue that needs to be considered is that cloud computing makes it possible for individuals and companies to access other people's information and personal details, without that person necessarily knowing that their information is being accessed. Conduct research on the internet about the ethical issues surrounding cloud storage. Identify three potential ethics issues surrounding cloud storage. What can you do to ensure your cloud-based information is secure?

6   Windows 10

What To Expect

After completing this lesson, you will be able to:

- define Windows 10, discuss its new features, and list the versions available;
- explain how to operate Windows 10;
- discuss the various ways of interacting with and customizing Windows 10.

Chapter Topics

- What Makes Windows 10 Stand Out
- Using Windows 10
- Navigating Windows 10
- Windows 10 Settings
- Live Tiles in Windows 10
- Working with Apps in Windows 10
- Working with Documents
- Windows 10 Control Panel
- Customizing Windows 10

Copyright © McGraw Hill   Anton Watman/Shutterstock.

Let's Explore Windows 10

The Evolution of Windows

Before we jump into the specifics of **Windows 10**, let's take a quick look back at the journey of Microsoft Windows. Windows is an operating system that has been a cornerstone of personal computing!

1. **The Birth of Windows.** Microsoft's foray into the operating system market began in 1985 with the release of Windows 1.0. Although it was a modest start, Windows was destined for greater things. It introduced a graphical user interface (GUI) that gradually replaced the text-based interfaces of the past.

2. **Windows 95: A Game-Changing Milestone.** The turning point came in 1995 with Windows 95. This version introduced the iconic Start button and Taskbar, setting a standard that would define Windows for years to come. Subsequent iterations, such as Windows 98 and Windows 2000, built upon this foundation.

3. **The Era of Windows XP.** In 2001, Windows XP emerged as a reliable and user-friendly operating system, earning a special place in the hearts of many users. Its stability made it one of the most beloved Windows versions, remaining in use for an extended period.

4. **The Transition from Windows 7 to Windows 8.** After the success of Windows 7, Microsoft ventured into the touch-centric interface with Windows 8. While it faced some resistance from traditional desktop users, it paved the way for what was to come.

5. **Windows 10: The Best of Both Worlds.** And now, we arrive at **Windows 10**, released in 2015. **Windows 10** embodies the best elements of Windows 7 and 8 while introducing new features. It marks a significant milestone in Microsoft's journey, offering users a versatile and seamless computing experience.

What Makes Windows 10 Stand Out

Windows 10 isn't just another iteration of Windows; it brings noteworthy improvements to the table. Here's what sets **Windows 10** apart:

- **The Return of the Start Menu. Windows 10** brings back the beloved Start menu. The iconic Start button is back, providing a familiar and efficient way to access apps and settings.

- **The Unified Store. Windows 10** introduces the Microsoft Store, your one-stop destination for apps, music, videos, and games. It simplifies the process of discovering and installing new software, streamlining customization.

- **Meet Cortana. Cortana**, your personal digital assistant in **Windows 10**, is a standout feature. Similar to Apple's Siri, **Cortana** can perform tasks, provide information, and assist you in various ways. Whether you need to find files, schedule meetings, or even tell a joke, **Cortana** is ready to help.

- **Microsoft Edge: A Modern Web Browser. Microsoft Edge** takes the place of Internet Explorer in **Windows 10**. It offers a faster, more secure, and more contemporary web browsing experience, enhancing your online activities.

Windows 10 Editions. Windows 10 is available in different editions, each tailored to specific user needs:

- **Windows 10 Home:** Designed for home and small business users.
- **Windows 10 Pro:** Geared towards business and tech professionals.
- **Windows 10 Enterprise:** Ideal for large organizations with dedicated IT personnel.
- **Windows 10 Mobile and Mobile Enterprise:** Optimized for smartphones and tablets.

As we explore **Windows 10** further, you'll discover its capabilities and features, which can empower you in various ways, whether you're a casual user or part of a professional organization. Let's dive into the world of **Windows 10** and see how it can enhance your computing experience.

Using Windows 10

Starting Windows 10

Here are the first things you'll need to do:

- **Create a user account:** Windows enables each user of a device to create a user account. User accounts identify which Windows resources, such as storage locations and apps, a user can access when working on the digital device.
- **Create a username:** A username should uniquely identify you and be easily remembered.
- **Create a password:** A password should include a variety of numbers, letters, and symbols and contain at least eight characters.

You can also sign into **Windows 10** using a Microsoft account. If you do this, the email address linked with your Microsoft account will be shown instead of a traditional username. You will need to provide the password you use to sign into your Microsoft account when you sign into **Windows 10**.

Sign Out of Windows 10

If you do not want to shut your device totally off, it is a good idea to sign out of **Windows 10**. This will help you protect your device and information.

To sign out of **Windows 10**:

1. Click the Start button to display the Start menu.
2. Click your name or account info associated with **Windows 10** to display a menu of options.
3. Click Sign Out and return to the lock screen. You will now see the lock screen with the password box displayed (not your password!).

Using Lock/Unlock in Windows 10

The Lock feature in **Windows 10** allows you to hide the desktop behind a logon screen. You can use this option if you need to walk away from your computer for a while.

When you lock the system in **Windows 10**, any open programs and files remain in memory, but the logon screen is displayed. You must enter the logon password before you can resume access to your files and programs.

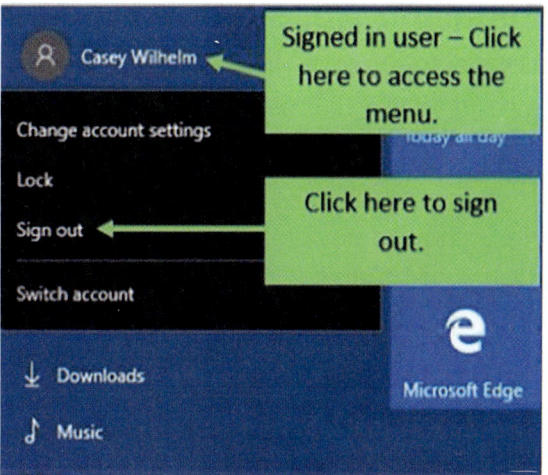

Here's how to lock/unlock your screen:

1. Hold down the Ctrl + Alt + Delete key combination and select Lock or click the Start button.

2. Select the appropriate username.

3. Select Lock.

To unlock your computer, click on the logon screen and enter your password. If no password is associated with the active account, simply click the image of the desired user.

Updating Windows 10

Making sure **Windows 10** has the most up-to-date information is important to maintain security and usability.

Here's how to check your Windows update settings:

1. Click the **Cortana** button or **Search box** button.

2. In the **Search box**, type "Windows Update" and select Windows Update Settings.

3. In Windows Update Settings, check the date for the last update.

4. You can also click Check for Updates to see if there are any needed Windows updates.

5. Click Advanced Options and make sure Automatic (Recommended) is selected.

6. When you are finished, close Windows Update Settings.

Installing Apps Using the Microsoft Store

Many different apps are available in the Microsoft Store in **Windows 10**. Many are free, while others require payment to download. Be cautious when providing any financial information online. Use a credit card or other third-party payment method and never use your debit account.

Here's how to download an app using the Microsoft Store:

1. First, click the Store App button on the taskbar or in the All Apps menu.

2. Next, type the app's name in the Search window.

3. Click on the app.

4. Finally, click the Install button or the Free button. If you are not logged into your Microsoft account, you will see the Free button. You will have to sign into a Microsoft account to complete the download.

Usernames and Passwords in Windows 10

A **password** is a secure combination of numbers, letters, and symbols associated with a username. The correct password allows access to the user's **Windows 10** account. **Picture passwords** in **Windows 10** allow the user to control access to a **Windows 10** account.

Shutting Down Windows 10

Shutting down a digital device running **Windows 10** means turning the device all the way off. Turning a device off allows you to save battery life and conserve energy.

Here's how to shut down a device running **Windows 10**:

1. Click the Start button to display the Start menu.

2. Click Power in the Start menu. Click Shut Down.

3. You can also select Restart or Sleep.

Navigating Windows 10

There are variety of ways to interact with **Windows 10**, including a touchscreen or touchpad, a mouse, an on-screen or virtual keyboard, scroll bars and scroll arrows, and even voice inputs.

Using a Touchscreen in Windows 10

You interact with a **Windows 10** touchscreen using gestures—motions of the fingers or hands. There are several types of gestures, most of which you probably use with your smartphone. One example is to tap or double-tap to run programs or apps. A double-tap gesture also can allow you to zoom in on objects. Another gesture is the finger drag or slide, which is accomplished by pressing and holding one finger then dragging or sliding. This is used for screen scrolling or for moving objects.

Other common gestures include the finger stretch and pinch, which are accomplished by placing two fingers on the screen and moving them apart or bringing them together. This is used to zoom in/out on areas of the screen.

Using the Task View in Windows 10

Task View allows you to view all your open windows and desktops with just one click. **The Task View** button is located on the taskbar next to your **Cortana** or Search button. It looks like a small vertical rectangle lying over a small horizontal rectangle. When you click the Task View button, all your open windows are displayed on the desktop, allowing you to quickly choose between each window.

Additionally, clicking the Task View icon displays the New Desktop button. The New Desktop button is a plus symbol located on the lower-right side of your desktop. A new desktop allows you to stay organized by having all your open work windows on one desktop and your open entertainment windows on another desktop.

Windows 10 Mouse Operations

There are a variety of mouse operations in **Windows 10**. Here are some of the most common:

- **Click or double-click:** Press and release the left mouse button one time to select or deselect items on the screen or run an app.
- **Triple-click:** Press and release the mouse button three consecutive times to select a paragraph.
- **Right-click:** Press and release the right mouse button to display shortcut menus.
- **Drag:** Point to an item and hold down theleft mouse button to move the item.
- **Wheel:** Rotate the wheel forward and backward to scroll through pages.

Using the Windows 10 On-Screen Touch Keyboard

Windows 10 has a built-in on-screen keyboard designed for use on touchscreen devices. You can use the on-screen keyboard to enter data using your fingers. The onscreen keyboard works much the same as a traditional keyboard. To enable the onscreen keyboard on touchscreen devices, click the Touch Keyboard button on the taskbar.

Windows 10 Keyboard Shortcuts

There are many different keyboard shortcuts on a traditional keyboard. When using a keyboard shortcut, hold down the first key indicated and then press the second key indicated. There is no need to enter the plus (+) sign. This is used only to indicate the action. Some popular keyboard shortcuts include holding down the Ctrl key and then pressing C to copy, holding down the Ctrl key and then pressing V to paste, and holding down the Ctrl key and pressing Z to undo an action.

Displaying the Windows Desktop

The easiest way to display the desktop in **Windows 10** is by clicking the Show Desktop button.

The Show Desktop button is in the extreme right corner of the taskbar by the notifications area.

Show Desktop can be useful when you want to quickly minimize all active windows to gain access to your desktop.

Cortana

According to Microsoft, "**Cortana** is your clever, new personal assistant. **Cortana** will help you find things on your PC, manage your calendar, find files, chat with you, and tell jokes. The more you use **Cortana**, the more personalized your experience will be."

To get started, type a question in the **Search box** on the taskbar. Or select the microphone icon (if available) and talk to **Cortana**. When you use **Cortana** for the first time, there is a brief setup process. To use **Cortana**, you must be signed into Windows with a Microsoft account.

Using Cortana

You can use **Cortana** for a variety of things, including finding files and apps, searching the web, getting weather and traffic updates, and many other tasks.

Simply type your file's name or app name in the **Search box** or, if you have a microphone, speak to **Cortana** to execute your search.

File Explorer in Windows 10

In **Windows 10**, **File Explorer** can be used to view files on your device.

File Explorer can also be used for organizing and naming files, putting files into folders, and moving or copying files from a USB drive or other external storage device.

It is a good idea to utilize file management techniques to keep files organized.

File Explorer Window

There are various ways to interact with the **File Explorer** window. Here are some common features:

- **File Location:** Displays the location of the file on the device and the folder in which the file is located
- **Back/Forward:** Enables you to navigate back or forward to previous views of files or folders
- **Search Box:** Provides an area into which you can enter criteria to search for a file or folder
- **Control Buttons:** Allow you to minimize, maximize, or close the screen
- **Navigation Pane:** Displays folders and drives as well as their contents

Opening File Explorer

To open **File Explorer** in **Windows 10**, either click the **File Explorer** button on the taskbar or press the Windows key + E on your keyboard. **File Explorer** automatically opens in Quick Access mode. Frequently used folders and files are listed there.

In **Windows 10**, My Computer is now called This PC. To display This PC in **Windows 10**, open **File Explorer** and select This PC in the left pane.

Jump Lists in Windows 10

Jump lists are lists of recently opened items, such as files, folders, or websites. Jump lists are organized by the program that you use to open them. You can use jump lists to open items and pin favorites for easy access.

Displaying a Jump List in Windows 10

The Start menu in **File Explorer** contains a jump list, which allows you to access different locations in **File Explorer**. Using a jump list makes accessing apps, file locations, and app features (settings) easier.

To access the jump list, right-click on the app's icon on the taskbar or in the Start menu; or click on an app's icon in the taskbar and drag upward.

Search Box

The **Search box** allows you to search for files on your devices, search the web for information, and interact with **Cortana**.

Searching the Web Using the Search Box

Here's how to search the web using the **Search box**:

1.   Click the **Cortana** button or the **Search box** to activate it.

2.   In the **Search box**, type "most in demand jobs 2016". Notice as you type that a list of suggested sites will appear.

3.   Click "most in demand jobs 2016" to display the search results in **Microsoft Edge**.

4.   Click the Close button to exit the browser.

Windows 10 Settings

Setting Reminders

Setting reminders can help you stay organized. You can set reminders for meetings, upcoming assignments, your work schedule, or basically anything that requires you to meet a deadline.

Here are the steps to set a reminder for an upcoming exam:

1.   Click the **Cortana** button or the Search button (if **Cortana** hasn't been activated) to access the **Search box**.

2.   Click the light bulb icon to set a new reminder.

3.   In the Remember To ... box, type "Introduction to Computers Exam 1".

4.   Click Time and select the time of the exam. Click the check mark.

5.   Click Tomorrow and select the date for the exam. Click the check mark.

6.   Click Remind, and you're all set!

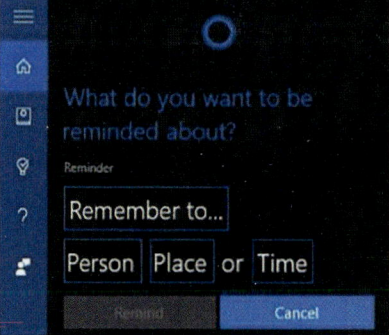

Changing System Settings Using the Search Box

To change settings, you can click the Windows Start button and then click the Settings button.

You can also use the **Search box** to find a variety of actions, including changing system settings and interacting with **Windows 10**. Here's how to use the **Search box** to change your power settings:

1. Type "settings" in the **Search box**.

2. Select Settings.

3. Select System, and then Power & Sleep.

4. Adjust your power and screen settings.

Live Tiles in Windows 10

Live Tiles in **Windows 10** displays useful information without opening an app. They are constantly updated when you are connected to the internet.

You can resize, rearrange, and move tiles based on your personal preferences. Examples of live tile apps include:

- Sports
- weather
- news
- money

Turning Live Tiles On and Off in Windows 10

You might want to turn on or off the information that is displayed in a Live Tile. Here are the steps:

1. If necessary, display the Start menu by clicking the Microsoft button.

2. Once the Start menu is displayed, right-click a Live Tile to display the Shortcut menu.

3. Click Turn Live Tile Off or Turn Live Tile On, based on your preference.

Working with Apps in Windows 10

You interact with **Windows 10** using a variety of apps. App is short for application. Application and computer program are synonymous. Two common ways to interact with apps in **Windows 10** are by clicking the Start button or by interacting with tiles.

A tile in **Windows 10** is a visual representation of an app that is easily accessed by touching or clicking. Tiles were developed to increase the interactivity and ease of use of apps in touchscreen devices.

Running an App from the Start Menu

Running an app from the Start menu is one of the most common methods to open an app. Here are the steps for using the Start menu to open the Microsoft Maps app:

1. Click the Start button to display the Start menu.

2. Use the scroll bar located to the right of the listed apps to find the Maps app.

3. Select Maps in the All Apps list to run the app.

Switching and Closing Apps in Windows 10

To switch between apps in **Windows 10**, click the desired app on the taskbar or press Alt + Tab to switch between open apps.

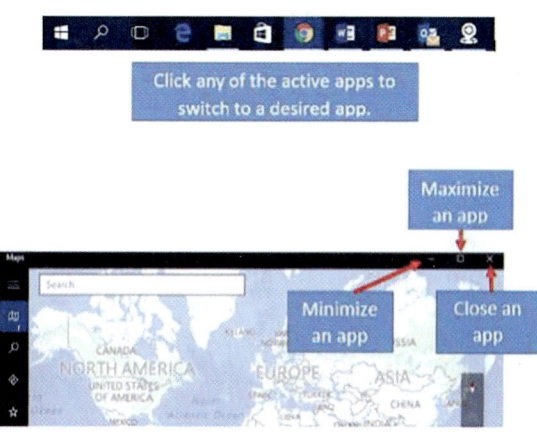

It is a good idea to close apps when you are finished working in them. This will free up system resources. Here's how to close an app:

1. Select the app you want to close.

2. Click the X in the upper-right corner of the app screen to close the app.

3. You can also right-click the app you want to close in the taskbar or press Alt + F4.

What Is a Pinned App in Windows 10?

Pinning an app adds a tile for it to the Start menu, allowing easy access to frequently used apps. Here's how to pin an app:

1. Click the Start button on the taskbar.

2. Select from the All Apps list the app you want to pin and right-click it.

3. With the shortcut menu displayed, click Pin to Start.

Unpinning an App from the Start Menu

You may want to customize the Start menu to best fit your preferences. Unpinning an app or tile allows you to keep the Start menu organized. Here's how to unpin an app:

1. Display the Start menu.

2. Right-click the tile or app you want to remove. This will display the Shortcut menu.

3. Click Unpin from Start.

4. Reorganize the remaining items.

Pinning an App to the Taskbar in Windows 10

Pinning an app to the taskbar makes it easy to open frequently used apps!

By default, Task View, **Microsoft Edge**, **File Explorer**, and Store buttons are pinned to the taskbar. If you want to pin another app, such as Outlook, to the taskbar, follow these steps:

1. Display the Start menu.

2. Right-click the Outlook tile to display the Shortcut menu.

3. Select More.

4. Click Pin to Taskbar.

Due to limited space, you may want to remove infrequently used apps from the taskbar. Here's how to remove a pinned app from the taskbar:

1. Right-click the app you want to remove.

2. With the Shortcut menu displayed, click Unpin from Taskbar. The app should now be removed from the taskbar.

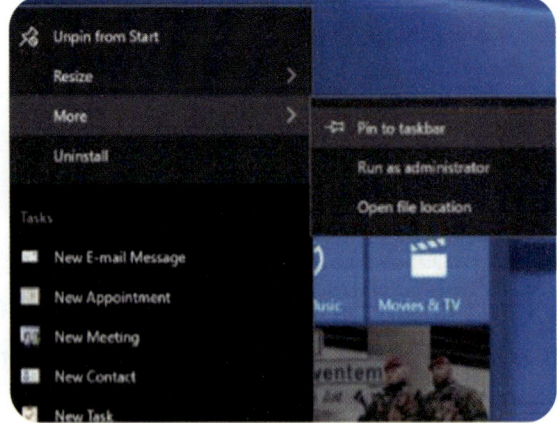

Running an App in Windows 10 Using the Search Box

Another quick way to search for an app or file is to use the **Search box**.

Here are the steps:

1. Click the **Search box** on the taskbar.

2. Type the name of the file or app you are looking for. You do not need to worry about capitalization when using the **Search box**.

3. Select the app or file you want to run/open.

Uninstalling Apps

It is important to use good file management practices. One key practice is to uninstall unused apps. Uninstalling unused apps will increase the amount of available hard drive space on your device.

Here's how to uninstall the Candy Crush Soda Saga app:

1. Click the Start button to display the Start menu.

2. Scroll to find the Candy Crush Soda Saga app.

3. Right-click the app to display the Shortcut menu.

4. Click Uninstall to remove the app.

5. A prompt box will be displayed. Click Uninstall again.

Installing Apps Using the Microsoft Store App

Many different apps are available in the Microsoft Store in **Windows 10**. Some are free, while others require a payment to download. Be cautious when providing any financial information online. Use a credit card or other third-party payment method—never use your debit account.

Here's how to download an app using the Microsoft Store:

1. Click the Store app button on the taskbar or in the All Apps menu.

2. Type the app's name in the Search window.

3. Click the app.

4. Click the Install button or the Free button. If you are not logged into your Microsoft account, you will see the Free button. You will have to sign into a Microsoft account to complete the download.

Working with Documents

Creating a Document in WordPad

WordPad is included with **Windows 10**. It is a scaled-down version of Microsoft Word.

If you don't have Microsoft Word installed on your device and you need to compose a document, WordPad is an excellent alternative.

Here's how to create a document using WordPad:

1. Click the **Cortana** button or the Search button to open the **Search box**.
2. Type "WordPad" and select the app.
3. Use WordPad to compose your document.
4. Save the document and close WordPad.

Creating a Document in the Documents Folder

Windows 10 has the capability to create WordPad documents in the Documents folder. Here's how:

1. Click the **File Explorer** button and then select Documents.
2. Right-click any open area of the Documents window to display the Shortcut menu.
3. Point to New on the Shortcut menu to display a list of options.
4. Click Text Document and name the file.
5. Right-click the newly created file to open the Shortcut menu, point to Open With, and select WordPad.

Saving a Document in the Documents Folder

When you create a document in **Windows 10**, by default the document is in random access memory (RAM). To permanently save the file (and prevent losing your work), you must save the document.

The Documents folder in **Windows 10** contains a specific user's documents and files. Here's how to save a WordPad file in the Documents folder:

1. Open WordPad and create a document.
2. Click **File**.
3. Click Save As.
4. Click Documents in the Navigation pane to specify Documents as the destination where the file will be saved.
5. Click Save.

What Are Files, Filenames, and File Extensions?

It is important to know the difference between files, filenames, and file extensions.

A **file** is data that has been identified by a filename. Examples of files include images, videos, documents, spreadsheets, and presentations.

A **filename** is the specific name given to a file to identify it. Filenames are usually pertinent to the information contained in the file.

A file extension is assigned by **Windows 10** and helps to associate a file with an app. **File extensions** are indicated by a period followed by three or more characters. WordPad files are saved in Rich Text Format. The file extension for Rich Text Format is ".rtf".

Printing a Document in Windows 10

There are times when you will be required to print a document in **Windows 10**. Here are the steps:

1. Open WordPad and compose a document or open an existing document.

2. Click **File** to display the **File** menu.

3. Click Print on the **File** menu to display the Print dialog box.

There are a variety of print options, including Select the Printer, Number of Pages, and Page Range.

You can also click Preferences to display other print options. Click the Print button to print the document.

Windows 10 Control Panel

Because Settings can be accessed by clicking the Start menu, the Control Panel in **Windows 10** has been greatly deemphasized. However, it may still be necessary to access the Control Panel to change certain settings.

Here are the steps to access the Control Panel:

1. Click the **Cortana** button or the **Search box** button.

2. Type "Control Panel" in the **Search box**.

3. Double-click the Control Panel desktop app.

4. View the various operations of the computer listed in the Control Panel.

5. Close the Control Panel desktop app.

Using the Windows 10 Control Panel to Change Time and Date Settings

Here are the steps to adjust time and date settings using the Control Panel:

1. Click the **Cortana** button or the **Search box** button.

2. Type "Control Panel" in the **Search box**. Double-click the Control Panel desktop app.

3. With the Control Panel displayed, select Clock, Language, and Region.

4. Select Change Date and Time.

5. Make the desired changes.

6. Select Apply and close the Control Panel.

Customizing Windows 10

You can customize a variety of settings in **Windows 10**, including display settings, personalization settings, and language settings.

Adjusting Display Settings in Windows 10

There are a variety of display settings you can adjust using the Control Panel. Here's how:

1. Click the **Cortana** button or the **Search box** button.

2. Type "Control Panel" in the **Search box**.

Control Panel
Desktop app

All Control Panel Items

↑ > Control Panel > All Control Panel Items

Adjust your computer's settings

Administrative Tools	AutoPlay
Color Management	Credential Manager
Dell Audio	Dell Command	Power Manager
Devices and Printers	Display
File History	Flash Player (32-bit)
HomeGroup	Indexing Options
Java (32-bit)	Keyboard
Mouse	Network and Sharing Center
Power Options	Programs and Features
Region	RemoteApp and Desktop Connections
Speech Recognition	Storage Spaces
Taskbar and Navigation	Troubleshooting
Windows Firewall	Windows Mobility Center

3. Double-click the Control Panel desktop app.

4. With the Control Panel displayed, select Appearance and **Personalization**.

5. Adjust the desired display settings and click Apply.

6. When you are finished, close the display settings on the Control Panel.

Customizing the Windows 10 Desktop: Desktop Display Options

Personalization allows you to change pictures, colors, and sounds on your computer. There are several ways to access the personalization settings in **Windows 10**.

One way is to right-click on any blank area of the desktop and select **Personalization**. Another method is to click the Start menu, then click the Settings button, then select **Personalization**. A third method is to access the Control Panel. Here are the steps:

1. Click the **Cortana** button.

2. Type "Control Panel".

3. Click the Control Panel desktop app.

4. Inside the Control Panel, select Appearance and **Personalization**.

5. Personalize your device and then close the app.

Customizing the Windows 10 Desktop: Language Options

When you purchase a new device, the language has been preset for you based on the location where the computer is purchased. **Windows 10** allows you to easily change languages and input options.

Here's how to change language options using the Control Panel:

1. Click the **Cortana** button or the **Search box** button.

2. In the **Search box** type "Control Panel".

3. Select the Control Panel desktop app.

4. In the Control Panel select Clock, Language, and Region.

You can then adjust the various language settings, including adding a language, keyboard input settings, and number format.

Customizing the Windows 10 Desktop: Accessibility Options

Windows 10 includes accessibility options and programs that make it easier for people with disabilities to see, hear, and use the computer. **Windows 10** supports adaptive devices for people with impairments.

Here is how to start the Ease of Access Center:

1. Click the **Cortana** button or the **Search box** button. In the search box, type "Ease of Access".

2. Select Ease of Access in the Control Panel. Adjust the following settings:

   - Magnifier: Enlarges portions of the screen, making it easier to view text and images and to see the whole screen more easily
   - Narrator: Reads aloud the text on the screen and describes actions
   - On-Screen Keyboard: Displays a visual keyboard you can use with a pointing device
   - High-Contrast Display Settings: Increases the contrast of colors to reduce eyestrain.

Chapter Review

1. What are the key differences between **Windows 10** and its previous versions?

2. How can you create and customize user accounts in **Windows 10**?

3. What options are available for signing into **Windows 10**, and why might you choose one over the other?

4. Why is it important to keep **Windows 10** up-to-date, and how can you check for updates?

5. What is the role of a password in **Windows 10**, and how can you create a secure password?

6. What precautions should you take when downloading apps from the Microsoft Store in **Windows 10**?

7. How can you install and uninstall apps using the Microsoft Store in **Windows 10**?

8. What are **Live Tiles** in **Windows 10**, and how can you customize their display?

9. How do you run, switch between, and close apps efficiently in **Windows 10**?

10. What are jump lists in **Windows 10**, and how can they enhance productivity?

11. How can you create and save documents using WordPad and the Documents folder in **Windows 10**?

12. What is the significance of filenames and file extensions in **Windows 10**?

13. What options are available for printing documents in **Windows 10**?

14. How can you access and utilize the **Windows 10** Control Panel for system settings?

15. What customization options does **Windows 10** offer, and how can you adjust display settings, personalize the desktop, and change language settings?

Ethics Discussion

Is it ethically responsible for Microsoft to collect user data for the purpose of improving **Windows 10**, and if so, what measures should be in place to ensure user privacy and consent in this data collection process?

7 Windows 11

What To Expect

After completing this lesson, you will be able to:

- describe what Windows 11 is, discuss its new features, and list the versions available;
- explain how to operate Windows 11;
- discuss the various ways of interacting with and customizing Windows 11.

Chapter Topics

- Running Windows 11
- Updating Windows 11
- Navigating Windows 11
- Microsoft Cortana
- Windows 11 File Explorer
- Working with Apps
- Creating and Managing Documents

Copyright © McGraw Hill   Rawf8/Alamy Stock Photo

Let's Explore Windows 11

Let's Review Windows' Evolution

As you explore the features of Windows 11, you'll discover how this new version caters to the modern computing needs of users. It's not just about a fresh look; Windows 11 is equipped with enhancements that elevate productivity and user satisfaction. Let's explore these changes. Let's take a moment to trace the evolution of Microsoft Windows:

The Beginning of a Revolution: In the mid-1980s, Microsoft introduced Windows 1.0, laying the foundation for a graphical user interface (GUI) that would eventually revolutionize personal computing. While this initial version was limited in functionality, it set the stage for a series of transformative releases.

Windows 95 - A Turning Point: The watershed moment arrived in 1995 with Windows 95. This release introduced the Start button and Taskbar, shaping the user experience for years to come. Subsequent versions like Windows 98 and Windows 2000 built upon this foundation.

The Era of Windows XP: The early 2000s saw the emergence of Windows XP, a beloved and enduring OS known for its stability and user-friendliness. Windows XP became a staple for many users and businesses, maintaining its presence for over a decade.

Transitioning through Windows 7 and 8: Following the success of Windows XP, Microsoft navigated through a transition phase with Windows 7 and Windows 8. While Windows 7 was celebrated for its reliability, Windows 8 introduced a touch-centric interface, anticipating changes in the computing landscape.

Windows 10 - Bridging the Gap: In 2015, Windows 10 arrived as a bridge between the traditional desktop experience and the touch-centric world. It combined the best elements of Windows 7 and 8 while introducing new features, offering a versatile and cohesive user experience.

Windows 11: A New Chapter Begins

The 2021 release is more than just an iteration; it's a leap forward in design and functionality, bringing exciting changes to the familiar Windows ecosystem.

A Fresh Look and Feel: One of the most immediately noticeable differences in Windows 11 is its refreshed user interface. The design draws inspiration from Chrome OS and macOS, creating a sleek and modern look that stands out. Here's what sets Windows 11 apart:

Centered Taskbar and Start Button: The Taskbar and Start button, traditionally located at the bottom of the screen, have been moved to the center, aligning them with the aesthetic seen in Chrome OS.

Rounded Corners and Elegance: Windows 11 introduces rounded corners for windows, resembling the design found in macOS. This design choice brings a touch of elegance and consistency to the interface.

Different Versions of Windows 11

Windows 11 is available in two main versions, or editions. These are **Windows 11 Home** and **Windows 11 Pro**. Both share many similar features, but there are some differences.

- **Windows 11 Home** is designed for consumer users who intend to use their devices at home or for personal use.
- **Windows 11 Pro** is designed for use in business and organizational environments. It has enhanced security features including BitLocker device encryption, Windows Information Protection (WIP), and enhanced business management and deployment options.

Additionally, **Windows 11 SE** is a cloud-first operating system that was built for the K–8 education market.

Running Windows 11

Initial Start-up of Windows 11

The initial start-up and setup process for **Windows 11** is fairly straightforward. You will navigate through a series of screens that guide you to proper setup. Following are the steps that cover the initial start-up procedure for **Windows 11**.

Step 1—Choose a Language

The first screen you will see during initial setup of **Windows 11** asks you to select a language. Next, you will be prompted to select the region where you are located and to select the keyboard layout you are using. You can change the time, language, and keyboard layout at any time by visiting Settings > Time & language or Settings > Language & region.

Step 2—Select a Network

The next screen prompts you to select a network for internet connectivity. If your device is not connected to the internet via an ethernet connection, **Windows 11** will display the Wi-Fi networks that are in range. Choose the network you want to connect to and then enter the network password.

It may take a little time after you have selected your network to access it, as your device will check for updates. After this step has been executed, you will be asked to name your **Windows 11** PC. This name will be used to identify the device on local networks.

Signing into Windows 11

PIN Code

You can sign into **Windows 11** using a PIN code. You can choose a PIN code during the initial setup process. Additionally, you can choose a PIN code using the Settings app. To set up a PIN code using the Settings app, select Settings > Accounts > Sign-in options. You can then set up a PIN code access for your device.

Password

From the sign-in screen, select your account and then type your password. You can click the arrow button next to the password box or hit Enter on the keyboard to enter the password.

Facial Recognition

You can use facial recognition (using your webcam) to sign into **Windows 11**.

Fingerprint Recognition

If your device has a fingerprint scanner (or if you have a fingerprint scanner peripheral), you can sign into **Windows 11** using your fingerprint.

Signing out of Windows 11

If you do not want to shut your device totally off, it is a good idea to sign out of **Windows 11**. This will help you protect your device and information. To sign out of **Windows 11**, select Start, select the Accounts icon (or picture), and then select Sign out. You should then be signed out of **Windows 11**.

Locking Windows 11

The **lock feature** in **Windows 11** allows you to hide the desktop behind a logon screen. You can use this option if you need to walk away from your computer. When you lock the system in **Windows 11**, any open programs and files remain in memory, but the logon screen is displayed. You must enter the log-in password before you can resume accessing your files and programs.

Here are some different ways to lock your screen:

- Hold down the Ctrl + Alt + Delete keys and select Lock.
- Press the Windows key, click the Start button, select the username, and select Lock. Press the
- Windows key + L simultaneously on your keyboard to lock your device.

Usernames and Passwords

A **username** is a unique grouping of numbers, letters, or symbols that identifies a specific user of **Windows 11**. A **password** is a secure combination of numbers, letters, and symbols associated with the username that allows access to a user's **Windows 11** account. **Picture Passwords** allow the user to control access to a **Windows 11** account via a picture.

Shutting Down

Shutting down a digital device running **Windows 11** means turning the device all the way off. Turning a device off allows you to save battery life and conserve energy.

There are several ways to shut down your **Windows 11** Device including:

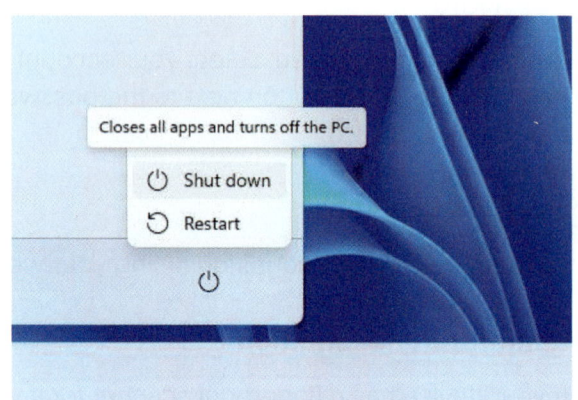

- Click Start (or press the Windows key), select the Power button, click **Shut down**. Hit the power button on your device to fully turn the power off.
- Click the Alt + F4 keys to fully power the device off.

Sleep Mode

If you don't want to fully power your device off, you can select Sleep mode. This mode uses very little power and allows you to start up faster than when the device is fully powered off. It uses an autosave feature that will save your work when the battery runs low and the device must be powered off. To access sleep mode, click Start (or press the Windows key), select the Power button, and click Sleep.

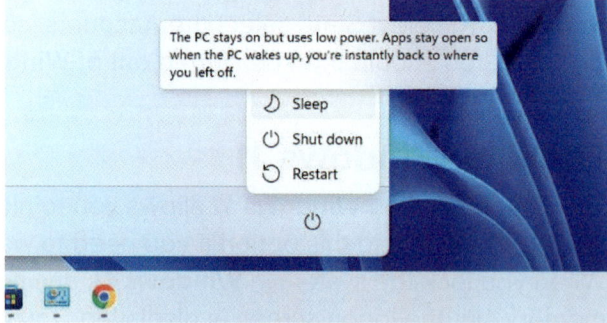

To change sleep settings for your device, select Start > Settings > System > Power & sleep > Additional power settings. Next, select Choose what the power setting does, then select "when I press the power button select Sleep". Finally, select Save changes.

Updating Windows 11

Ensuring **Windows 11** has the most up-to-date information is important for maintaining security and usability.

To check your Windows update settings, select Start > Settings. Alternatively, you can open the Settings app by pressing the Windows key + I simultaneously. Then click on the Windows Update category.

The **Windows Update page (app)** displays the current update status as well as other features, such as pause updates, update history, advanced options, and the Windows Insider Program. If you want to see if there are updates, you can select the Check for updates button, which will then scan the system and provide any information about needed updates (if there are any).

You can choose to delay updates by selecting Pause updates and selecting the amount of time you want to pause **Windows 11** updates. It is not advised to pause updates for too long as Microsoft often provides updates to security. If you turn off the update feature, you may not have the most current security features protecting your device.

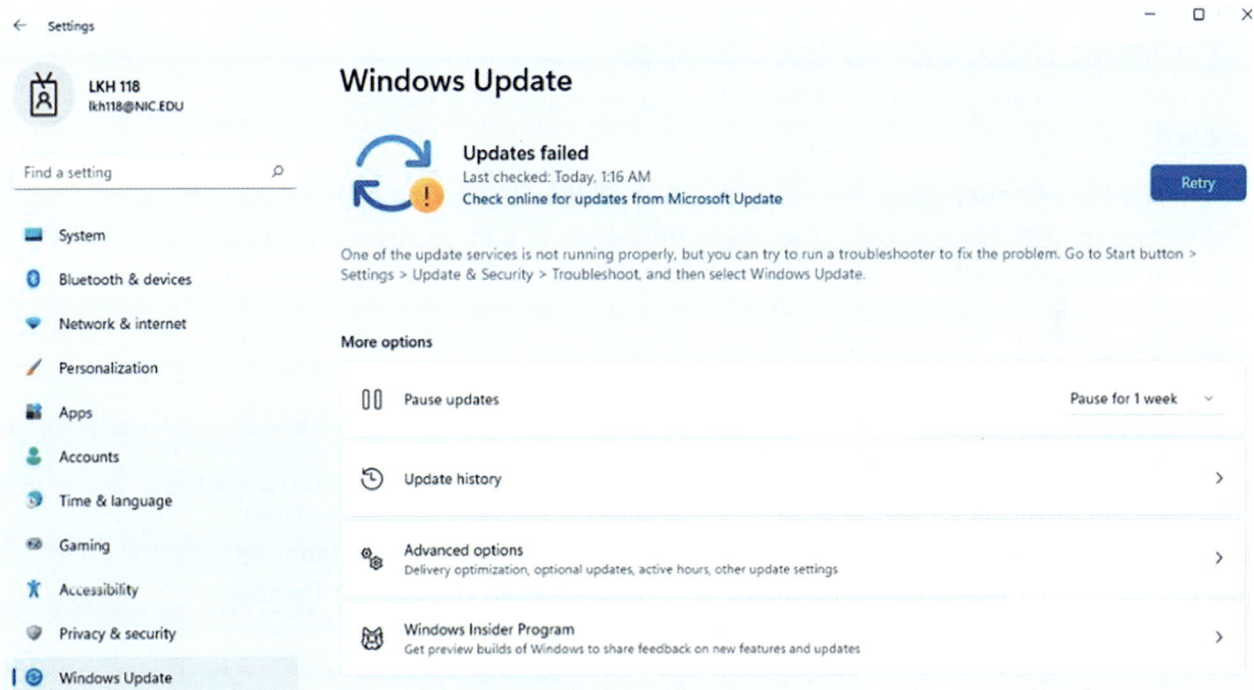

Navigating Windows 11

The **taskbar** in **Windows 11** displays the most used apps on your device. It is an easy way for you to access and visualize the apps you use most. By default, the taskbar is visible at the bottom of the screen, In **Windows 11** you can change the color, pin your favorite apps, and move or rearrange taskbar buttons.

It is a good idea to get comfortable using various input methods to interact with **Windows 11**. There are variety of ways to interact with **Windows 11** including using a mouse, a touchpad or touchscreen, the traditional and on-screen keyboard, scroll bars and scroll arrows, and voice inputs.

Adjusting Taskbar Settings

To change your **Taskbar** settings, press and hold or right-click any empty space on the taskbar and then select **Taskbar** settings. In the **Taskbar** settings, you can adjust the alignment of the taskbar, rearrange taskbar buttons, use the taskbar to show the desktop, and automatically hide the taskbar. When adjusting the alignment of the taskbar, you can align taskbar icons to the center or left of the screen. There is no option to move the taskbar to the top or right of the screen.

Changing the Color of the Taskbar

This option will change the color of the taskbar to the color of the theme you have applied.

To change the color of the taskbar, select Start > Settings > Personalization. Select Colors and scroll to Accent color. Then select the option to turn on Show accent color on Start and taskbar.

Windows 11 Search Button

In **Windows 11**, it is likely that the **Search button** will not be displayed by default. If you'd like to use the **Search button**, you'll need to enable it.

The **Search button** allows you to search for files and folders in Windows, find and open installed apps, search the web for information, and interact with **Cortana**. To enable the search button, right-click anywhere on the taskbar. Then select **Taskbar** settings. With **Taskbar** settings displayed, toggle the **Search button** on.

Windows 11 Focus Assist

Focus is a feature in **Windows 11** that helps to minimize distractions when working. To help you stay on track, it integrates with the Clock app and includes features such as a focus timer and music integration.

When you're in a focus session, a focus timer will appear on your screen and a Do not disturb sign will turn on. Apps in the taskbar won't flash to alert you, and badge notifications on apps in the taskbar will turn off. This can help you stay on task without distractions from the screen.

How to Start Focus

To start **Focus**, select notification center in the taskbar. Choose an amount of time for your focus session. Then select **Focus** to start your session

To Set Do Not Disturb During Focus

To set Do Not Disturb during **Focus**, select Start > Settings > System > Notifications. In Do not disturb, you can automatically turn off notifications outside of your work hours and set priority notifications so reminders, calls, and specific apps can alert you even when "Do not disturb" is turned on.

Windows 11 Snap

Snap is a new feature in **Windows 11** that allows users to execute window snapping. You can access window snap layouts by hovering your mouse over an open window's maximize button or by pressing the Windows button + Z.

Once this feature has been activated, a menu will be displayed that shows available layouts. Click on a zone in a layout to snap a window to a particular zone. Then use **Snap** Assist to complete the buildout of the window layout.

Windows Hello

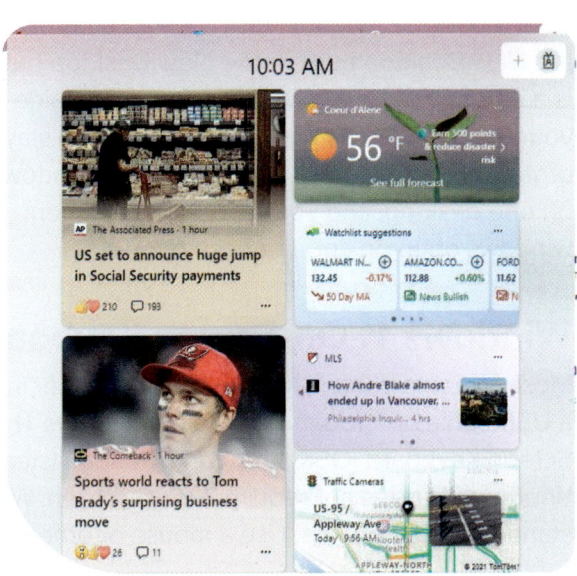

Windows Hello is a feature in Windows that allows you to create more personal and secure methods for accessing a **Windows 11** device. To access **Windows Hello**, select Start > Settings > Accounts > Sign-in options. You will then be given the options to set up facial recognition, fingerprint recognition (requires a fingerprint reader), or PIN.

Widgets

Widgets are cards displaying frequently updated content from apps and services that you select on your Windows desktop. Content displayed in the widgets is frequently updated and requires an active Internet connection.

Widgets are displayed on the widgets board. Here you can discover, pin, unpin, arrange, resize, and customize widgets based on your preferences and usage habits. Popular widgets include weather, sports, the Microsoft Outlook calendar, Microsoft OneDrive photos, and third-party apps.

Using a Touchscreen in Windows 11

You interact with a **Windows 11** touchscreen using gestures. **Gestures** are motions of the fingers or hands that are used to interact with a touchscreen.

Here are some examples of gestures to use when operating a touchscreen:

- You can use a finger tap or double tap to run programs or apps. A double-tap gesture also can allow you to zoom in on objects.

- The finger drag or slide is accomplished by pressing and holding one finger and then dragging. This is used for scrolling or for moving objects.

- Other common gestures include the finger stretch and pinch, which are accomplished by placing two fingers on the screen and moving them apart or bringing them together. This is used to zoom in or out on areas of the screen.

Using the Task View in Windows 11

The **Task View** allows you to view all your open Windows and desktops with just one click. It is considered a virtual desktop system, also known as a task switcher. The **Task View** button is located on the taskbar to the right of the **Search button**. When you click the **Task View** button, all your open windows are displayed on the desktop, allowing you to quickly choose between each window.

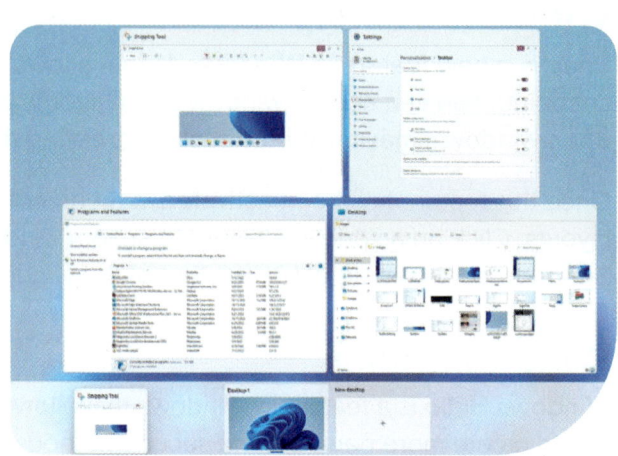

Additionally, clicking the **Task View** icon displays the New Desktop button. The New Desktop button is a plus symbol located on the lower right side of your desktop. A new desktop allows you to stay organized by having all your open work windows on one desktop, and your open entertainment windows open on another desktop.

Windows 11 On-Screen Keyboard

Most computers use a keyboard that is either installed on the device or attached with a USB cable or through Bluetooth. **Windows 11** has a built-in Accessibility tool, the On-Screen Keyboard (OSK), that can be used instead of a physical keyboard. The OSK in **Windows 11** does not require a touchscreen. When activated, a visual keyboard with standard keys appears. Use a mouse or other pointing device to select the keys.

To activate the OSK, go to Start > Settings > Accessibility > Keyboard. Then turn on the On-Screen Keyboard. The OSK can be moved around the screen to make it more convenient.

Windows 11 Action Center

The **Action Center in Windows 11** is an area that displays notifications and quick actions you can take on your device. Notifications include email messages and news alerts. Quick actions include several different tasks including battery, network, screen brightness, and settings.

To access the Action Center, press the Windows key + A. You can adjust the Quick Settings that are displayed in the Action Center. To adjust Quick Settings from the Action Center, click the pencil icon (Edit quick settings). From this area, you can add or remove the Quick Settings that are displayed.

Windows 11 Shortcuts

There are many different keyboard shortcuts that may be used when typing on a traditional keyboard. When using a keyboard shortcut, you need to hold down the first key indicated and then hit the second key indicated. There is no need to enter the "+" sign. This is used here only to indicate the action.

Here are some of the most popular shortcuts in **Windows 11**:

- Copy: Ctrl + C Paste: Ctrl + V Undo: Ctrl + Z
- Switch between apps: Alt + Tab
- Lock PC: Windows button + L
- Open the Action Center: Windows button + A
- Open Search: Windows button + S

Displaying the Windows 11 Desktop

The easiest way to display the desktop in **Windows 11** is by clicking the Show desktop button. The Show desktop button is in the extreme right corner of the taskbar by the notifications area.

If the Show desktop feature isn't working, you can troubleshoot the issue. Select Setting > Personalization. Then select **Taskbar** > **Taskbar** behaviors and click on Select the far corner of the taskbar to show the desktop.

Show desktop can be useful when you want to quickly minimize all active windows to gain access to your device's desktop.

Microsoft Cortana

According to Microsoft, "**Cortana** is your clever personal assistant. **Cortana** will help you find things on your PC, manage your calendar, find files, chat with you, and tell jokes. The more you use **Cortana**, the more personalized your experience will be." **Cortana** is similar to Amazon's Alexa and Apple's Siri.

To get started, type a question in the Search box on the taskbar or select the microphone icon and talk to **Cortana**. Typing works on most Windows devices, but you need a microphone to talk.

When you use **Cortana** for the first time, there is a brief setup process.

Using Cortona

Cortana has many uses including finding files and apps, searching the web, and getting weather and traffic updates.

To open **Cortana**, you can either click the **Cortana** button on the taskbar or navigate to **Cortana** by clicking the Start button and the All apps button.

Some devices may not have **Cortana** activated. To activate **Cortana** on your device, from the Start menu, open **Cortana** and select Settings, and then Talk to **Cortana**. Under Hey **Cortana**, switch the toggle to On.

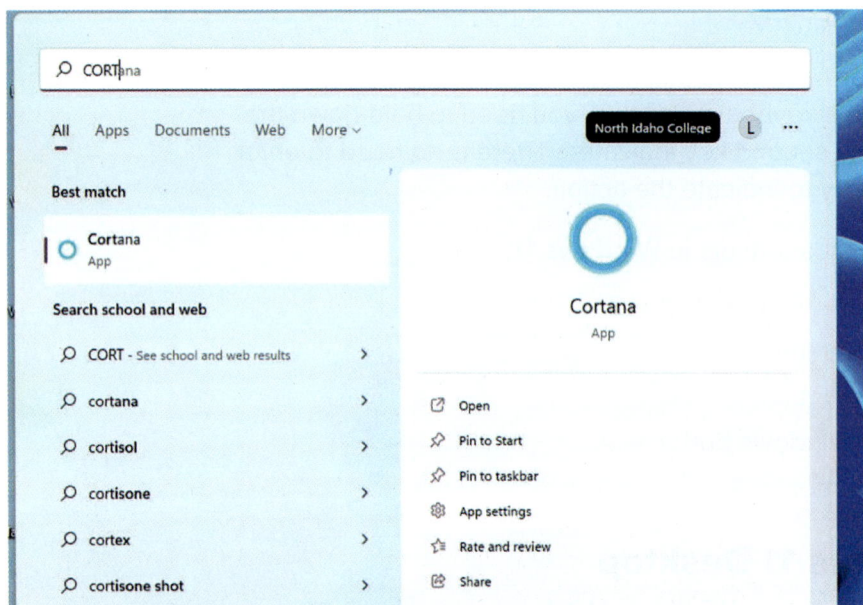

To use **Cortana**, you can either type your file or app name in the Search box or use a microphone and speak to **Cortana**.

Windows 11 File Explorer

In **Windows 11**, the **File Explorer** can be used to view files on your device. Use efficient file management techniques to keep files organized.

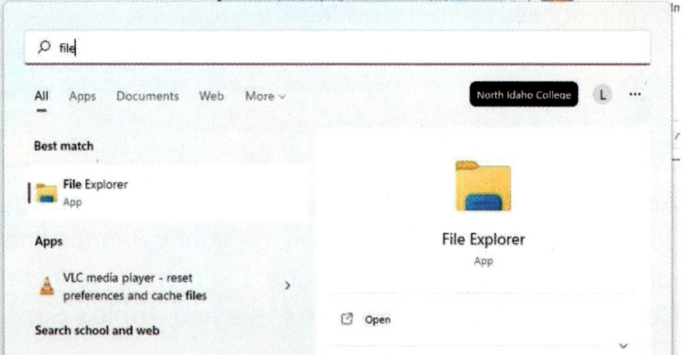

File Explorer is used to open and view files, to name files, to move files into folders, and for moving or copying files from a USB drive or other external storage device.

The **Windows 11 File Explorer** has some new features. It now includes tabs to help you organize your **File Explorer** sessions like you do in Microsoft Edge. It includes Suggested Actions for items that you copy. And you can now share to more devices using network sharing.

The File Explorer Window

There are various ways to interact with the **File Explorer** Window. Some common features include the following:

- **File** Location—This displays the location of the file on the device and the folder the file is located in.
- Back/Forward—This is used to navigate back or forward to display previous views of files or folders.

- Search Box—This provides an area into which you can enter criteria to search for a file or folder.
- Control Buttons—These allow you to minimize, maximize, or close the screen.
- Navigation Pane—This displays folders and drives that you can double-click to see their contents.

Opening File Explorer

To open **File Explorer**, click the **File Explorer** icon on the desktop or use the keyboard shortcut Windows key + E. In **Windows 11**, My Computer is called This PC. Just open **File Explorer** and click This PC to open. **File Explorer** automatically opens in Quick access. Frequently used folders and files are listed there.

Jump Lists

Jump Lists are lists of recently opened items, such as files, folders, or websites. Use **Jump Lists** to open items and pin favorites for easy access. They are organized by the program that you use to open them.

Displaying Jump Lists

The **File Explorer** on the Start menu contains a Jump List. The Jump List allows you to access different locations in **File Explorer**. Using the Jump List makes accessing apps, file locations, and app features easier.

To access the Jump List, right click on the app's icon either on the taskbar or on the Start menu. Additionally, clicking on an app's icon in the taskbar and dragging upward will open its Jump List. To access the Jump List with an app active on the taskbar, right-click on the app's icon or click on the app's icon and drag upward. To access the Jump List from the Start menu, simply right-click on the app name.

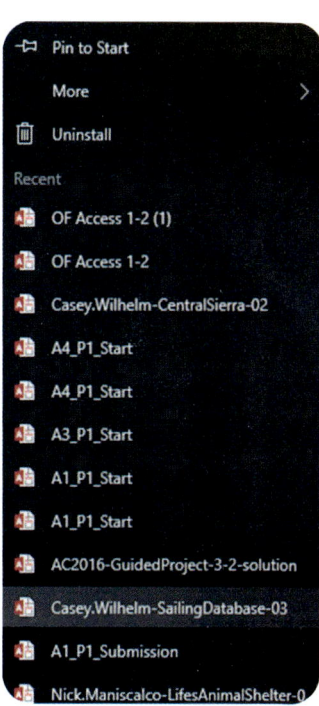

Working with Apps

You interact with **Windows 11** using a variety of apps. To open the apps list in **Windows 11**, click the Start or Windows button to open the Start Menu and then click All apps in the top right corner of the screen. A list of apps installed on your device will be displayed. Scroll through the list and select the app you want to use.

The apps list in **Windows 11** includes a speed dial feature. If you click a letter in the apps list (for example the letter *B*), you will be taken to the speed dial. Select a letter in the speed dial to be taken to a list of apps that begin with that letter.

Pinning Apps

Pinning an app to the Start menu is a convenient feature that is designed to save time when accessing commonly used apps in **Windows 11**.

Pinning an app to the Start menu makes it appear as a tile (a large icon) on the right side of the Start Menu. This makes the application quick and easy to select, rather than scrolling through all your apps to find the right one.

To pin an app, click the Start button on the taskbar, right-click on the app to display the menu, and click Pin to Start.

Pinning and Unpinning Apps from the Windows Taskbar

Pinning an app to the **Taskbar** is a convenient feature designed to save time when accessing commonly used apps. In addition to apps, you can pin other **Windows 11** features including files, folders, drives, and even websites.

These apps have been Pinned to the Taskbar

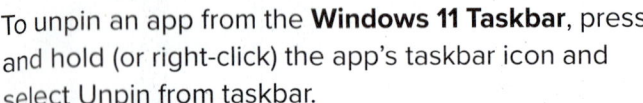

To pin an app to the **Windows 11 Taskbar**, click the Start button (the Windows button) on the taskbar and then click the arrow next to All apps. Then right-click the app and select: More > Pin to taskbar.

To unpin an app from the **Windows 11 Taskbar**, press and hold (or right-click) the app's taskbar icon and select Unpin from taskbar.

Installing Apps Using the Microsoft Store App

There are numerous apps available in the Microsoft Store in **Windows 11**. Many are free, while others require a payment to download. Be cautious when providing any financial information online. Use a credit card or other third-party payment method and never use your debit account.

For example, follow these steps to download and install the WhatsApp application from the Microsoft Store:

1. Select the Start button and then from the apps list, select Microsoft Store.

2. Select the Apps tab or type "WhatsApp" in the Search apps, games, movies, and more in the search box in the Microsoft Store.

3. Select WhatsApp and select Get.

Uninstalling Apps in Windows 11

It is important to use good file management practices. One key file management practice is to uninstall unused apps. Uninstalling unused apps will increase the amount of available hard drive space on your device. You can uninstall an app or program from the Start menu, the Settings page, or the Control Panel.

To uninstall an app from the **Windows 11** Start menu, select Start and find the app or program you want to uninstall. Then press and hold (or right-click) on the app and select Uninstall.

To uninstall an app from the **Windows 11** Settings page, select Start > Settings > Apps > Apps & features. Then select the app you want to remove and select Uninstall.

To uninstall an app from the **Windows 11** Control Panel, in the search box on the taskbar, type "Control Panel" and select it from the results. Then select Programs > Programs and Features. Press and hold (or right-click) on the program you want to remove and select Uninstall or Uninstall/Change. Follow the directions on the screen to complete the uninstall process.

Creating and Managing Documents

Creating Documents in WordPad

WordPad is included with **Windows 11**. **WordPad** is a scaled-down version of Microsoft Word. If you do not have Microsoft Word installed on your device and you need to compose a document, **WordPad** is an excellent alternative.

WordPad files are saved in Rich Text Format. The file extension for Rich Text Format is .rtf.

To create a document using **WordPad**, click the **Cortana** button or the **Search button** to open the Search box. Type "**WordPad**" and select the app. Use **WordPad** to compose your document. Then save the document and close **WordPad**.

Creating Documents in the Documents Folder

To create a **WordPad** document in the Documents folder, click **File Explorer** and select Documents. Right-click any open area of the Documents windows to display the shortcut menu, click New > Text Document, and name the file. Right-click the newly created file to open the shortcut menu, point to Open with, and select **WordPad**.

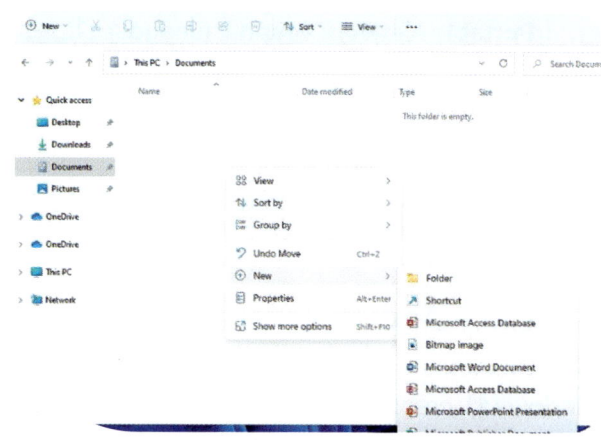

Saving Documents in the Documents Folder

When you create a document in **Windows 11**, the document, by default, is in random access memory (RAM). To permanently save the file (and prevent losing your work), you must save the document. The Documents Folder in **Windows 11** is a place where a specific user can save documents and files.

To save a **WordPad** file in the Documents Folder, open **WordPad** and create a document. Then click **File** to open the **File** menu. Click Save As to open the Save As dialog box. Then click Documents in the Navigation pane, which will specify Documents as the destination folder. Then click Save.

Files, File Names, and File Extensions

A **file** is data that have been identified by a filename. Examples of files include images, videos, documents, spreadsheets, and presentations. A **file name** is the specific name given to a file that identifies the file. **File** names are usually pertinent to the information contained in the file.

A **file extension** is assigned by **Windows 11** and helps to associate a file with an app. **File** extensions are indicated by a period followed by three or more characters.

Common File Extensions	
Microsoft Excel	.xlsx
Microsoft Word	.docx
Microsoft PowerPoint	.pptx
Microsoft Access	.accdb
Portable Document Format (created by Adobe)	.pdf
Joint Photographic Exports Group (images)	.jpg
Portable Network Graphic (images)	.png
Rich Text Format (text)	.rtf
Audio, Video, Image format	.mp4

Printing Documents in Windows 11

There are times when you will be required to print a document. To print a document in **Windows 11**, create a document or open an existing document. Click **File** to display the **File** Menu. Here you can click Print on the **File** menu to display the print dialog box.

The print dialog box will give you a variety of print options. You will need to select the correct printer, enter the number of pages to print, and the Page Range. You can also click Preferences to display other print options. Then click the Print button to print the document. You can use the keyboard shortcut Ctrl + P to skip some of those steps.

Desktop Language Options

The display language you select changes the default language used by **Windows 11** features including Settings and **File Explorer**.

If you want to change the language used on your **Windows 11** device, select Start > Settings > Time & language > Language & region. Then choose a language from the Windows display language menu.

Windows 11 Accessibility Settings

The Ease of Access Center in **Windows 11** allows you to adjust the settings on your device to make it more accessible. Some features available include the following:

- A narrator feature that provides audio descriptions of elements on the screen
- High contrast mode and Magnifier, which make information on the screen appear in a larger format
- Keyboard features including Sticky Keys, Toggle Keys, and the On-Screen Keyboard (OSK)
- The option to change the pointer size of your mouse and Mouse Keys which allow you to control the mouse using your keyboard

To open the Ease of Access Center, select the Start button. Then select Settings > Ease of Access.

Adjusting Power Settings

To save battery life, **Windows 11** sets the power setting to Balanced, by default. However, you may want to adjust power settings to maximize the performance of your device or to increase the amount of time you can use your device while running off the internal battery.

To adjust the power settings on a **Windows 11** device, select the Microsoft button. Next, select Settings > System > Power and battery. In the Power and battery screen, select Power mode. Here you can select Best power efficiency, Balanced, or Best performance.

Changing your Default Web Browser

Windows 11 comes with the Microsoft Edge web browser preinstalled and is likely the default web browser on your Windows device. You may want to change the default browser to a browser that is more secure or one you are more familiar with.

To change the default browser on a **Windows 11** device, click the Windows button and select Settings. Click Apps > Default Apps. Search for the web browser you'd like to use and select Set default to make the newly selected browser your default browser.

Turning Off Notifications in Windows 11

Some notifications contain pertinent information, but too many notifications can be annoying. To adjust notifications in **Windows 11**, select the Windows button. Then select Settings > System > Notifications. In the Notifications screen, turn off the areas you don't want to receive notifications from.

Consider turning off notifications for Offer suggestions on how I can set up my device and Get tips and suggestions when I use Windows.

Adjusting Screen Refresh Rate and Brightness

The **refresh rate**—measured in Hertz (Hz)—of a display is the number of times per second a new image is drawn on the display. Displays with faster refresh rates can display content rapidly and smoothly. This is especially important for gamers and viewing of fast-motion activities like sports.

Common refresh rates include 60 Hz, 120 Hz, 144 Hz, and 240 Hz.

Windows 11 allows you to adjust the refresh rate of your device. To adjust the refresh rate of your display, click the Windows button and select Settings. Select System > Display > Advanced Display and select the Choose a refresh rate option.

Chapter Review

1. What are the key differences between **Windows 11** and its predecessor, Windows 10?

2. How can you create and customize user accounts in **Windows 11**?

3. What sign-in options are available in **Windows 11**, and what factors should influence your choice among them?

4. Why is it crucial to keep **Windows 11** up-to-date, and how can you check for updates?

5. What role does a password play in **Windows 11**, and how can you establish a secure password?

6. What precautions should you exercise when downloading apps from the Microsoft Store in **Windows 11**?

7. How do you install and uninstall apps using the Microsoft Store in **Windows 11**?

8. What are Live Tiles in **Windows 11**, and how can you personalize their appearance?

9. How can you efficiently run, switch between, and close apps in **Windows 11**?

10. What are **Jump Lists** in **Windows 11**, and how can they boost productivity?

11. How can you create and save documents using **WordPad** and the Documents folder in **Windows 11**?

12. What is the significance of file names and file extensions in **Windows 11**?

13. What options are available for printing documents in **Windows 11**?

14. How can you access and utilize the **Windows 11** Control Panel for system settings?

15. What customization features does **Windows 11** provide, and how can you adjust display settings, personalize the desktop, and change language settings?

Ethics Discussion

Is it ethically responsible for Microsoft to collect user data for the purpose of improving **Windows 11**, and if so, what measures should be in place to ensure user privacy and consent in this data collection process?

8 Mac OS

What To Expect

After completing this lesson, you will be able to:

- turn the Mac on and off and navigate through the desktop;
- use the keyboard, trackpad, and gestures;
- describe how to find and use apps;
- explain Mac security settings;
- describe how to configure a Mac's appearance;
- explain file management in a Mac;
- describe Mac utility features

Chapter Topics

- Mac OS Basics
- Customizing Mac Settings
- Mac OS File Management

Let's Explore the Mac OS

A Journey Through Apple's Computing History

To fully appreciate the Mac OS and understand what sets it apart from its Windows counterpart, let's take a journey through Apple's rich history of computing.

The Genesis of Apple

Our story begins in the late 1970s when two visionaries, Steve Jobs and Steve Wozniak, founded Apple Computer, Inc. Their mission was clear: to bring personal computing to the masses with an emphasis on simplicity and elegance. In 1984, they introduced the Macintosh, a groundbreaking computer that featured the graphical user interface (GUI) and a mouse – innovations that would forever change the way we interacted with computers.

The Macintosh and Mac OS: A Revolution in Computing

With the Macintosh came the birth of the Mac OS. The original Mac OS (Macintosh System Software) offered a radical departure from the command-line interfaces of the time. Its iconic desktop metaphor, complete with icons, folders, and a trash can, made computing more intuitive and accessible to a broader audience. Apple's commitment to user-friendly design was evident from the start.

Mac OS X: The Power of Unix, the Elegance of Apple

In 2001, Apple unveiled Mac OS X (pronounced "ten"), a complete overhaul of its operating system. This version combined the elegance of the Macintosh with the power of Unix, creating a robust and stable platform. Mac OS X introduced features like the Dock, Exposé, and the Aqua user interface, enhancing both aesthetics and functionality.

The Intel Transition and Mac OS on Windows Hardware

Another pivotal moment came in 2006 when Apple announced its transition from PowerPC processors to Intel processors. This change not only improved performance but also opened up the possibility of running Mac OS on non-Apple hardware, although Apple's licensing agreements generally prohibited this.

Mac OS: Modern, Robust, and Privacy-Focused

Fast forward to the present, and we find ourselves in the era of macOS. Apple has continued to refine its operating system, focusing on security, privacy, and integration with its ecosystem of devices and services.

As we journey deeper into the world of macOS, we'll explore its features, capabilities, and the unique experiences it offers to users. Whether you're a long-time Mac enthusiast or new to the platform, there's always something special to discover in the realm of the Mac OS.

Turning on a Mac

To turn on your Mac desktop, simply press its power button. The MacBook Pro turns on automatically when the lid is raised. If the MacBook is in Sleep Mode, press the Touch ID button, located on the far-right side of the touch bar.

How to Turn Off a Mac

The best way to turn off a Mac is to open the Apple menu (in the upper-left corner of the screen) and select Shut Down. This method initiates a shutdown process that closes all open programs and provides opportunities to save any work before powering off.

Another method is to hold down the Touch ID button at the top-right corner of the keyboard for approximately 5 seconds. This method is sometimes called a "force shutdown," as it closes all programs without providing any option to save files.

The Mac OS Desktop

There are many useful features on the Mac OS Desktop

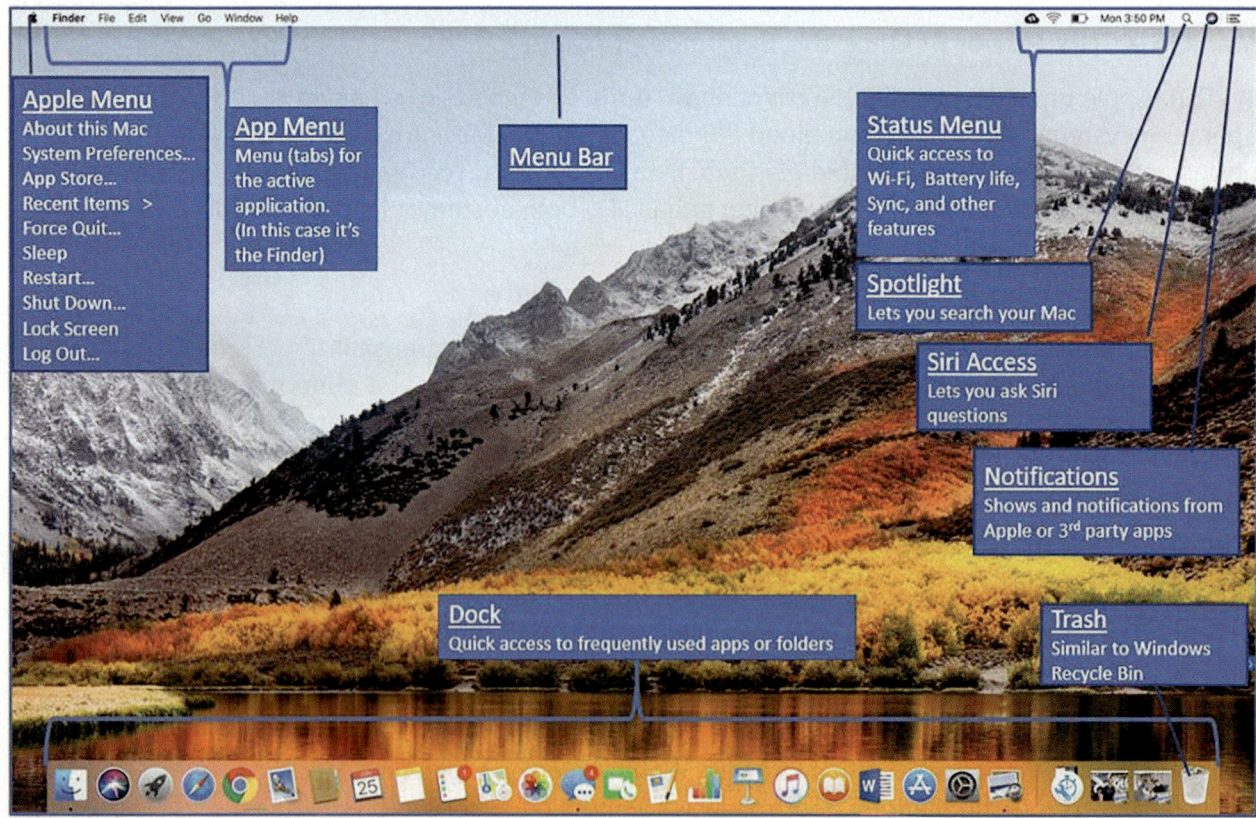

The Mac Keyboard

The Mac keyboard is different from a PC keyboard in several respects.

The major differences are the keys to the left and right of the Space bar, and the Delete key in the upper-right corner. To the immediate right and left of the Mac's Space bar is the Command key.

The **Command** key is used for many Mac shortcuts, much like the **Ctrl** key on a PC. For example, just as **Ctrl + S** is a shortcut for **Save** on a PC, **Command + S** is a shortcut for **Save** on a Mac. The **Control** key on the Mac keyboardis also used very often.

Control + touchpad or **Control + mouse** click often opens menu options in the same way that a right-click on a PC touchpad or mouse does.

The **Delete** key on a Mac keyboard, located in the place that usually contains a Backspace key on a PC keyboard, deletes the character to the left of the insertion point. **Function (fn) + Delete** removes content to the right of the insertion point.

The MacBook Pro Trackpad

The MacBook has a multitouch, force-sensitive trackpad. Multitouch means that you can use multiple fingers on the touchpad. For example, if you spread your fingers when an image is displayed you can zoom in on the image.

Force-sensitive means that when you press down on the trackpad, it makes a click sound that indicates you have pressed it. Just putting your finger or fingers on the trackpad allows you to move the pointer and manipulate icons on the screen. Pressing down on the trackpad (clicking) allows you to use other features. For example, if you click and hold down a link in Safari, a preview of the webpage is displayed.

The Control-Click Gesture

Many PC users are accustomed to using the right-click method of opening menu options. Because the Macintosh mouse was designed with one button, this step is accomplished by holding the **Control** key and clicking the mouse button or trackpad.

Most Mac users prefer to use the two-finger click method instead. Clicking the trackpad with two fingers performs the same function as the **Control + Click** method.

The Touch Bar

Located on the keyboard above the numbers, the **touch bar** provides quick access to many useful tools and adapts to the applications being used at the time. When you first open the Mac, the touch bar displays Touch ID, giving the user the option of logging into the computer with their finger.

After login, the touch bar displays an escape button (**esc**), as well as speaker volume control, brightness control, and access to Siri. When using an app, the touch bar displays commonly used features in the app. Additionally, pressing the Function button (**fn**) displays the function buttons on the touch bar.

The Finder

The Finder menu opens when the MacBook Pro turns on. Clicking on the Finder app allows you to see the applications on the Mac, recent files that have been opened, as well as documents and items on the user's iCloud or other cloud-based accounts such as Google Drive. Basically, the Finder provides an easy place to search or to find files or programs.

The Launchpad

The **Mac Launchpad**, denoted by a grey icon with a rocket, allows users to start apps from a single location. Like the Windows Start Menu, the Launchpad displays icons for every app on the MacBook Pro. It provides an interface that looks like an iPhone screen.

The Dashboard

The **Dashboard** allows users to get to many of the commonly used mini apps (such as the clock, calculator, calendar, and memo pad) as well as small apps, called widgets, that can be used to view weather, news, or business information. To open the **Dashboard**, click on its icon in the Launchpad or use the Finder to open it. To add widgets to the **Dashboard**, click the small plus sign in the lower-left corner of the **Dashboard**.

Mission Control

Mission Control displays all open apps in one screen so users can quickly switch between different apps. To open the **Mission Control** screen, swipe up with three fingers or click the **Mission Control** icon in the Launchpad. You can also open **Mission Control** by using the Finder.

Sleep Mode and Power Nap

Sleep mode allows the Mac to save battery life during periods of inactivity.

To set the Mac to sleep mode or to adjust the specific time and wake settings, open the **Apple** menu (in the upper-left corner of the screen), select **System Preferences**, and then open the **Energy Saver** menu. This menu enables the user to choose the length of inactivity time before the Mac enters **Sleep mode** or to keep the Mac from ever entering **Sleep mode**. It also provides the Power Nap option.

With **Power Nap, the Mac**, while in **Sleep mode**, checks for new email and updates calendar events.

Customizing Mac Settings

How to Add a User to Your Mac

Sharing a Mac can save money and be far more efficient than having everyone in a household using their own computer. The Mac OS makes it relatively easy for multiple users to have their own login information, their own desktop, and their own internet access settings. This is particularly useful for households with younger users.

To add a user to a Mac, the computer's owner, or administrator, first goes to the Apple menu or the Launchpad and opens the System Preferences menu, then selects System Preferences, then Users & Groups. After unlocking the settings, the administrator can click the plus symbol to add new users.

The administrator can set parental controls, share files, and even determine which apps the other users can access.

The MacBook's Security Setting Options

The Mac OS includes several programs to keep malicious software from attacking the computer.

FireVault encrypts your computer's hard drive so that if someone else gets possession of your Mac, your data are still secure. **FireVault** also has an Instant Wipe function that removes the **FireVault** encryption key and empties the hard drive.

The Mac OS also gives you the option of turning off Location Services, so that apps cannot access your location.

Additionally, the Mac OS has a strong built-in firewall program to limit incoming connections. All these security setting options are in the Security & Privacy menu in System Preferences.

How to Create a Backup Using the Time Machine

The Mac OS includes an application called Time Machine that makes it easy to back up or restore your computer.

To use Time Machine, connect your computer to an external hard drive or a storage server on a network. Apple sells an external storage device called a **Time Capsule** that interfaces easily with Macintosh computers, but nearly any external storage device will also work with Time Machine.

When you initially connect the storage device to a Macintosh computer, the Mac OS should recognize the device, and you should be given the option to use it as a backup disc.

If you do not see this option, use the Apple menu or the Launchpad to navigate to System Preferences, then select Time Machine. In the Time Machine menu, select Backup Disk, choose your disk from the list, and click Use Disk.

To restore files that have been accidentally deleted, navigate to the Time Machine menu and open the **Documents** folder. Scroll through the dates on the right side of the window to find the file and then click Restore.

How to Change the Wallpaper

Changing your MacBook's wallpaper, also known as the desktop picture, allows you to customize the look of your computer. There are several ways to select an image as wallpaper.

One method is to choose the image, hold down the Control key, and click the trackpad or mouse button to open a shortcut menu. From this menu, select Share, and then select Set Desktop Picture.

Another method is to open the Finder and navigate to the desired image. Then, Control -> Click the image and select Set Desktop Picture from the menu.

Finally, another way to set an image as wallpaper is to use the Apple menu or the LaunchPad to open the System Preferences menu and select Desktop & Screen Saver. Click the Desktop tab, navigate to the desired image, and click on it to set the wallpaper.

How to Adjust Screen Savers and Hot Corners

A **screen saver** is an image or images that the operating system displays when the computer is in **Sleep mode**. Changing your MacBook's screen saver allows you to customize the look of your computer and adjust when it goes into **Sleep mode**.

The Screen Saver menu includes many options, including displaying a slide show and having a message on the screen. The Mac OS screen saver options also provide a tool called **Hot Corners**, which has several controls that can be activated by clicking your cursor on a corner of the screen.

Hot Corners options include entering **Sleep mode** and opening the **Dashboard**, **Mission Control**, the Launchpad, and other choices. To adjust the screen saver or to set **Hot Corners**, use the Apple menu or the Launchpad to navigate to System Preferences, select Desktop & Screen Saver, and then select the Screen Saver tab.

How to Adjust the Force Touch Trackpad

The MacBook Pro comes with what Apple calls a Force Touch trackpad. Using this trackpad's capabilities allows users to operate the MacBook more efficiently. To view these capabilities, use the Apple menu or the Launchpad to navigate to System Preferences, then select Trackpad.

The Trackpad menu has three tabs: Point & Click, Scroll & Zoom, and More Gestures. Under Point & Click, you can select what occurs when you tap the touchpad (without clicking), click the touch pad, and click or tap the touchpad with two fingers.

You can also adjust the force needed to click and how fast the cursor tracks as you move your finger across the trackpad. Scroll & Zoom allows you to choose which direction you scroll and allows you to "pinch" to zoom in on an image.

More Gestures provides options for using three fingers or fingers and the thumb.

Mac OS File Management

How to Create a Folder

Files and documents on a MacBook are kept in folders. Creating folders helps users keep files organized. To create a new folder, first use the Finder to navigate to where you would like the new folder to be stored, or just create a new folder directly on the desktop. Once at the site of the new folder, there are three methods for creating the folder.

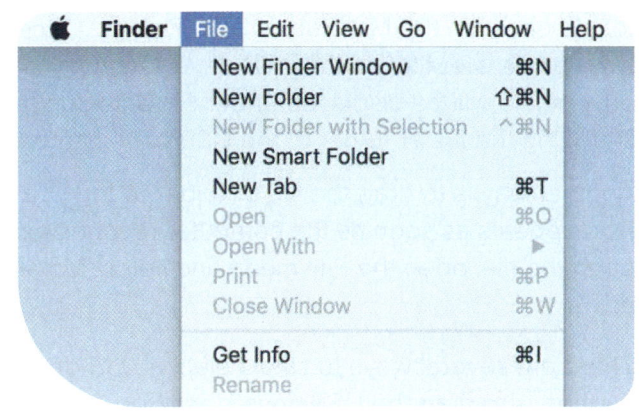

One method is to open the **Finder**, open the **File** menu, and select **New Folder**. Another method is to hold down the **Control** key and click the **trackpad**. A third method is to press **Shift + Command + N**. Once the new folder is created, enter a name for it and press **Return**.

How to Save a File in a Folder

There are several ways to save a file in a folder. When you first create a document, you name it and save it. At this point, you have the option to select which folder to save it in. The default folder is called **Documents**. If you have already created a file, such as an image or a document, you can drag it into the folder. Alternatively, you can open the File menu and then select Move To.... If you wish to save a file to a completely different disk, hold down the Command key and then drag the file to the disk.

How to Copy a File

There are several ways to copy a file. If you have an opened document, such as a Pages document, then you can open the File menu and select Duplicate. With the file open, you can also use the shortcut of Shift + Command + S (recall that a shortcut for Save is Command + S).

If the file is closed, you can simply hold down the Option key and drag the file's icon to a new folder. This places a copy of the file into the new folder while leaving the original in its folder. To create a copy of the file in the same folder, select the file and click Command + D.

How to Copy a File to a USB Drive

Storing files on USB drives provides an easy way to back up files and transfer them to other devices.

Because many USB drives have USB-A connectors, you may need an adapter to use one with the MacBook's USB-C ports. Once connected, there are several ways to transfer files to the USB drive. One way is to open the Finder. The USB drive name will be displayed under Favorites on the left side of the Finder window. Dragging the file or folder to the USB drive name copies the file onto the USB drive.

Another way is to drag the file or folder to the USB drive icon on the desktop. The icon appears as soon as the computer recognizes the USB drive. A third way is to open the file, open the File menu and select Move To..., and then select the USB drive.

There are several ways to safely eject a USB drive from your computer. A common method is to drag the USB drive's desktop icon to the Trash. When you drag the drive's icon, the Trash icon changes to an Eject icon. The USB drive icon that appeared on the desktop when the drive was inserted will disappear, indicating that it is safe to remove the drive.

Another way is to open a Finder window. On the left side of the window, under Favorites, locate the USB drive's name. Next to the name is a small Eject symbol. Clicking this Eject symbol closes the Finder window and removes the USB icon from the desktop, indicating that it is safe to remove the drive.

Another method is to click on the USB drive icon on the desktop, open the Finder, and then expand the File menu. The option to eject the USB drive will be toward the bottom of the menu.

How to Rename a File

There are several ways to rename files and folders.

One way is to select the file or folder by clicking on it. Then press Return and enter the new name.

Alternatively, for a file such as a document, you can open the File menu and then select Rename....

Unlike in Windows, the Mac OS allows you to use nearly every symbol when naming files or folders except for the colon (:). This is because the colon, and sometimes the slash (/), are used to separate directories in paths or file addresses.

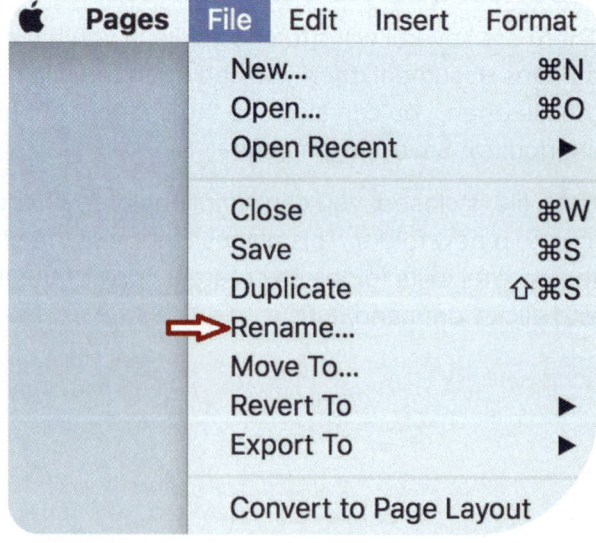

How to Delete a File

To delete a file or folder from a MacBook, first drag it to the Trash. Trash is a folder located on the far-right side of the dock. The icon for the Trash is easily identified because it looks like a trash can.

Like the Recycle Bin in the Windows OS, files placed in the Trash folder are not deleted until the Trash is emptied.

Another method is to hold down the Control key while selecting the file. This opens a menu from which you can choose Move to Trash.

A shortcut for deleting a file or folder is to select it by clicking on it, and then pressing Command + Delete.

How to Empty the Trash

Trash is a folder that the Mac OS uses to store files and folders before permanently deleting them. To clear the computer of unneeded files, drag them to Trash, located in the far-right side of the dock. The files can be retrieved at any time by clicking on the Trash and dragging the desired file out.

Empty Trash means that the Mac OS can rewrite over the hard-drive space that previously held the discarded files. Essentially the files are irretrievable.

To use Empty Trash, select Finder, then open the Finder menu and click Empty Trash. A shortcut to Empty Trash is Shift + Command + Delete. Alternatively, holding down the Control key and clicking on Trash allows you to select Empty Trash.

How to Take a Screenshot

There is no PrtScr button on a Mac keyboard. The shortcut for a screenshot on the Mac is Command + Shift + 3. This pastes a .png file of a screenshot of the entire screen onto the desktop.

To select a rectangular area of the screen and create a screenshot, use Command + Shift + 4. This pastes a .png file of the selected area onto the desktop.

If you wish to copy the screen and then paste it into a document, like using the PrtScr button on a PC keyboard, press Command + Control + Shift + 3. In the same way, Command + Control + Shift + 4 copies a selected area of the screen.

How to Use Grab

The Grab utility in the Macintosh OS works similarly to the Snipping Tool found in the Windows OS. To use Grab, open the Finder and select Applications. In Applications, open Utilities and select Grab.

Dragging the Grab icon to the dock allows you to open the utility with one click.

To "grab" a portion of a document or image, open the Capture menu and click Selection. The Grab tool has many interesting options such as capturing timed actions and sounds.

Chapter Review

1. How do you turn on a MacBook Pro, and what is the alternative method if it's in Sleep Mode?

2. What are the steps to properly turn off a Mac, and how does this process differ from a force shutdown?

3. What are the major differences between a Mac keyboard and a PC keyboard, and how are Command and Control keys used?

4. How is the Control-Click gesture used on a Mac, and what is its equivalent for PC users?

5. Explain the functionality of the touch bar on a MacBook Pro and its adaptability to different applications.

6. What steps should be followed to add a user to a Mac, and what privileges can the administrator control for these users?

7. Why is it beneficial to create folders in Mac OS, and what are the methods to create a new folder?

8. Describe the steps to copy a file or folder to a USB drive using the Finder in Mac OS.

9. How can you transfer files to a USB drive by dragging them to the USB drive icon on the desktop, and when does this icon appear?

10. How does dragging the USB drive's desktop icon to the Trash indicate that it's safe to eject the drive, and what happens to the icon afterward?

11. Describe the process of using the Finder window to eject a USB drive, including locating the drive's name and the eject symbol.

12. Explain the process of permanently deleting files from the Trash and how it compares to the Recycle Bin in Windows.

13. What is the keyboard shortcut for moving a file to the Trash when you select it, and why might this be a convenient option for users?

Ethics Discussion

Is it ethically responsible for technology companies like Apple to design their operating systems in a way that makes it easy for users to store and transfer personal data, while also maintaining robust security features to protect that data from unauthorized access?

9 Networking and Connecting to the Internet

What To Expect

After completing this chapter, you will be able to:

- describe how the Internet works, what is meant by an Internet Service Provider and how to choose an ISP;
- describe how to create and effectively use a home network;
- describe the World Wide Web and how to safely browse the web;
- identify different types of networks;
- identify Cloud computing and its characteristics;
- describe how Cloud computing affects the business landscape;
- describe the benefits and limitations of Cloud computing;
- describe various Cloud deployment models.

Chapter Topics

- How the Internet Works
- The Web
- Internet Transmission Basics
- Internet Communication Devices
- Connecting to the Internet
- Determining Your Internet Speeds
- Setting Up a Wireless Network at Home
- Intranet vs. Extranet
- Networks
- Cloud Computing
- File Transfer Utilities
- Edge Computing
- Net Neutrality

Copyright © McGraw Hill   spainter_vfx/Shutterstock

Let's Explore the Internet

Imagine an Internet City

The internet is a global network of interconnected computers and systems that share and exchange information. But for now, let's think about the internet like a massive city. Instead of buildings and interconnected roads, it's made of computers and networks linked together. Like a city, it has different departments and rules that address its infrastructure, but it doesn't have a single controller governing all the roads and routes. Each connected part of the internet decides how it wants to move and send information.

In our "internet city," let's think about a basketball team as a piece of information that must get from one place to another. The information could be an email, a photograph, or a computer program you want to download. Imagine there's a basketball tournament in the city, and teams from every part of the nation are walking through the sprawling metropolis to eat, rest, and most importantly, reach their games on time.

Initially, the teams (our pieces of information) walk or drive from place to place together, but they often find themselves stuck in traffic jams or just have trouble moving together as a group. Can you imagine a full team of basketball players trying to walk down a crowded street? It's impressive, but slow. The players need to stick to larger sidewalks and always make sure everyone is together.

Tired of the delays, one coach has an idea. Each player, with their agile movements, could navigate the city streets faster alone. So, they all leave their hotel separately, and run through the city individually. The players zip on electric scooters, cruise down avenues on bicycles, run through parks, and even get into cabs on main roads whenever things seem clear.

The idea works great! Players begin arriving at the stadium from different directions. Some are early, others barely make it, but enough players are there to start the game. This nimble navigation becomes their new game-plan off the court, proving sometimes it's faster to divide and conquer.

So, what can this basketball saga teach us about the internet? First, it's faster to move smaller bits of information along different paths. Second, if the information is moving along different paths, it's unlikely that you'll lose all the information. And, even if you lose some of the information, you'll likely still have enough to use.

What's on the Internet?

In our "internet city" analogy, the internet is the city's infrastructure—that's the solid stuff that makes up the city. Everything else in the city—the cars, trucks, shops, and services, is being hosted on the internet. So, what exactly can we find in this vast digital metropolis?

World Wide Web (WWW): The WWW consists of websites and browsers. When you open a browser and type in a website address, you're visiting a specific location on the WWW. It's like window-shopping in a mall, where each store is a different website.

Email (Electronic Mail): Email is one of the oldest and most essential parts of the internet. It's the digital equivalent of a postal system, where you can send and receive letters (emails) to and from anyone in the world.

File Transfer Protocol (FTP): The internet isn't just about browsing websites or sending emails. FTP services allow you to send or receive large files, almost like sending packages in the mail.

Voice and Video Calls (VoIP): Services like Skype, Zoom, and WhatsApp allow us to make voice or video calls over the internet. It's like having a global telephone network without the traditional phone lines.

Social Media Platforms: Places like Facebook, Twitter, and Instagram are digital social hubs where people connect, share, and interact with friends, family, and even strangers.

Streaming and Entertainment: Platforms like Netflix, YouTube, and Spotify let us watch movies, listen to music, or even attend virtual concerts and events.

Online Gaming: Digital arenas are where players from around the world can compete, collaborate, and immerse themselves in virtual worlds.

Cloud Services: Think of this as renting storage lockers or spaces, but digitally. Services like Google Drive or Dropbox allow you to store, share, and access your files from anywhere.

E-commerce and Online Marketplaces: Platforms like Amazon and eBay are the bustling marketplaces of the digital city, where you can buy and sell almost anything.

Learning and Information Platforms: From digital libraries like Wikipedia to online courses on platforms like Coursera, the internet is a treasure trove of knowledge.

This list is just the tip of the iceberg. As technology evolves, new "districts" and "neighborhoods" continue to emerge in our digital city, enriching the ways we connect, learn, play, and work on the internet.

How the Internet Works

Networks: Networks are like our basketball teams, each with its own game plan. Every network that's part of the **Internet** City chooses its own way of operating. If one network, or team, faces a problem, it doesn't mean the whole tournament stops.

ICANN: This acronym stands for "The **Internet** Corporation for Assigned Names and Numbers." Just like a basketball tournament needs an organizer, the internet has ICANN. It's responsible for making sure every team, or in this case, every network, has a unique name and number.

The Backbone: Picture the main highways in our city. That's like the internet backbone. It's the fastest route made of large cables that connect different parts of the city.

Nodes: The major crossroads on the streets are the nodes.

Servers: Just like stadiums host the games and store records, servers are the computers that store the web pages we visit.

Routers: Think of routers as city guides, coaches, or the navigation system that tells the players where to go. Routers know the routes, what the traffic looks like, and decide the quickest way for information to reach its destination.

Packets: Packets are the pieces of information that need get from place to place. In our analogy, the critical piece of information is the team, and the packets are the players. But here's where our analogy breaks down a bit. The packets are agile and small, but they don't decide where to go on their own—the overall network conditions and routers control them.

Packet Switching: Think of breaking a big message into smaller parts and sending each part on its own quick journey. All these small parts can take different roads, and when all the parts get to their destination, they come back together as the full message. In our analogy, packet switching is when each player finds their own way to the game.

Circuit Switching: Circuit switching is when you make a special path just for two people to talk, and no one else can use that path until they're done talking. The public telephone system that runs on traditional land lines uses circuit switches.

In our basketball tournament analogy, you might have been thinking—why don't they take a bus? Circuit switching is a bit like taking a bus and booking a private road. It's secure and, if we're really talking about moving people though a city—it would be the saftest wa to travel. But what if the bus is delayed? You lose the whole team. Likewise, in circuit switching, if the path breaks, you lose the connection.

History and Techniques of Internet Transfer

SAGE

In the 1950s, the United States set up a large radar system called Semi-Automatic Ground Environment (SAGE). Think of it like a network of lookouts. But there was a big problem: if one lookout got knocked out, the whole system might break. Just like in our city, if the bus taking the basketball players to the game broke down, it'd be game over for that team.

That's when Paul Baran stepped in. He was like the basketball coach in our story who wanted a foolproof way to get his team to the game. He thought of a clever way to send messages. Instead of sending one big message on one path, he asked, "Why not break the message into small pieces and send each piece a different way?" That way, even if one path was blocked, the message could still get through on other paths. And, even if the whole message didn't get through, enough of the message could get through to make sense of it.

ARPAnet

In the 1960s, using Paul Baran's cool idea about sending messages in bits and pieces, the US Defense Department made ARPAnet. They hooked up computers from four big schools so they could share stuff and chat. The best part? Even if one computer stopped working, the rest kept going. This was a big step toward creating today's internet.

The Web

In a transportation analogy, the internet is the infrastructure, the network of "roads." Now, imagine the World Wide Web (or just "the web") as all the buildings, houses, stores, and places you can visit on these roads. Each building or house is like a website or web page. Just like you would travel on roads to visit different places, you use the internet to access different parts of the web.

History of the Web

The Internet without the Web

In the late 1970s and early 1980s, even though the concept of the internet existed, it wasn't exactly user-friendly. If you wanted to access information from another computer on the other side of the world, you'd have to go through a tedious process, kind of like texting commands. Imagine having to type a long string of commands every time you wanted to open a single photo or read a short document. It was powerful, but for most people, it was like trying to navigate a maze blindfolded.

Steve Jobs and the GUI

Now, around the same time, some people started experimenting with a graphical user interface (GUI). That's a mouthful, but you can just say "gooey." It's simply the idea of interacting with your computer using visuals—like clicking on folders or dragging icons. You do it all the time nowadays.

The person who really pushed this idea into homes around the world was Steve Jobs, the founder of Apple. In 1984, his Apple Macintosh computer brought this visual way of computing to everyday folks, and computers suddenly felt less intimidating.

THE ARPA NETWORK

DEC 1969

4 NODES

Tim Berners-Lee and Hyperlinking

Seeing this visual revolution, Tim Berners-Lee, a British computer scientist, had a eureka moment: What if the internet could be as visual and intuitive as this GUI thing? He wanted to make the internet feel like flipping through a magazine, where you could just tap a word or image and instantly zip over to a new page or topic. This idea of making connections so effortlessly was called "hypertext," and it was groundbreaking. Tim worked with Robert Cailliau, a Belgian computer scientist to make it happen.

Web Evolution: From 1.0 to 3.0

Thanks to hypertext, we got the web. Instead of a maze, the internet started to feel like a vast, interconnected library, with billions of books and every page linked to another. The US government saw the web's incredible potential and wanted to ensure it stayed organized and always improved. The US government asked Tim, Robert, and other bright minds to form the World Wide Web Consortium (W3C). Their job? Keep the web evolving, organized, and, most of all, fun to use for everyone. And the web, evolved over time.

Web 1.0: In the early days, mostly you'd just visit websites and read them, much like flipping through a digital magazine.

Web 2.0: At this stage, the web became interactive! You could write on blogs, share photos, and interact on platforms like social media. It was all about creating and sharing content. Faster internet and stronger computers made this possible.

Web 3.0 (Semantic Web): Picture the web having its own brain. It starts to understand and connect information in smarter ways. So, when you search for something, it knows just how to help, giving you better and more relevant results.

URLS

A Uniform Resource Locator **(URL)** is an address for a website or a web page. Think of it like your home address but for a site. It has three main parts:

- **Protocol:** This is often "http," and it's like the rules for how to open a web page.
- **Domain Name:** This is the main part of the web address, like "google.com."
- **Path:** This leads you to a specific page or image on that website.

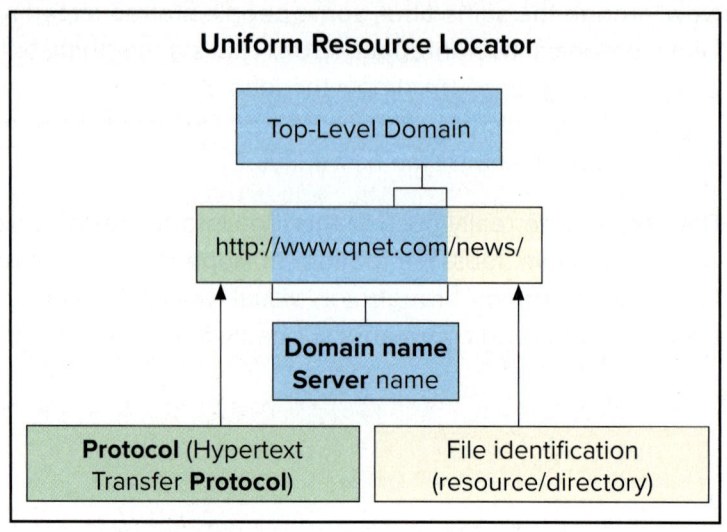

Understanding Protocols

A protocol is like a set of rules for how to do something on the internet. For example, http shows webpages, and ftp helps move files from one computer to another.

Domain Names and Top-level Domains

The domain name points to a specific place on the internet. Take "www.google.com" as an example:

- "google" is the domain name.
- ".com" is the top-level domain. Other top-level domains are ".edu" and ".org."

Web Browsers

Browsers are like doorways to the web. They help you visit and explore websites. Popular web browsers include

- Chrome;
- Microsoft Edge (previously **Internet** Explorer);
- Safari;
- Firefox.

Choosing a Browser

Choosing a browser depends on what you like and your computer type. But it's smart to have more than one, just in case!

- If you're working in Windows, Microsoft Edge may be the smoothest and fastest browser because it is made by Microsoft.
- With Macintosh computers, Safari works fastest and smoothest because they are both Apple products.
- Chrome is the most comfortable for those who like using the Google search engine and for those who use Gmail.
- Firefox is a very agile browser and has a reputation of being great with most plug-ins.

Whichever you choose, be sure to have at least two browsers on your computer. That way if one has a glitch, you'll have another means of accessing the web.

Browser Add-ons: Plug-ins

Plug-ins are like tools or add-ons for your browser. They let you do more things, like watch videos or play games. Examples include the following:

- Adobe Flash Player for watching certain videos;
- Java for some interactive content.

Staying Safe: Browser Filters

Brower filters are settings in your browser that block certain websites or content. Parents might use them to keep kids safe online. Some countries use them to limit what their people can see.

Examples of Top-Level Domains

Original

.com = commercial

.gov = government

.mil = military

.edu = education

.org = non-profits

Generic

.menu = restaurants

.bar = pubs

.surf = surfing

.gripes = opinion sites

.cab = taxis

Country code

.ca = Canada

.uk = United Kingdom

.il = Israel

.au = Australia

.ie = Ireland

Internet Transmission Basics

Transmission Media

This refers to the different pathways or channels that internet signals use to reach devices. Different types of transmission media have different capabilities. Let's explore some of the main ones.

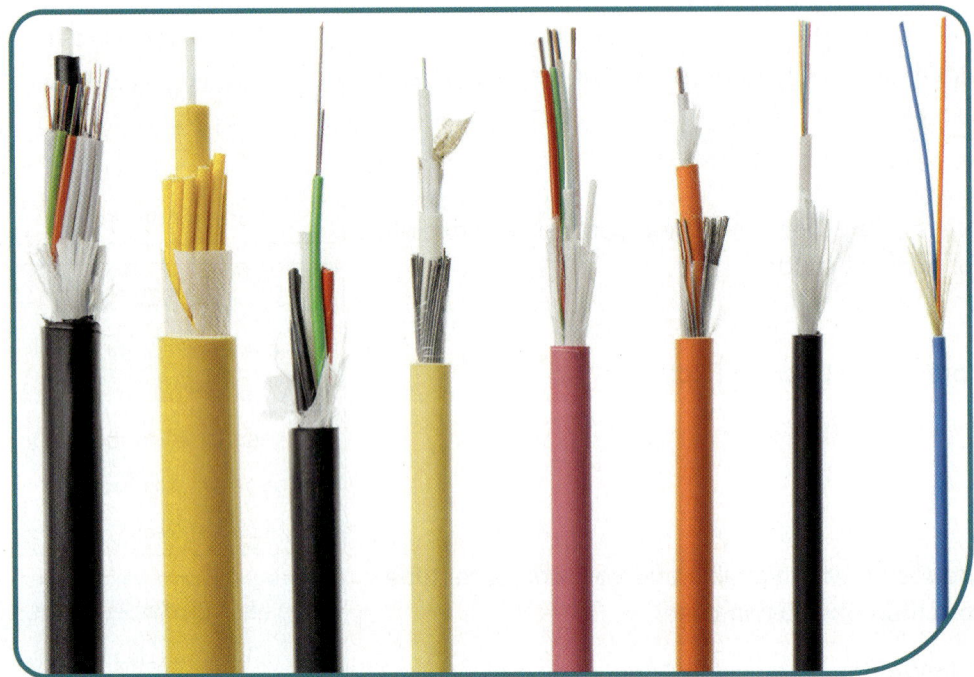

Fiber-Optic Cable: Fiber-optic cable uses light waves to carry internet signals at extremely high speeds. It is composed of a thin glass wire wrapped in cladding and insulation and is particularly useful for transmitting signals over long distances. Because light waves rather than electricity carry the signal, there's minimal electromagnetic interference. Fiber-optic cables are even used to connect continents with cables running across the ocean's floor.

Coaxial Cable: Known as TV cable, it's widely used for home internet connections.

Twisted Pair: Common telephone wires connect homes to local stations.

Ethernet Cable: This type of cable is specifically designed for connecting computers and digital devices.

Wi-Fi: Wi-Fi provides wireless access to the internet. The term *Wi-Fi* is generally thought to mean "wireless fidelity." But it's just a snappy trademarked name that has no meaning. And it gets the job done. With Wi-Fi, multiple devices can use the same Wi-Fi signal. Wi-Fi sources are often called "hotspots."

Cellular: Mobile devices use cellular signals to access the internet. Smartphones can also turn cellular signals into Wi-Fi.

Microwave: Useful in remote areas, microwaves can send signals over huge distances, including to satellites.

Bandwidth

This refers to the amount of data that can a transmission medium can carry in a certain amount of time, usually a second. Different transmission media can handle different amounts of data. For instance, fiber-optic cable can carry way more data at once than a coaxial cable. That's why some internet connections are faster than others.

Broadband

This term describes internet services that are always on and fast. The kind of transmission media used often determines how "broad" or fast the broadband is. For example, a fiber-optic service usually offers higher broadband speeds than a DSL service using twisted pair. The US government even has standards for what counts as broadband, based on how much data the connection can carry.

Internet Communication Devices

Once you know the means through which you will be able to connect to the internet, you need the right device to facilitate the connection. Modems, routers, switches, and MAC addresses all have a role to play in this process.

Modems

The word *modem* is a contraction for *modulation-demodulation*. Modems are nearly always provided by the internet service provider (ISP). However, if you want to save rental fees, you can buy a router/modem. If your internet connection seems to be slow, it may be time to get a new router/modem.

Internet signals traveling over long distances travel in waves. Waves have height (also called amplitude) and frequency. This height plus frequency signal is called an analog signal or a modulated signal.

Computers store and process only on/off signals. These on/off signals are called digital signals or demodulated signals. A modem changes the incoming modulated signals into digital signals and changes outgoing digital signals into modulated signals.

Routers

The router is the most critical part of a home network. A router connects two networks, for example, a home network to an internet nternet service provider's network. Because of this, a router can serve as a gate that protects a home network. This protection is called a hardware firewall.

Routers can be wired only, or wired-wireless. With a wired only router, Ethernet cables connect the router to computers. With a wired-wireless router, the router also transmits Wi-Fi signals. Routers include a device called a switch. **Switches** allow multiple computers to use the same router.

Dual-Band Routers

Dual-band routers are Wi-Fi–capable routers that use both shorter- and higher-frequency signals to transmit data. With the same power, lower-frequency signals travel farther but are subject to more interference and transmit less data. Higher-frequency signals transmit more data and have less interference, but they travel shorter distances. Also, microwave ovens, car alarms, and Bluetooth devices send signals in lower frequencies, creating more interference there.

Wi-Fi Protocol	Frequency	Potential Speeds
802.11a	2.4GHz	54Mbps
802.11b	2.4GHz	54Mbps
802.11g	2.4GHz	54Mbps
802.11n	2.4 or 5GHz	600Mbps
802.11ac	2.4 and 5GHz	~1Gbps

Dual-band routers

Also note: microwave ovens, car alarms, and bluetooth devices transmit waves in the 2.4GHz range, often causing interference.

Switches

Switches connect devices that are within the same network. Most home routers come with built-in switches that identify the different computers, printers, game consoles, and other devices that are connected to the router. **Switches** ensure that the correct data are sent to and from the correct device within the network.

MAC Addresses

Media access control addresses, known as MAC addresses, allow switches to distinguish between different devices in the same network. Each device in a network has a network interface card, or NIC. Each NIC has a unique MAC address. Because of MAC addresses, switches can ensure that the correct data are sent to and from the correct device in the network.

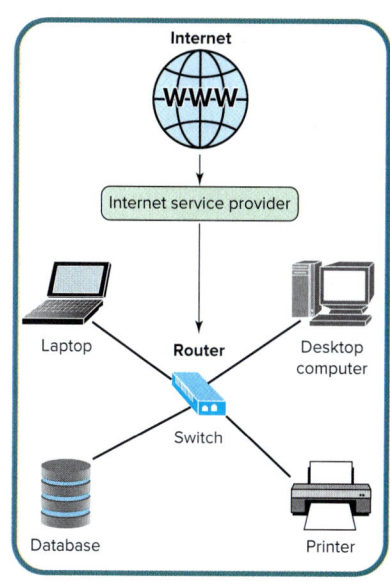

IP Addresses

When you log onto the internet from your home computer, your ISP assigns an internet protocol (IP) address to your router. This IP address allows other computers on the internet to send data to your home computer. All devices that connect through the same ISP router use the same public IP address. When you log off the internet, your ISP may assign the IP address to another subscriber. This is called dynamic addressing. Some organizations have permanently assigned IP addresses. This is called a static IP address.

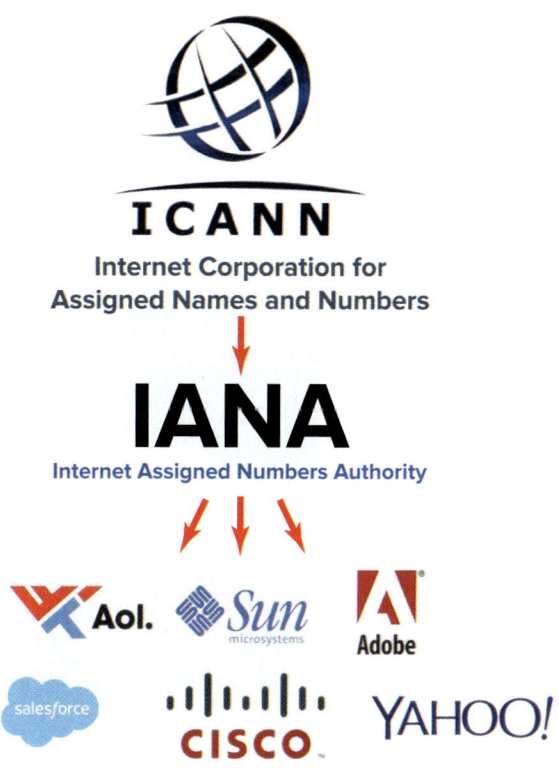

Portions of an IP Address

An IP address identifies the network on which the host resides and the host on a network. To identify these elements, an IP address is broken into two portions. The network portion, or network identifier, is indicated by a certain number of bits, starting with the leftmost bit.

The host portion is indicated by the remaining bits that are present after the network identifier and identifies the host of the network.

Networking devices use the network and host portions of an IP address to determine on which network the host resides and to determine whether the host is local or remote.

How are IP addresses assigned?

IP addresses are assigned by the **Internet** Corporation for Assigned Names and Numbers (ICANN). ICANN assigns blocks of IP addresses, which in turn assign the addresses to their customers.

When you purchase your internet service from an ISP, you receive the right to use their block of assigned addresses.

How are IP addresses determined?

An IP address is determined by the network on which it is hosted. All hosts on the same network share the same established network address but are assigned a unique host number. For example, the two nodes on the 262.168.147 network might be identified by 262.168.147.240 and 262.168.147.242.

Often IP addresses are structured and assigned by a service called a **Dynamic Host Configuration Protocol (DHCP)** or by a network administrator. A DHCP service will automatically assign IP addresses to a host that leases the IP addresses from the service provider.

Connecting to the Internet

To get on the internet, you need an ISP. The ISP allows you to surf the web, watch videos, and send emails. Just like there are different types of roads, there are different ways ISPs connect you to the internet—through fiber-optic cables, regular cables, DSL, dial-up, or even using your cell phone signal. Remember, if you want to get online, you'll need an ISP.

Finding an ISP

So, how do you pick the best ISP for you? First, check what's available in your area. An easy way to do this is to search online. If you don't have internet yet, you can use a computer at a local library. A couple of handy sites to help you are www.dslreports.com and www.broadbandmap.gov. Type in your address or zip code, and you'll see a list of companies that can help you get connected.

DSL

Digital Subscriber Line, or DSL, provides internet service with signals that are carried over local telephone lines. DSL is available in most areas depending on the home's distance from the main telephone exchange.

Because individual wires, which are not shared, carry the signal, DSL download and upload speeds are very consistent. The bandwidth of telephone wires isn't as large as television (coaxial) cable bandwidth, so DSL's top speeds are usually less than cable internet's top speeds.

Many DSL providers include wireless routers with their DSL modems, simplifying the installation process.

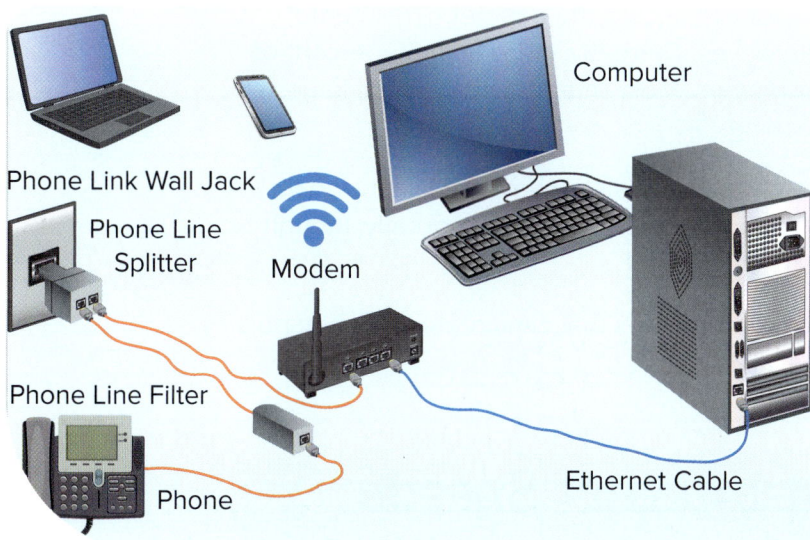

Cable Internet Service

Cable internet service providers send internet signals to your home using television (coaxial) cables. Cable internet service is available wherever cable television service is available.

Top cable internet upload and download speeds are usually faster than top DSL speeds, but because the cable's bandwidth is shared with your neighbors, your speeds can be inconsistent, particularly during peak usage hours.

Sometimes cable modems don't include routers, so you may need to purchase one separately.

Let's Explore Search Engines

Imagining the internet as an enormous, interconnected, rapidly changing city is a great way to think about search engines. Navigating through an urban maze to find exactly what you're looking for can be a daunting task. Your search engine is your city guide. And this guide doesn't just know every part of the city; it knows how to get you there quickly!

The Role of Your Guide. First, your guide listens to your question, or (in search engine terms) your query. Then it consults its extensive map, checks its database to find where this topic might be discussed or displayed, and points you to the places where you can find what you're seeking.

Well-Known Guides. Some city guides (like Google, Bing, or Yahoo) have become famous for their expertise. They all employ intricate maps—a database of URLs—to take you exactly where you need to go.

Navigational Mastery: Boolean Operators. When asking for directions, specific phrasing matters. For example, if you're looking for a restaurant that serves "burgers OR pizza AND dessert," Boolean operators help refine your inquiry to find the perfect spot.

Advanced Exploration Tools. Just as a real city guide might offer specialized tours, search engines offer advanced search tools. For example, you can search for the exact phrase "chocolate chip cookie recipe" in quotation marks to find recipes that include those exact words.

Behind the Map: The Database. Just like city officials keep updated maps, search engine companies maintain extensive databases. Think of it as a comprehensive travel guidebook that's continually updated with new attractions and places to explore.

The Art of Attraction: SEO. Imagine a shopkeeper who wants more foot traffic. In our digital city, this is known as Search Engine Optimization (SEO). For example, a bookstore might use the keyword *bestsellers* on its website to attract more visitors searching for popular books.

Vetting the Locales: Website Evaluation. Not all streets in a city are safe or reliable. If a site about health advice has numerous spelling errors and no medical credentials, it's like a sketchy, poorly maintained area you might want to avoid.

Expert-Level Exploration. For those interested in truly diving into the depths of our digital city, advanced search techniques can provide an entirely new level of detail in your explorations. Using HTML objects in your search could be like using a specialized tour guide who knows the architectural details of each building.

With search engines, you're not just a visitor; you're a traveler in the ever-expanding digital city that is the internet. Equip yourself with these skills, and you'll never be lost again. Happy exploring!

Search Engines

Though they all aim to make the world's information easily accessible and useful, not all search engines are created equal. Three of the most used search engines are Google, Bing, and Yahoo!.

Common Features

Before diving into what sets these search engines apart, let's talk about what they have in common. Most search engines, including these big three, use a similar underlying technology. They rely on databases filled with internet Uniform Resource Locators (URLs). Each record in this database includes a variety of information such as the site's URL, description, title, and even the keywords contained within the site. When you type a query into the search bar, the search engine quickly sifts through this database to provide the most relevant results.

What Makes Them Different?

Google

Google is the most widely used search engine and is known for its simple, clean interface. It offers various services integrated into its platform, such as Google Maps, Google Scholar, and Google Images, among others. Google's algorithms are highly sophisticated, often providing the most relevant search results, which is why many people prefer using it.

Bing

Bing, developed by Microsoft, offers a visually rich interface with a new background image every day. Some users find Bing's image and video searches to be superior to Google's. Bing is also integrated into Microsoft's suite of services and products, including Windows OS and Microsoft Office, offering a more seamless experience for users in the Microsoft ecosystem.

Yahoo!

Yahoo! Search, while not as popular as Google or Bing, still has a dedicated user base. Yahoo! offers an integrated experience with its other services such as Yahoo! Mail and Yahoo! Finance. Its interface is a bit more cluttered, packed with news, trending topics, and more, which some users find more informative right off the bat.

User Interface and Preferences

The user interface plays a big role in why someone might prefer one search engine over another. Google offers a minimalist design focused on speed and efficiency. Some people prefer its no-nonsense, straightforward approach that lets them find what they're looking for without distraction. Bing provides a more visually engaging experience, aiming to inspire and entertain as you search. Some users appreciate these aesthetic touches and the deeper integration with Microsoft's services. Yahoo! serves up a content-rich interface that tries to offer something for everyone. It appeals to those who like to have a variety of information at their fingertips right from the homepage.

In summary, while search engines may share a lot of similarities, the user experience, interface, and integrated services often guide user preferences. Whether it's the efficiency of Google, the visual appeal of Bing, or the content variety of Yahoo!, each search engine has unique offerings that attract different users for different reasons.

Search Engine Databases

The information contained in a search engine database is determined by how each specific company records URL information. It is common practice for the website owner to submit their URL and associated information to the search engine company (SEC). SECs use indexes to speed up the search process. These indexes include information about the URL and the content contained on the site, including directories, links, and even frequently asked question (FAQ) pages.

What's the Deal with Hashtags?

You've probably seen hashtags—those words or phrases that start with the pound sign (#)—on social media. But what do they do? Hashtags are like labels for posts. They help you find all the posts about the same topic. So if lots of people are sharing pictures of their vacations and using the hashtag #StudyBreak, you can click or search that hashtag and see everyone's vacation pictures in one place!

The cool thing is that the hashtag isn't anything fancy or technical. It's just a regular pound sign followed by a word or phrase. But when lots of people use the same hashtag, it makes it super easy to search for a topic. So the next time you see a hashtag, you'll know it's a way to group things together and make them easier to find!

SEO

Search engine optimization (SEO) is a set of strategies, techniques, and practices used to increase the number of visitors to a specific website by getting a high-ranking placement in the search engine.

Studies show that 87 percent of web searches end at the first page of results, so a higher-ranking page helps to get more website visitors. SEO helps to ensure that a website is known and reachable by search engines, thus increasing the likelihood of more site hits.

SEO Best Practices

- **Choose the right words.** Think of words that people would use to find your website. These are your "keywords." Make sure to use these words in your website's content so it shows up when people search for those terms.

- **Keep content helpful and relevant.** Make sure the information on your website is useful to the people you want to visit it. If you're making a website about pets, for example, include tips on pet care or fun pet facts.

- **Make keywords and pictures work together.** Use your keywords in the titles and descriptions on your website. For pictures, add captions and alt text (text that describes the picture) so search engines know what the picture is about.

- **Make your website user-friendly.** Make sure your website loads quickly and is easy to read. Keep things neat and organized so visitors can find what they're looking for.

Boolean Operators

Boolean operators allow searchers to find more accurate search results. Some of these characters include the terms *AND*, *NOT*, and *OR* (all capitalized). Using BOOLEAN operators can either reduce or expand the number of search results returned to a user when conducting a web search. Their use allows for more targeted search results and can limit the number of inappropriate results that are returned.

- Connecting keywords with *AND* searches for multiple items in the same web page or article.
- *NOT* excludes web pages or articles with the designated keyword.
- *OR* searches for multiple keywords on different pages.

The Boolean Operators *OR* and *NOT*

When you're searching for information and want to include more than one term, you can use *OR* in your search. For example, let's say you're interested in learning more about puppies and kittens. You could type "puppies OR kittens" in the search bar. This will give you all the results that talk about puppies, kittens, or both. It's a cool way to cast a wider net and get more info on the things you're curious about! And, just so you know, you can also use a special symbol like "/^" to do the same kind of *OR* search.

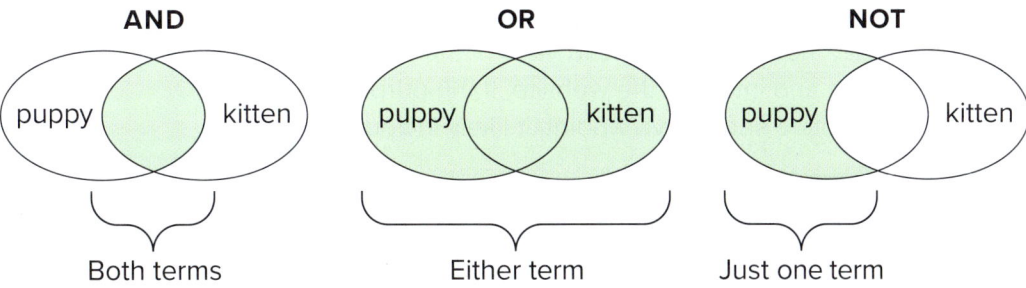

Advanced Search Tools

To be more specific when conducting a search on the web, tools such as parentheses quotation marks can help narrow the search results.

- Quotation marks (" ") allow the searcher to search for an entire phrase.
- *AND* (all capitalized) between two keywords tells the search engine that the results must contain both keywords.
- *OR* (all capitalized) between two keywords will display all results that have either keyword.

Parentheses are used in conjunction with *AND* and *OR* to further specify a result. For example, a search of recipe AND ("banana pudding" OR "Nilla wafers") will show recipes for banana pudding as well as recipes that use Nilla wafers.

Evaluating Websites

It is important to evaluate websites to ensure you are visiting reliable sites with accurate information. In today's world, finding information is easier than ever, thanks to search engines like Google and Bing. But not everything you find online is trustworthy. That's why it's important to check the quality of a website before you use its info for your homework or anything else. When you find a website, look at who wrote it and what they're trying to say. If the website is trying to sell you something or make you believe a certain way, it might not be the best source for unbiased info.

Even with fancy AI and search engines, it's up to you to make sure information is reliable, fair, and up to date. Some websites will have errors or old information. You can use special fact-checking websites to double-check content. Remember, just because a website shows up first in the search results, that doesn't mean it's the best one. So use your detective skills to make sure you're getting the best and most accurate info.

Evaluating Websites—Objectivity and Bias

Pay attention to the objectivity and bias of the information on the website. It is nearly impossible to find information without any bias, but that does not mean these sources are not credible. When viewing a website, investigate the background of the author and the site. Try to find information that demonstrates as little bias as possible, as this is the most constructive. Use sites that include multiple perspectives. This usually indicates a balanced view of the topic being discussed. Determine the purpose and intent of the website. If the purpose or intent is to convince, sell, or promote a specific viewpoint, it likely includes bias.

Evaluating Websites—Reliability and Relevance

It is important to investigate the reliability and relevance of the information contained on a website. When evaluating a website, look for errors in facts, spelling, or language use. If many errors are present, you should question the validity of a site. Additionally, using fact-checking sites such as FactCheck.org or Fact Check from Duke Reporters' Lab can help to determine the reliability of information found on a website. Check to see how current the information on a website is. Out-of-date information can be an indicator that the site may not be the best resource. Consider whether the website is relevant to specific information needs. Evaluate the depth and breadth of the content to ensure it covers the specific questions that need to be answered.

Evaluating Websites—Validity and Authenticity

Validity and authenticity are important factors to consider when evaluating a website. If you are using a website for research purposes, you should examine the publisher's qualifications. Check to see the author's background, who published the author's work, and what the relationship is between the author and publisher. You can find most of this information by visiting the About Us, Corporate Profile, and Background sections of the site.

You should also check to see what references are included on the site. Extensively documented references are a good sign that the site is using viable information. Investigating the URL can also indicate validity. For example, if the top-level domain includes ".gov," you can be fairly certain this is a site sponsored by the US government. If you are using blogs or wikis, be aware that the author does not always need to meet certain qualifications to post content.

It is a good idea to use these validity checks as a starting point for your research. If you decide to use the blogs or wikis, make certain to verify the author and information.

Whois Lookup

It is often useful to determine the owner or sponsor of a specific website or to determine whether a domain name is already being used. A Whois lookup can provide that information.

Specific information contained in a Whois lookup includes:

- the registered domain name of the website you are searching;
- the name, address, email, and phone number of the organization responsible for the domain, as well as inception and expiration dates for the domain;
- status of the domain, including whether it is active, suspended, or pending deletion.

Several sites, such as www.easyWhois.com or www.Whois.net can quickly show users who owns the domain name. Websites often provide links labeled "About" that describe who owns the website.

More Advanced Searching Techniques

If the methods already covered in this chapter don't work out you, try these more advanced, in-depth techniques to find the information you need.

HTML Objects

There are a variety of tactics you can use to find information through internet searches, including using Hypertext Markup Language (**HTML**) objects as part of the search criteria. **HTML** is used to structure and present information and content on the web. **HTML** objects are components of a webpage that are defined using **HTML**. **HTML** objects are considered foundational elements of web pages. **HTML** objects can be used as part of a web search and may decrease the time needed to find relevant information by returning results that are related to specific elements found on a web page. For example, a user can use **HTML** objects to return only web results that include images.

Here are some **HTML** objects that are commonly used:

- anchor—finds web pages with hyperlinks (e.g., jungle tours anchor: Costa Rica)
- image—finds web pages containing images (e.g., sloth image: Costa Rica)
- link—displays web pages about the search topic with links included (e.g., McGraw-Hill Connect link: MH.com)
- site—displays web pages about search criteria on a specific site (e.g., PlayStation site: Sony.com)

Keywords and Hashtags

Keywords and hashtags are used to tell a search engine what the internet user is looking for on the web. For example, if a user wanted to find information on the Dallas Cowboys, she might type "Cowboys" or "Dallas Cowboys" into her search window. Alternatively, she could type "#Cowboys" or "#DallasCowboys" into the search window because Twitter and Instagram feeds use hashtags to identify topics. For this reason, keywords and hashtags are used extensively by marketers to help ensure that the products and services they are promoting are found when potential buyers are browsing.

Note that using "#Cowboys" and "Cowboys" as keywords will return different results. "#Cowboys" will display social media feeds, but because not all news outlets post on social media, some articles would not be displayed. Typing "Cowboys" as a keyword would display both social media and those news articles without Twitter or Instagram feeds.

Google Image Searches

Every major search engine has tools that allow users to search the web for images or videos as well as articles on subjects. For most search engines, just type the subject of the image and then click the Images button. For example, in Google, Bing, or Yahoo!, typing "pepperoni pizza" in the search box and clicking the Images button will bring up numerous images of pepperoni pizza.

Internet Research

Sources of Data for Internet Research

In the past, when conducting research for a paper, college students would go to the library. Today, nearly all scholars first go to the internet. The challenge for today's students is finding reliable sources of information.

The best sources fall into these categories:

- Peer-reviewed academic journal articles: These tend to be reliable because they were checked and reviewed by other academics (professors) before they were published. One issue, however, is that occasionally the author will turn to like-minded academics for their reviews, leading to the problem of groupthink.

- Governmental sites: Most .gov sites tend to be reliable sources of information, particularly regarding statistical data such as employment figures or census data.

- Books by respected authors: These tend to be credible because the authors are very motivated to maintain their reputations. As a result, responsible authors conduct significant research before making any claim in a text.

The CRAAP Test

You may have heard the term *fact-checking* a lot in the news lately. **Fact-checking** is an often-politicized concept that involves verifying publicly posted documents, news articles, and statements for truth, accuracy, and validity. When working on academic assignments such as case studies, group projects, research papers, and even discussion posts that require references, it is important to ensure the information you are using can be verified and is factual. One method to accomplish verification is the CRAAP test. *CRAAP* is an acronym for Currency, Relevance, Authority, Accuracy, and Purpose.

Here's how to perform the CRAAP test:

1. **Currency:** Evaluate how current the information is. Is it up-to-date for your topic or research? Look for a publication date on the source. Some topics require very recent information, while others can be more timeless. Check if the source has been updated or revised. Outdated information may not be reliable.

2. **Relevance:** Consider whether the information is relevant to your research or topic. Determine if the source provides all of the information you need. Analyze the source's content to ensure it aligns with your research question or purpose.

3. **Authority:** Assess the credentials and qualifications of the author, organization, or publisher. Look for information about the author's expertise and affiliations. Consider whether the source is published by a reputable institution, organization, or publisher.

4. **Accuracy:** Verify the accuracy of the information presented in the source. Check for errors or inconsistencies. Look for citations or references to other reputable sources that support the information provided. Cross-reference information with other sources to confirm its accuracy.

5. **Purpose:** Analyze the purpose of the source. Is it intended to inform, persuade, entertain, or sell a product or service? Consider potential biases in the source, including political, commercial, or ideological biases. Determine if the source is credible and objective or if it has a particular agenda.

Forms of Plagiarism and How to Avoid It

Plagiarism results when someone uses the work of someone else and does not give proper credit (or credit at all) to the person or organization that created the work. **Plagiarism** can be intentional or unintentional. A lack of intent or knowledge about plagiarism does not excuse the offense. Most colleges and universities have detailed policies on plagiarism. You are encouraged to research and understand your institution's plagiarism policies to ensure you do not violate them.

Types of Plagiarism and How to Avoid Them

Type of Plagiarism	Explanation	How to Avoid Plagiarism
Complete/Direct	• Submitting someone else's work as your own. For example, paying someone to write a paper for you or stealing someone else's work and using it as your own are examples of complete/direct plagiarism.	• Do not use someone else's work without giving proper credit for the work. • Do not steal or "borrow" someone else's work and submit it as your own.
Paraphrasing	• Using someone else's work and making subtle changes to words and phrases to make the work appear different from the source.	• If you use thoughts, ideas, or work from someone else, give proper credit to them for their work. • Paraphrasing still constitutes plagiarism if a citation does not accompany it.
Patchwork	• Intertwining your own original work with the work of someone else. For example, writing a research paper using most of your own thoughts, but using the work of others to bolster your work (without proper citation) constitutes patchwork plagiarism.	• If you are using someone else's work to augment your analysis, make sure to properly cite any sources you use.
Source-Based	• Correctly citing a used source but misrepresenting the content in the source. For example, if you cite a reference but paraphrase or change the structure of the content, you are committing source-based plagiarism.	• When citing sources, make sure not to change the structure of the information. • Using direct quotes helps to avoid this.
Accidental	• Unknowingly using the work from someone else without proper citation. Common types of accidental plagiarism include • overlooking the inclusion of citations in your work; • incorrect citation; • forgetting to include direct quotes when appropriate.	• Use proper citation techniques (APA, MLA, Chicago Style, etc.). • If you use a source, make sure to cite it correctly. • Proofread your work to ensure you have included citations when appropriate.

Credible and Noncredible Digital Sources

Suppose you have recently been assigned a research assignment in your computer class. The professor has asked you to research one of the following topics: blockchain and cryptocurrency, artificial intelligence, cloud computing, or emerging technologies.

Unfortunately, you know very little about these topics and know you must conduct research to determine which topic you will cover. You begin your research using Google but find it difficult to decide which references are credible and noncredible.

Here are some ways to distinguish credible and noncredible resources:

Credible Sources

- peer-reviewed journal articles
- most articles found in scholastic research databases, including Google Scholar, Academic Search Premier, and Lexis Nexus Academic Government agencies
- professional organizations Major newspapers and magazine
- PhD or graduate-level dissertations

Noncredible Sources

- Wikipedia: Anyone can edit Wikipedia.
- Blog posts: These tend to be subjective and difficult to verify.
- Articles without citations: Avoid the use of articles without proper citations.
- Unknown authors: Make sure to research the author of the work.
- Forums and personal websites: These may include subjective information.
- Opinionated articles or posts: Many include biased information.

Deepfake and Altered Images

Photographers began altering pictures shortly after the invention of the camera. Similarly, the art of editing and enhancing video began quickly once moving pictures were developed. **Deepfake** uses artificial intelligence and machine learning to take image and video editing to an entirely new level. Now amateur videographers, using apps such as Faceswap, can change photos and films with relative ease. This technology offers many benefits. Using deepfake techniques, filmmakers can bring the deceased back to life on film. If audio and sufficient images exist, children can watch their parents speak to them in 3D.

Like all technology, deepfake is subject to abuse. Hospitals have had to enhance cybersecurity when it was shown that deepfake technology could be used to add or delete tumors on X-rays, CAT scans, and MRIs.

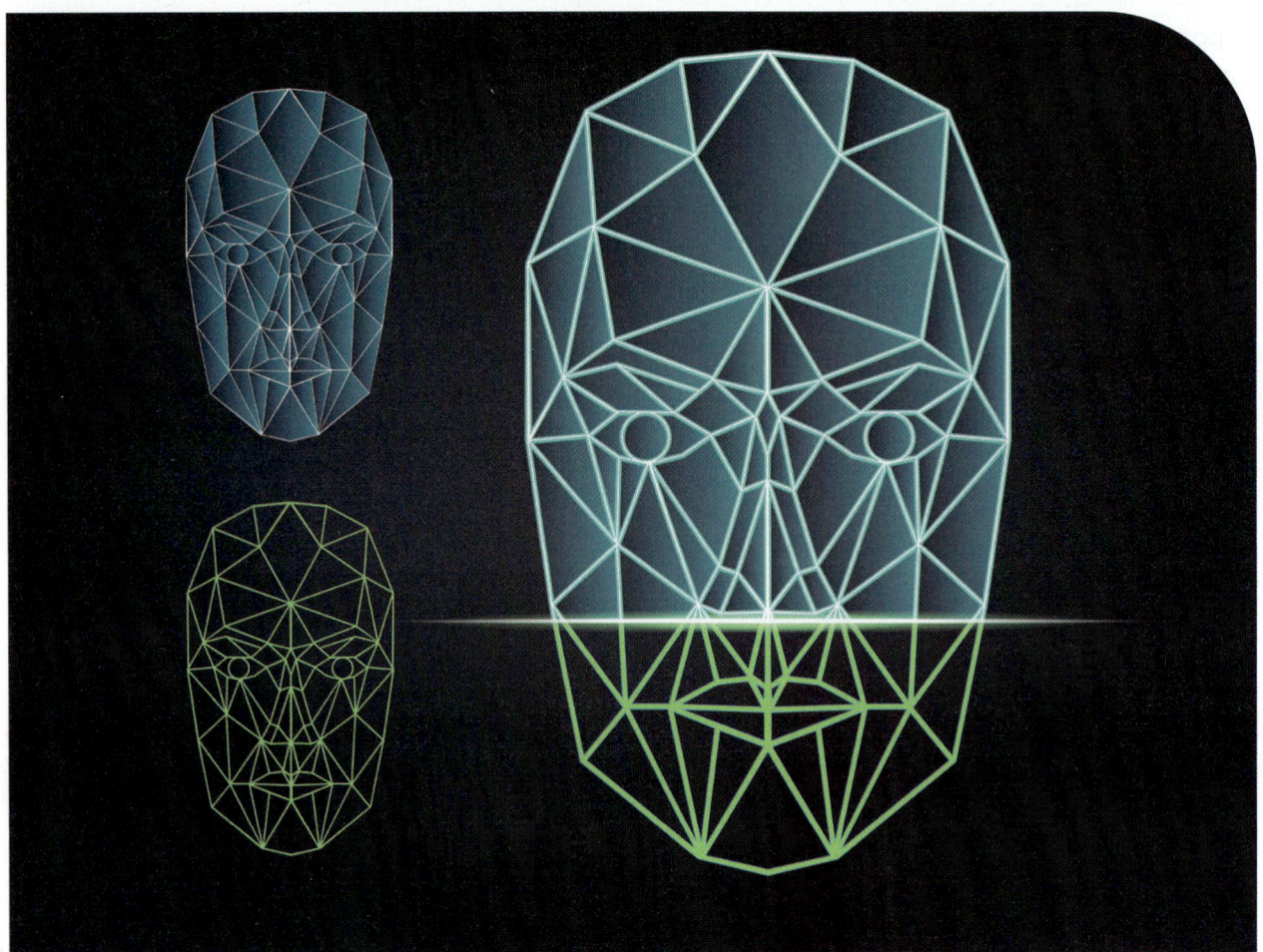

Chapter Review

1. Describe the three basic tasks of a search engine.

2. Research the term *web spider* or *web crawler*, listing alternative names and describing their purpose.

3. Give examples of three commonly used search engines.

4. Describe the differences in the user interface of each search engine. Do you prefer one search engine over another? Give reasons for your preference.

5. Define search engine optimization (SEO) and list three SEOs.

6. Give several examples of the use of **Boolean operators** *AND, OR,* and *NOT* and explain how they work.

7. Explain how parentheses and quotation marks help you to make advanced web searches.

8. The module suggests that you evaluate websites using six criteria. Select a website and discuss how you would apply each criterion to that website.

9. When you use Whois lookup to determine the owner of a specific website, you enter the second-level domain name and the top-level domain name (TLD) separated by a period. An example is "microsoft.com." "Microsoft" is a second-level domain name and "com" is a TLD. Go to one of the Whois lookup sites described in the module and research the ownership or sponsor of a specific website.

10. Do a search on a keyword using a hashtag (#), and then do a search on the same keyword without using a hashtag. Describe the differences you observe in the results.

Ethics Discussion

Research and discuss search engine optimization (SEO) ethics. Find and list at least three examples of unethical SEO practices.

11 Digital Citizenship

What To Expect

After completing this chapter, you will be able to:

- explore and practice good online behavior;
- identify what "digital identity" is and ways to protect it;
- discuss how to keep your online information private;
- identify and learn ways to protect online intellectual property;
- analyze the right ways to acquire and use online content;
- describe the AUP and its role in online safety;
- reflect on how social media influences your identity;
- identify and manage online bullying;
- explore the fundamentals of e-commerce and online shopping;
- communicate online with respect and responsibility;
- understand how branding can shape a positive online image.

Chapter Topics

- Ethical Behavior
- Digital Identity
- Internet Privacy
- Intellectual Property Rights
- Acquiring Digital Content
- Acceptable Use Policy
- Social Media
- Cyberbullying
- E-commerce
- Online Games
- Electronic Communications etiquette
- Branding

Let's Explore Digital Citizenship

What's a citizen?

A citizen is someone who belongs to and participates in a community, whether it's a small town, a big city, or even a whole country. It's not just about living in a place; it's about being a part of it. Citizens have rights, like the freedom to express their opinions or to vote in elections. But they also have responsibilities, like following laws and looking out for their neighbors. It means taking part in making the community a better place, helping others, and being accountable for one's actions. The idea of citizenship is all about the balance of enjoying the benefits of a community while also contributing to its well-being.

Leaving Tracks

The internet might seem vast and ever-changing, but our actions on it don't simply disappear. Think of it like a bustling city where every move we make leaves a trace. Just like in a real city, where we're citizens with responsibilities to our community, in the digital city, we are digital citizens. Being a digital citizen means understanding that our actions, whether positive or negative, impact others. So, just as we're thoughtful and responsible in our daily lives, we should bring that same care to our online world.

If the internet were a sprawling metropolis, think of each click, like, and share as steps you take down its bustling streets. Every website you visit, every post you make, and every interaction you have leaves a mark, like footprints on a sandy path. But take care . . . those footprints aren't in sand—in the digital world, it's more like leaving tracks in concrete!

In our digital city, picture everyone wearing a unique pair of sneakers, leaving distinctive tracks wherever they tread. Jackie, for instance, wore bright blue sneakers, always the same pair, to every place she went: the park where she played soccer, the library where they studied, the diner where she ate her favorite burger. Soon, everyone knew those blue sneaker prints. Everyone knew Jackie loved mystery novels because of the prints near the library's mystery section. They knew her favorite park bench, and they even had a guess of where Jackie lived based on the direction of her tracks.

One day, Jackie realized her tracks were giving away more than she intended. She decided it was time to switch things up. For the park, she wore running shoes. For her library trips, comfy slip-ons. And for her diner trips? Fun blue boots. This way, she could still enjoy all her favorite activities without giving away every detail of her life.

Just as Jackie learned the importance of wearing the right shoes for the right occasion, in the digital metropolis, we too should be mindful of the "footprints" we leave. Different activities, different platforms, different choices—all require their unique approach. By choosing our "shoes" wisely, we can enjoy the digital world while still preserving our privacy and safety. And remember, just as concrete sets and hardens, our actions online have lasting impressions. Make sure they're the right ones.

Stepping into the Digital World

Just like Jackie carefully considered which sneakers to wear for different occasions, we need to think about how we "step" into different online spaces. The "shoes" we choose—our actions, behaviors, and the information we share—leave impressions. And each part of the digital city has its own "dress code." Let's explore how to make the best choices in various digital "neighborhoods."

- **Ethical Behavior:** In the digital city, just like in any community, there's a way to act that's right and fair. By knowing what is and what isn't okay, you can be a good digital citizen.

- **Digital Identity:** Think of the way you dress or the places you go in the city. These choices reflect who you are. Similarly, your online actions present an image of yourself to the world. Making positive choices can lead to great online experiences.

- **Internet Privacy:** Sometimes in the city, you want to stand out; other times, you'd prefer to blend in. Online, you get to choose what to share and what to keep to yourself. Picking the right privacy settings ensures you maintain control over your information.

- **Intellectual Property Rights:** The digital city has art, music, and more. But remember, liking something doesn't mean you can just take it. By understanding these rights, you show respect to creators.

- **Acquiring Digital Content:** Street vendors in the city might offer freebies, but be cautious. Online, ensure you're downloading or purchasing from trusted sources to avoid surprises.

- **Acceptable Use Policy (AUP):** Just like there are rules in a city park, there are guidelines for how to behave online. Knowing them makes sure you and everyone else can enjoy the digital space.

- **Social Media:** Think of it as the bustling city square with graffiti walls, newsstands, and art displays. Some choose to leave beautiful art or inspirational messages on this wall, while others might scribble thoughtlessly. Positive contributions brighten the park, while negative ones can cast shadows. Your choices determine the kind of park it becomes for all.

- **Cyberbullying:** Just as you would stand up against bullying in the city, online bullying needs to be addressed. Knowing the signs and how to respond makes the digital world safer for everyone.

- **E-commerce:** Shopping streets in the city have both known stores and new ones. Online shopping offers variety and convenience, but it's essential to know how to navigate it safely.

- **Online Games:** Playgrounds and arcades are full of fun and challenges. But there are risks to your wallet, your identity, and your time.

- **Electronic Communications:** Saying "please" and "thank you" matters both in the city and online. Being polite in your online interactions builds trust and respect.

- **Branding:** In the digital city, branding is like the unique style or signature you carry. Understanding how to create a positive image for yourself sets you apart in a good way.

As you traverse the digital city, remember that each action, like footprints, tells a story about you. Choosing wisely not only enhances your online journey but also benefits the entire digital community. So, stride with confidence and make choices you're proud of.

Ethical Behavior

Five Components for Ethical Behavior

For many years, various organizations have put together teams that focus on doing the right thing. Even though what's "right" can change depending on where you are or whom you ask, there are some guidelines that most people can agree on. As we dive into this chapter, let's consider how these ethical principles apply in our digital lives.

(1) Accessing Information

- We all deserve to keep our personal details private.
- If we want someone's information, we should always ask first.

(2) Protecting Creative Works

- When someone crafts a piece of art, writes a song, or tells a unique story, it's their own special creation known as intellectual property.
- Taking or copying someone else's intellectual property without saying "may I?" or giving something in return isn't ethical. Imagine if someone took your painting from the web without asking you—it wouldn't feel good, right?

(3) Being Accountable

- People should be answerable for their actions online, especially if they break rules about privacy or other important things.
- Just like in the real world, when mistakes are made online, it's important to take responsibility and correct them.
- With technology evolving rapidly, new challenges arise, making it even more crucial for users to stay informed and act responsibly.

(4) Setting Standards

- Every community, whether it's online or offline, has rules to ensure everyone gets along and things run smoothly.
- Just like how schools have specific guidelines about behavior, online spaces often have their own set of rules to keep everyone safe and respectful.
- Deciding on the best rules, especially for online platforms like **Instagram** or TikTok, can be challenging because the digital world changes so fast.

(5) Balancing Freedom and Safety

- The internet is a place to share thoughts, but it's essential to think about our safety and the feelings of others.
- It's a balancing act—ideally, we should be able to speak our minds but in ways that don't harm people or put ourselves in danger.

Digital Identity

Digital Identity: When you use a phone, computer, or even some video games, you're creating a digital identity. It's like having an online version of yourself. This identity can show up in many places, like social media apps, game platforms, and even school websites.

Different Faces for Different Spaces: You might act one way with your friends and another way with your family. Online, it's the same thing! You could have one username for a game, another for an educational site, and yet another for a chat app. Each one is a part of your digital identity, kind of like wearing different clothes for different occasions!

Protecting Your Digital Self: Just like you wouldn't want everyone knowing everything about your day-to-day life, you must be careful with your digital identity. Here's where privacy and security come in. You can protect your digital identity by keeping the following in mind:

- **Usernames and Passwords:** Keep these secret and make sure they're strong and unique.

- **Online Searches and Sites:** Remember, what you search for or the sites you visit can be tracked. It's like leaving a trail of breadcrumbs behind you.

- **Online Purchases:** If you buy something (or even if your parents do it for you), that's another part of your footprints. It's like leaving a receipt behind.

- **Personal Info:** This is super important! Things like your full name, where you live, or your birthday are private. Just like you wouldn't shout out your home address in the middle of a crowded mall, don't share it online.

Your digital footprints shape the story of who you are in the online world. It's like a digital storybook of your adventures. So, as you explore, chat, and play, always think about the footprints you're leaving behind. Make sure it's a story you'd be proud to share!

Internet Privacy

In the online world, just as in real life, you have the right to keep your personal stuff, well, personal. This is called internet privacy. Think of it like your bedroom door. Sometimes you want to keep it open, and sometimes you want to close it to keep out nosy siblings. The same idea applies online, but it gets a bit more complex.

Let's simplify it:

- **Internet Privacy:** This is your right to decide who gets to know what about you when you're online. So, when you're playing a game, posting a photo, or even doing homework, you have some choices about what gets shared and with whom.

- **Data Collection:** This is when apps or websites gather information about what you do on them. It's like when a store remembers your shoe size.

- **Data Security:** Make sure the information that apps and websites collect is safe and protected, kind of like locking it in a treasure chest.

- **Consent and Control:** Always have a say in what information you share. Think of this like choosing to show or hide your game scores from friends.

- **Internet Privacy around the World:** Just like different schools have different rules, countries have their own sets of privacy laws. So, when you travel or chat with friends in other countries, the rules might be different.

- **Protecting Your Privacy in the United States:** The US federal government has rules to help keep your online life private and safe. Here are some important ones.
  - **The Red Flags Rule:** Think of this as an online neighborhood watch. Businesses look out for red flags, or signs, that someone might be trying to steal your identity.
  - **The Gramm-Leach-Bliley Act:** This one's like a promise ring. Financial places (like banks) promise to keep your super private information (like account numbers) safe and sound.
  - **COPPA (The Children's Online Privacy Protection Act):** This one's for the younger crowd. It gives parents and guardians the power to decide what information websites can know about their kids. If you're under 13, this is like having a digital guardian.

Intellectual Property Rights

As we defined earlier, intellectual property is any original idea, creation, or invention that comes from someone's mind. It could be a story you wrote, a drawing you made, an invention you thought of, or even a special way of doing something. It's not something you can hold, like a new pair of shoes, but it's something valuable because it's unique and belongs to you. That makes it a special kind of property: intellectual property!

Types of Intellectual Property

- **Patents:** A patent is special certificate you get when you invent something new. It stops others from making or selling your invention without asking you first.

- **Trademarks:** Have you seen special symbols or logos that brands use? Those are trademarks. They show that a product belongs to a particular company.

- **Trade Secrets:** Companies use a private recipe or method that's so special they keep it hidden so others can't copy it.

- **Copyrights:** These are for the creators. If you write a story, draw a picture, or make music, copyright helps make sure others don't just copy and sell it as their own.

US Copyright Protection

The US Copyright Office provides protection through copyright for original works of authorship. The following types of creations included are

- literary
- dramatic
- musical
- architectural
- cartographic
- choreographic
- pantomimic
- pictorial
- graphic
- sculptural
- audiovisual

Open Educational Resources

Open Educational Resources (OER) are learning materials available online without any cost. From digital textbooks to educational videos, assignments, and more, OER offers a variety of tools for both learners and educators. These resources come with an open license, allowing you to use and share them. However, they have some restrictions. Most OERs are intended purely for educational purposes. So, a teacher can use it for instruction, but businesses cannot sell it. Even teachers can't take content from OERs and sell it. For example, a teacher could use a video of a chemical reaction on a website, but she couldn't take the same video and put it on a site that requires payment.

OERs are all about promoting, learning, and sharing knowledge without commercial interests.

Acquiring Digital Content

Downloading Content

Being a good digital citizen means understanding the impact of how your online actions affect you and others, including when downloading. It's important to remember that, just like in the real world, there are rules to follow and consequences for breaking them.

More and more, when you want something on your computer, you download it. **Downloading** is a way of transferring a file or piece of content from the internet to your computer or device. If you want a new game, a song, an e-book, a map, or even a picture of a sunset, you're essentially pulling it from a server on the web directly to your personal space. Just like in the real world, there are pros and cons of having something in your personal space.

Reasons to Download

- **Convenience:** There's no need to visit a physical store; you can get what you want from the comfort of your home.
- **Accessibility:** Some resources, whether they're educational materials or rare tunes, might be exclusive to the online world.
- **Offline Access:** Once you've downloaded content, you can access it even without an internet connection.
- **Cost Savings:** The digital realm is full of free resources that are both useful and legal.

- **Customization:** You're in control. Download only what appeals to you, making your device a reflection of your tastes.
- **Reduced Internet Usage:** Download once and use repeatedly without eating into your internet data.
- **Optimal Performance:** Downloaded content often runs smoother, without the hiccups that sometimes come with streaming.
- **Longevity:** Web content can vanish or move. If you've downloaded something, you've got it for good, even if it disappears online.

Reasons Not to Download

- Downloaded content takes up space on your device.
- Your downloaded content may not be as up to date as online content. This is especially relevant when you're working with software.
- Not all available content is safe or even legal to download. You might come across "pirated software"—this is basically a knock-off, much like someone trying to sell you a photocopied book as if it were the real deal. Using this kind of software isn't just risky for your device; it's also unfair to the original creators who put in the work. Be ethical by ensuring you're not stealing someone else's work.

Acquiring Content Responsibly

Acquiring content goes beyond just downloading. When we make a purchase or even when accessing free content, it's crucial to be aware of the origin and the rights associated with that content. Here are some steps to keep in mind when acquiring content:

- **Verify Sources:** Always download from official or trusted sources. If it's a known brand or creator, go directly to their website.
- **Check Reviews:** Look for user reviews or comments about the software or content you're about to download. If many users report issues, think twice.
- **Avoid "Too Good to Be True" Deals:** If a site offers a pricey software for free, it's a red flag.
- **Use Security Software:** Have reliable antivirus and anti-malware software installed. They can scan and alert you about potential threats.
- **Read the License Agreement:** Though it might be long and tedious, it tells you how you're allowed to use the software or content.

Purchasing Online Content

When we talk about acquiring digital content, it doesn't always mean you're getting it for free. Often, you'll need to purchase content, much like buying a physical item from a store, but in this case, everything happens online.

Understanding Online Purchases

When you buy content online, what you're typically purchasing is a license to use that content. This could be a game, a movie, a piece of software, or an eBook. While it might feel like you own that content, you're technically just owning the rights to access and use it. Therefore it's very important to read the terms of service when making online purchases.

Why purchase content online?

- **Instant Access:** Once your payment is confirmed, you often get immediate access to your content.
- **Wide Selection:** Online platforms often have a more extensive selection than physical stores.
- **Digital Extras:** Sometimes, digital versions come with extra features or bonuses that physical copies might not have.
- **Green Choice: Downloading** digital content can be more environmentally friendly than producing and shipping physical copies.

Safe Online Purchasing Tips

- **Trusted Platforms:** Stick to well-known platforms or stores when buying content. Think of places like Amazon, Apple's App Store, or Steam for games.
- **Secure Connection:** Ensure the website's address starts with "https://" when making a purchase, indicating a secure connection.
- **Payment Options:** Using credit cards or trusted online payment systems like PayPal can offer added protection.
- **Receipts:** Always save or print a copy of your online purchase receipt. It's proof of your transaction and useful if you face any issues later.
- **Refunds:** Know the refund policy. Sometimes digital content might not work as expected or might not be what you thought it was. Familiarize yourself with the platform's return and refund policies before buying.

The Value of Genuine Content

Just as you appreciate when someone acknowledges your hard work, artists, software developers, and content creators appreciate when you buy their genuine content. Every purchase supports them, encouraging more fantastic content in the future. Plus, genuine content often comes with updates, support, and the peace of mind that you're getting the real deal.

In the end, being a good digital citizen in the realm of content means valuing and respecting the hard work of others. Whether downloading free content or making a purchase, always ensure it is ethical, safe, and from a trusted source. Remember, in the sprawling digital city, every choice leaves a footprint. Make sure yours leads in the right direction.

Software Piracy

Imagine if you painted a picture, and someone took a photo of it, then sold prints without asking you. That's stealing, right? Similarly, software piracy is when someone copies, shares, or sells a software program without the creator's permission.

Some famous programs that lots of people want to use are Microsoft Office and Adobe Photoshop. Due to their popularity , these programs often get pirated. It's essential always to make sure the software you're getting is from a trustworthy place.

If you ever see or hear about someone using or selling software that looks suspicious, there's a group dedicated to ensuring everyone follows the rules: **BSA**—The Software Alliance. They're like the guardians of the software world, making sure everyone plays fair.

Acceptable Use Policy

Think of an Acceptable Use Policy (AUP) like the rules for driving—like using stop signs and obeying speed limits. But instead of the rules of the road, these rules are for using computers, websites, and things like printers. An AUP is set of rules tells you what you may and may not do when you're on a computer or using the internet. For instance, not visiting certain websites or not printing a hundred photos of your favorite movie star. Businesses have these rules too! If you work at a place with computers, they might have an AUP that says you can't visit social media sites during work or use the internet in ways that could be unsafe.

You might also hear about something called an end-user license agreement (EULA) when you get a new app or game. It's like an AUP, but just for that one thing, while an AUP is for everything that's related to computers.

Social Media

What is Social media?

The internet-based software applications and interfaces that are designed to allow individuals, businesses, and other entities to interact with one another make up social media. **Social media** integrates a variety of multimedia, including images, videos, and messages, that make the sharing of information easy, interactive, and immediate.

Billions of people around the world use social media platforms. For perspective, **Facebook** alone has almost three billion users, which is close to the entire population of India and China combined! Meanwhile, YouTube, **Instagram**, and WhatsApp each boast over two billion users. Just think about that: with so many people connecting online, the world of social media is vast, making it a huge part of daily life for countless individuals globally.

Digital Citizenship and Social Media

Just like in any town, there are rules, customs, and potential pitfalls to watch out for. As you make your way through the social media landscape, keep these pointers in mind.

Digital Footprint and Reputation

Every post, like, share, and comment you make leaves a mark—this is your digital footprint. And on social media, your digital footprint is in concrete, not sand. You need to think about how you're representing yourself now and in the future. Ask yourself: Would you like a future co-worker to see this? What about someone interviewing you for a job?

So, think twice before posting . . . and then think again. Once it's out there, it's challenging to erase completely. Someone can take a screenshot of anything you put out there and share it with the world. Even if the person who shared it feels badly and stops, once they've shared it with other people, there's no way of getting it back.

Privacy and Security

Not everyone needs to know everything about you. Adjust your privacy settings to control who sees your posts and personal details. Be wary of sharing overly personal information. Oversharing can lead to unwanted attention or even identity theft.

Digital Relationships and Communication

It's easy to misinterpret tone online. Be clear in your communications, and always aim for kindness. Remember, there's a real person on the other side of the screen. Treat online interactions as you would face-to-face conversations.

Internet Safety

Watch out for suspicious links or requests from strangers. **Social media** can be a fantastic tool for connection, learning, and fun. By keeping these guidelines in mind, you can enjoy the digital realm while staying safe and respectful.

Social Networking

Today, people practice social networking by using social media to expand their business and social contacts.

While social networking has been a practice for thousands of years, the use of social media has significantly eased it.

The concept of social networking is rooted in the six degrees of separation, which asserts that you can connect with anyone through a chain of no more than six introductions or personal links.

Facebook

Facebook, with almost three billion monthly users, is a social networking site that allows you to share web content, images, videos, and commentary with friends, family, and acquaintances. **Facebook** also allows businesses a means to connect with their customers and even acts as a supplement to a website.

Once you create a personal **Facebook** account, the information is stored in your profile. A profile is your collection of the photos, stories, and experiences that tell people about you. You connect with friends and family by sending friend requests. Once you accept a friend request, you have access to that person's personal **Facebook** page. You connect with businesses by liking their profile. Pages are for businesses, brands, and organizations to share their stories and connect with people.

Risks of Using Facebook

There are some issues that should be considered when using **Facebook**, including the following:

- Anything posted on **Facebook** is not private once it's shared with others. Any posts or shared items can be easily sent to others.
- Photo tagging creates a link to your **Facebook** timeline. Cyberstalking and online predators are real problems.
- Every time you post something on **Facebook** or another social media site, you are creating your own permanent online brand identity.
- Third-party apps (software provided by a vendor that is different from the site that is offering it) can be designed to harvest email addresses and user data.
- **Social media** addiction is a valid concern.
- **Facebook** Marketplace scams including fake items, illegitimate payment methods, and requests for early payments or shipping are things to be mindful of.

Metaverse

In October 2021, **Facebook** officially changed its name to Meta. That led many to ask: what does meta, and the metaverse, mean? Because it's still in its infancy, the metaverse remains difficult to define. Most consider it a highly immersive internet experience, usually involving virtual reality (VR) or augmented reality (AR). Instead of simply using a smartphone or laptop to purchase shoes online, shoppers could wear VR headgear and enter virtual stores where the shoes would be displayed. In this virtual world, shoppers could try on the shoes and see how they look with different outfits. Visitors to this metaverse could virtually meet and talk with each other through avatars. Proponents believe that this could take the place of Zoom-type meetings or even traditional phone calls. Some critics contend that this is mostly hype, and that the metaverse will never be used for much more than immersive video gaming.

Instagram

Instagram is a photo-sharing app that is available for free to download to a variety of devices, including iPhones and Android devices. **Instagram** is a fun way to share your life with friends through pictures. **Instagram** includes many filters that let you easily edit your photos. You also can specify photo locations.

Pinterest

Pinterest is an online "pinboard" that uses visuals and images as posted content. You cannot post content without an image. A user's home feed is where to find Pins, people, and businesses the **Pinterest** algorithm surfaces based on your activity. Pins from the people and boards the user chooses to follow are also displayed. When you personally share an image, it is known as a Pin. A Pin is a visual bookmark. When you click a Pin, you are brought to the site or source that posted it. When you share someone else's Pin, it is called a Repin. Pins are grouped together by topic and organized onto Pinboards. **Pinterest** has become very popular for recipes, crafts, do-it-yourself (DIY) projects, and travel.

LinkedIn

LinkedIn, a free social networking site for business professionals, also offers a subscription service that increases access levels. People utilize **LinkedIn** to connect with other professionals, including business contacts, colleagues, and peers from college. Users employ this site to exchange contact information, share knowledge and ideas, and explore employment opportunities. To create your professional brand identity, make sure to integrate **LinkedIn** into your strategy. Therefore, it's essential to maintain professionalism in your **LinkedIn** posts and consider posting minimal personal information.

Best LinkedIn Practices

There are a variety of tips and techniques that can be used to maximize **LinkedIn**. Some of the best include the following:

- Edit and remove endorsements. It's a good idea to edit or remove endorsements that do not accurately reflect your skill set or experience.

- Use **LinkedIn** to apply for jobs. You should respond to job postings as quickly as possible to avoid missing closing dates.

- Use LinkedIn's résumé-building tool to create a professional résumé. Creating a résumé on **LinkedIn** will allow you to quickly respond to job postings with a properly formatted résumé.

- Update your profile. Use up-to-date information and images.

- Use keywords in your profile. It is important to use job/career-specific keywords in your profile. This will help attract recruiters to your page.

Snapchat

Snapchat is an image- and video-sharing application. Snaps are picture or video messages that are taken and shared with friends on **Snapchat**. The sender of the Snap sets the length of time the Snap can be viewed before it is no longer visible. Originally Snaps could be viewed for up to ten seconds, but they can now be set for infinite viewing. By using the screenshot feature, users can save Snaps in the photo gallery of their phone. Be aware that Snaps are not entirely temporary and can be saved and forwarded to others!

TikTok

TikTok allows users to create and share 15- to 60- second videos that are often edited to include music, sound effects, and stickers that are designed to increase engagement. TikTok is widely used for "infotainment," or content created by social media influencers that offer short snippets of advice and tips covering beauty, finance, fashion, cooking, and many more topics. TikTok has increasingly been used to promote and sell products to the over one billion active monthly users of the app. In the United States, TikTok has over 140 million active users, and 40 percent of them are between the ages of 18 to 24 years old. Users spend an average of 46 minutes per day on the app. TikTok has faced its share of controversy, especially from the U.S. and European Union (EU) governments. The app has faced increased scrutiny due to the alleged potential of the Chinese government's ability to mine data from U.S. and EU app users. The focus of this scrutiny is on the "For You" function and algorithm within the app. Some opponents fear the Chinese government could control data collection or the recommendation algorithm that could compromise digital devices that have the app installed.

Meet-up Apps

Meet-up apps have reshaped the way we connect with people around us. Whether you're searching for a local study group, hoping to join a book club, or searching for a partner for a weekend board game, there's probably an app designed to help you. Many adults also use similar platforms, like online dating apps, to pre-screen potential romantic interests before deciding to meet in person. These tools can be super handy, but they come with a note of caution. Not everyone online has the best intentions, so it's essential to be careful about how much personal information you share. Plus, if you or someone you know decides to meet someone from an app, it's always a good idea to choose a public place and tell someone you trust about it. While the digital age brings many conveniences, staying safe should always be a priority.

Ethical Issues with Social Media

Marketing

Imagine every time you walk into a sporting goods store, you're likely to glance at mountain bikes. Even if you don't buy one, you always take a moment to look. Think of it like this: if the grocery store knew you loved mountain bikes, wouldn't they place mountain biking magazines in your checkout lane? This is exactly what happens online. When you show interest in mountain biking content, that becomes a part of your digital footprint, and online retailers take note.

It's kind of amazing how closely our online actions are monitored. Have you ever searched for something, like mountain biking, and then suddenly see ads about it everywhere? Try it out: look up gear, clothing, or cool places to go mountain biking. After a bit, you'll probably notice mountain biking ads popping up on different websites or apps. What's even wilder? Your brother, sister, or anyone else in your house might start seeing those ads too, even if they've never searched for rock climbing themselves! Why? Because companies are excellent at figuring out your interests and showing you ads about it. And if your family chats, shares, or likes posts on social media apps, these companies think, "Hey, they might all be interested!" Plus, if everyone's using the same Wi-Fi at home, it's like the whole family's online actions are connected. It's interesting to think about, but it also makes you wonder: is that okay?

So, marketers, businesses, and any entity needing online behavior information find value in the data collected. However, the collection of this data raises many ethical concerns because consumers have little control over what data companies collect and how they use it.

Misinformation

Social media is a powerful tool, but it's also a place where false information can spread quickly. Sometimes, people might post things that aren't accurate about themselves, others, or even current events. This can be a photo that's edited, an exaggerated story, or completely made-up news.

False stories or "fake news" mislead people and cause confusion. Imagine reading a headline that sounds shocking, but later you discover it was just made up. It's essential to question what we see and read online. Remember, just because something is shared a lot doesn't mean it's true.

We all have a responsibility to be truthful online and to double-check before we share information. Spreading fake news or posting false things about someone can damage their reputation and misinform others. Being a smart digital citizen means knowing the difference between fact and fiction.

Self-Image and Identity

Social media often showcases the highlight reel of people's lives, which may not always reflect reality. It's okay if your everyday life looks different; everyone has their unique path. Be true to yourself. You don't have to craft an image that isn't genuinely you.

Mental Health Impacts

Constant comparison with others, the desire for validation through likes and comments, and cyberbullying lead to negative mental health outcomes.

Addiction

Imagine spending so much time playing your favorite video game that you forget to do your homework, miss out on hanging with friends, or even skip meals. Similarly, spending too much time scrolling through social media can become like that game—you might get so hooked that you lose track of time. This is called social media addiction. It means that you're using these apps so much that they start to interfere with your daily life and tasks. It's not just about wasting time; being constantly plugged in can also make you feel anxious or down. It's essential to strike a balance just like with anything fun—too much of anything isn't good.

Influence of Algorithms

Algorithms are programs that social media platforms use to decide which posts to show you. If you often click on "prevent pollution" content, the algorithm determines you only want to see that and keeps showing you more of the same. This can create an echo chamber, where you only see or hear things you already know or agree with. While it might seem cool to only see what you like, it also means you might miss out on different views or new information. So, it's good to sometimes step out of your comfort zone and explore diverse topics!

Social Media and Screen Time

Social media platforms offer a window into the lives of friends, celebrities, and the world at large. They provide a space to share, connect, and keep informed. However, the endless scroll of updates, stories, and notifications can become an addictive cycle, compelling users to spend more and more time online. Overindulgence in social media can lead to symptoms like sleep deprivation, decreased face-to-face social interaction, and even feelings of inadequacy or anxiety from constant comparisons. The lure of likes, retweets, and new followers can make it hard to disconnect. To combat this, users should set specific times to check social media, turn off non-essential notifications, and prioritize real-world interactions to ensure a well-rounded, healthy lifestyle.

Cyberbullying

Online Bullying

Cyberbullying is an unfortunate reality in our digital age, occurring when individuals use the internet to belittle, threaten, or harass someone. Like its offline counterpart, this form of bullying can lead to severe emotional distress for the victim. Though the digital realm provides anonymity, it doesn't strip away the pain caused by such actions.

Often, cyberbullying happens on platforms like social media, where it's easy to hide behind a screen. Some people might think it's just harmless teasing, but words can leave lasting scars, even if they're typed instead of spoken.

Think Before You Act: Navigating Cyberbullying

Using the internet is a lot like being in a big community. Before we act online, we need to think about our choices to help keep it a friendly and safe place for everyone. Here are some ways you can prevent cyberbullying and help create a safe place for everyone.

Consider the Impact

- Ask yourself: Would my post/comment hurt someone's feelings?
- Remember: Words have weight. What seems like a light-hearted joke to one person might be deeply hurtful to another.

Privacy Is Paramount

- Ask yourself: Am I sharing too much personal information?
- Remember: Protecting your private details can reduce your vulnerability and prevent potential bullying.

Engaging with Bullies

- Ask yourself: Will responding make the situation better or worse?
- Remember: Bullies often seek attention. Don't feed into their desire for drama. Block and report.

Sharing Content

- Ask yourself: Is this content potentially harmful or spreading rumors about someone else?
- Remember: Sharing or liking malicious content can be seen as an endorsement. Contribute to a positive online environment.

Reflect on Anonymity

- Ask yourself: Just because I'm behind a screen, does it make it okay?
- Remember: Online actions have real-world consequences. Always act as if you're face-to-face with the person.

See Something, Say Something

- Ask yourself: If someone were bullying my friend or sibling, what would I want others to do?
- Remember: Standing by silently can be just as harmful. Support victims and report malicious behavior.

Encourage Positive Interactions

- Ask yourself: How can I make someone's day better online?
- Remember: A kind word, compliment, or positive reaction can go a long way in counteracting negativity.

We all have the power to create a safer, more inclusive online community. By being vigilant about our actions and standing up against negative behavior, we can curb cyberbullying and foster a digital environment where everyone feels respected and valued.

E-commerce

In today's digital world, it's important to know how buying and selling merchandise online works. Just like we learn how to be safe and smart on social media, we also need to understand online shopping and money stuff. This online world lets us quickly buy cool stuff, sell things we make, and even use new tools to handle our money better. Let's explore this online marketplace and learn how to use it the right way.

Types of Electronic Commerce

E-commerce (short for electronic commerce) is the selling of products and services via the web. E-commerce activities are broken into three categories:

- B2C (business to consumer)
- C2C (consumer to consumer)
- B2B (business to business)

The consumer e-commerce market is divided into multiple categories including electronics, fashion, toys and hobbies, furniture, food, and media.

Electronic Funds Transfer

Electronic funds transfer (EFT) is the technical term for electronic or online banking. EFT uses computers and various technologies in place of checks and other paper transactions to transfer funds from one account to another. In the United States, the federal Electronic Fund Transfer Act (EFT Act) covers most consumer-oriented electronic transactions. One example of an EFT transaction is the use of an automated teller machine (ATM).

Electronic Data Interchange

Electronic Data Interchange (EDI) allows for the electronic communication of business-oriented transactions and information between people and organizations. EDI allows the exchange of data and information including purchase orders and invoices via any electronic means. EDI allows organizations and people in different countries and across the United States to participate in ecommerce.

Fintech

Fintech, short for financial technology, uses technology to offer financial services and solutions. PayPal is a prime fintech example, showcasing how technology can change how we manage our finances.

Imagine buying concert tickets online. Instead of using cash or sharing your credit card details directly, PayPal acts as a secure digital wallet. It protects your transaction by acting as a safeguard between your bank and the website. Additionally, PayPal provides extra protection—if your tickets don't arrive or aren't as expected, PayPal helps resolve the issue.

In today's tech-driven world, fintech tools like PayPal are reshaping how we handle money and transactions, making the tools more intelligent and secure for everyone.

Online Games

When you jump into the world of online games, including massive online role-playing games (MORPGs), there's so much fun to have, if you're careful. With online games, there are social and technical risks.

Technical Risks

- **Bad Downloads:** Sometimes, games can bring along unwanted buddies like viruses or malware.
- **Shaky Servers:** If the place (called a server) where the game lives online is not secure, anyone connecting to it might face issues.
- **Buggy Games:** Not all games are made perfectly. Some have coding problems, which means they might not work right and could be an open door for hackers.

Social-Emotional Risks

- Who are you, really? Be careful with your personal information. Some people might try to pretend to be you, which is called identity theft.
- In some games, players might try to "rob" you or bully you into giving them game items or real money.
- Games are meant to be fun, not take over your life. Some people find games, especially MORPGs, especially hard to quit. This could affect school, friends, and even family time.
- Remember, games are a blast, but always stay alert and know when it's time to hit the pause button!

Screen Time Risks

Online gaming can be an exhilarating escape, transporting players to imaginative worlds and allowing them to interact with peers across the globe. However, the captivating nature of these games means players can often lose track of time, leading to excessive screen time. Extended periods in front of the screen can cause physical strain on the eyes, disrupt sleep patterns, and limit physical activity, which is essential for overall health. Additionally, excessive gaming can lead to social isolation if players prioritize virtual interactions over real-life relationships. Finding a balance and setting time limits are crucial. It's essential to take regular breaks, engage in physical activities, and ensure that gaming doesn't overshadow other important aspects of life.

Electronic Communications Etiquette

Etiquette is the standard code for respectful behavior in society or within specific professions or groups. **Electronic etiquette**, also known as netiquette, governs how we should behave when using electronic communication tools. These communication tools include text messages, email, social media, and various digital communication software and hardware.

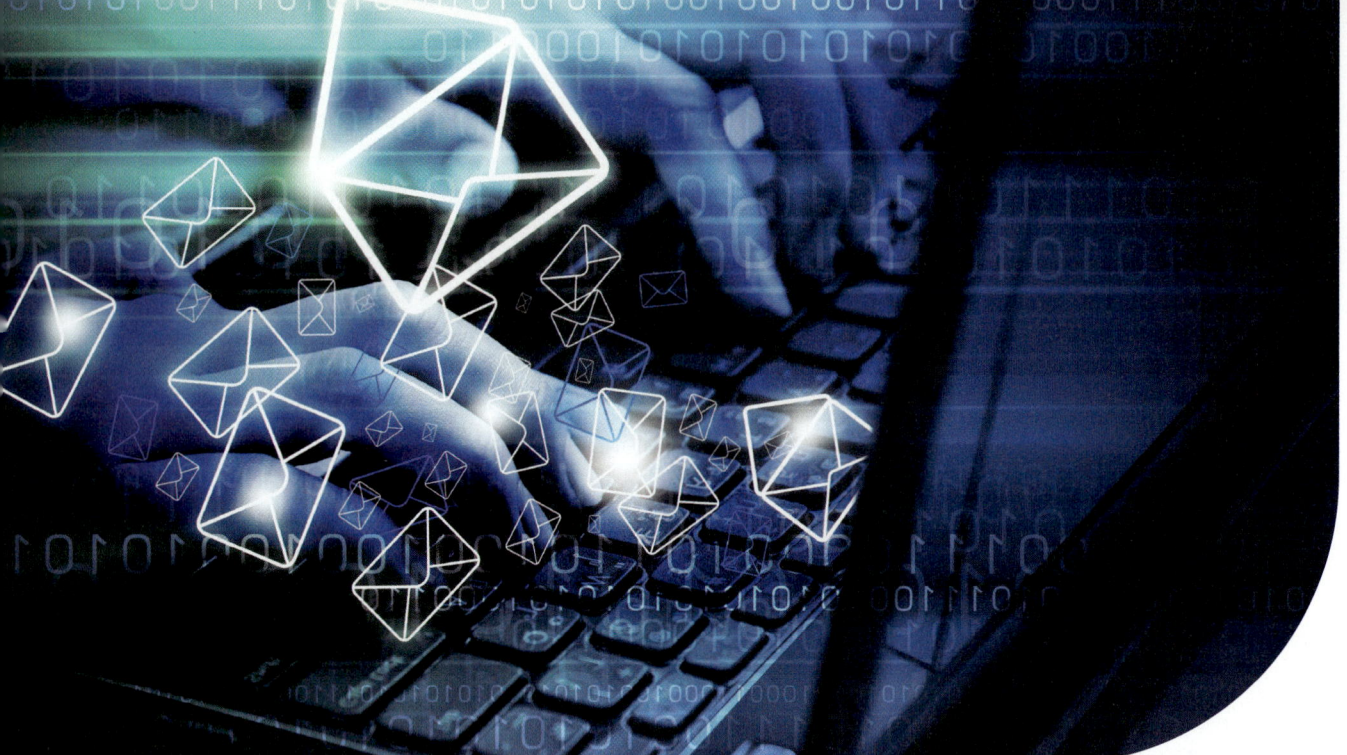

Email Etiquette

Write your messages carefully. Once you send a message, you cannot take it back. Sent messages are, for the most part, permanent and are saved for a very long time on email servers. You should also compose your message as if you were going to have to read it face to face to the person who receives it.

- Use proper spelling and grammar in your messages. It is important to avoid the use of all uppercase letters. Using all uppercase letters in an email is inappropriate because it indicates anger, even if you may simply intend to emphasize a word. In email etiquette, this is known as flaming and is considered rude.

- Keep your messages short and to the point.

- Make sure your message is clear. In the subject line, say what your message is about, and then talk about it in the body of the message.

- In professional messages, avoid the use of emoticons or acronyms. Using them is fine in communications with friends, but they shouldn't be used with people you don't know personally.

Proper Email Composition

Properly formatted professional email messages should have five main components:

1. **Subject:** The subject tells the recipient the context and importance of the message.

2. **Salutation:** A salutation is a greeting to the recipient.

3. **Body:** The body of the message should clearly and concisely state the information needing to be conveyed.

4. **Complimentary close:** A complimentary close immediately precedes the signature line. Common complimentary closes include the following: *Sincerely*, *All the best*, *Cordially*, and *Thank you*.

5. **Signature line:** The signature can be your first name for more informal communications or your full name for more formal messages.

What are internet acronyms?

An acronym is an abbreviation formed from the initial letters of other words. Internet acronyms are used specifically in email, chat, and text messages as a type of informal correspondence and should generally not be used in business messages or business email.

People use a wide variety of internet acronyms, so it is important to avoid using internet acronyms in formal messages and with people you are not familiar with, such as professors and employers.

Common Internet acronyms include

- BTW, or by the way;
- FWIW, or for what it's worth;
- LOL, or laughing out loud;
- TBH, or to be honest

New Internet acronyms seem to pop up daily!

Emoticons

Emoticon is a blending of the words *emotion* and *icon*. It refers to keyboard characters that are used to represent facial expressions and objects. Emoticons help to better convey feelings without the use of words.

Emojis

Emojis are pictures or graphics that are used in text and email messages. **Emojis** are different from emoticons in that emoticons are actual keyboard characters. **Emojis** are extensions to the coding/character used by many operating systems called Unicode. In Unicode computers, emojis are treated as letters from a nonwestern language in much the same way Japanese or Chinese characters are.

Employee-monitoring Software

Employee-monitoring software is designed to track and record the use of network and hardware resources in the workplace by employees. It can be used to track email, monitor network and internet usage, and set web filters. The use of employee-monitoring software should be explicitly stated in the AUP of the organization.

Branding

What is branding?

Branding is like giving something a special style or symbol to make it stand out. Just like a special soccer team jersey makes players recognizable, businesses and people use branding to be unique.

Personal Branding

Just like businesses, people can also have a brand. It's about showing what's unique about you that makes you stand out from everyone else.

Key Methods

- **Social Media:** Share cool stuff about what you like or do. But remember to always think before you post!
- **Talking to People:** How you talk and behave with others also shapes your brand.
- **Business Cards:** They're not just for adults! You can have one too.
- **Texts and Emails:** Even a simple "thank you" can make a difference.

Social Media Tips

- Reply fast if someone mentions your name.
- Look the same on all platforms, like having the same profile picture everywhere.
- Post regularly. Try to share something at least four times a week.
- Join groups or chats about things you love.

Creating Personal Websites and Portfolios

- **Your Online Home:** Think of a personal website like your space on the internet. It's where people can visit to learn all about you.
- **Showcase Your Skills:** A personal website can include a portfolio—It's like a digital showcase of what you can do!
- **More Than Just Words:** While a paper résumé can tell people about you, a website with a portfolio can display them. Photos, videos, art, or any project can come alive here.
- **Impress the Experts:** Many professionals believe that a personal website, can provide a better picture of who you are than just a written document. It's a step ahead in the digital world!

Building Your Brand

- Check what the internet says about you. Google your name and see.
- Have your own website.
- Share wisely. Every picture or post tells a story about you.
- Connect with friends, your school, and other groups you are part of.

Chapter Review

1. What are the five main components of ethical behavior discussed in this chapter, and why are they important in our digital lives?

2. Why is it essential to manage your digital footprint, and what steps can you take to ensure your online identity is secure and accurately represents you?

3. What is internet privacy, and why is it important to be cautious about data collection by online platforms?

4. What are the various types of intellectual property, and why is it essential to respect these rights in the online world?

5. What are some best practices to ensure you maintain a positive self-image and identity on social media while avoiding the pitfalls of comparison?

6. Describe how algorithms influence the content you see on social media and the potential implications of an echo chamber.

7. How can individuals contribute to a positive online environment and counteract the effects of cyberbullying?

8. What are the technological and social risks associated with online gaming, and how can individuals ensure they enjoy games responsibly?

9. What is a digital identity? How is your digital identity created? What steps can you take to protect your digital identity?

10. Define *e-commerce*. What are the three main types of e-commerce?

11. What is meant by the term *social media*?

12. What are some of the security risks associated with social media?

13. What is **LinkedIn**? How is **LinkedIn** used in business?

14. What is electronic etiquette? Why is it important? Describe proper email etiquette and email composition.

Ethics Discussion

Conduct research on the internet about copyrights and intellectual property. What is a copyright? What is intellectual property? How are copyrights used to protect intellectual property? What are some of the ethical considerations surrounding intellectual property and copyrights? Do you think it is ethical to use software or intellectual property without paying for it?

12 Computer Security

What To Expect

After you complete this chapter, you will be able to:

- describe basic cybersecurity concepts and safety measures;
- explain hacker types, motivations, and protection methods;
- identify malware types, transmission methods, and prevention strategies;
- navigate online nuisances and find protection against disruptions;
- explain about cookies and how to manage them in online browsing;
- describe how firewalls guard against digital threats;
- discuss how strong password security protects digital devices;
- practice safe browsing to shield personal information;
- explain how antivirus software fights threats and the pros of free vs. paid options.

Chapter Topics

- Protective Layers and Potential Threats of the Internet
- Cybercrime
- Hackers and Hacked Systems
- Malware: Types and Protection
- Online Nuisances and Protection
- Cookies and Online Browsing
- Firewalls: Protection and Function
- Password Security
- Safe Browsing Practices
- Antivirus Software: Role and Options

Let's Explore Computer Security

In today's interconnected world, *computer security* is more than just a technical term—it's a fundamental requirement for navigating our digital lives. Every click, download, and interaction carry potential risks, but with the right knowledge and precautions, we can guard against many of the threats that lurk in the digital shadows. To truly grasp the complexities and intricacies of computer security, let's envision it through a creative lens. Imagine our digital landscape as a vast, medieval kingdom and our computers as its fortified castles. Let's explore this analogy to better understand the protective measures and potential dangers in the realm of computer security before exploring the specifics.

Imagine the vast, sprawling world of the internet as a mighty kingdom, with its own landscapes, fortresses, and threats. In this kingdom, your computer or digital device is your own personal castle—a place of refuge, information, and entertainment. But, as with any castle, there are vulnerabilities and there are defenses. Let's embark on a journey to understand these protective layers and the potential threats that loom in dark places.

Protective Layers and Potential Threats of the Internet

Cybercrime: The vast and varied threats to a kingdom are many. From large armies laying siege to sly thieves sneaking in, understanding these threats helps fortify the castle. In the digital realm, recognizing and understanding the nature of cybercrime ensures that you're better prepared to defend against them.

Hackers and Hacked Systems: A castle's drawbridge is its main gateway. While the drawbridge serves as a strong barrier when raised, skilled adversaries might find covert ways to lower it. In the same vein, learning about hackers and the signs of a compromised system allows you to detect when your digital drawbridge has been tampered with, ensuring swift action.

Malware: Various threats seek to breach the castle, from giant battering rams to sneaky invaders. Just as a castle has defenses for each, understanding different malware types and their remedies keeps your digital fortress secure.

Online Nuisances: A kingdom doesn't only face threats from enemies. Sometimes, flocks of birds settle atop the towers or traders crowd the entrance, causing disruption. Similarly, online nuisances might not always be malicious, but they can still disrupt your digital experience. Knowing how to manage these ensures a smoother journey online.

Cookies and Online Browsing: Messengers approach the castle bearing seals of approval, signifying their trustworthiness. Similarly, cookies act as digital seals, signifying a mutual understanding between websites and their visitors. Recognizing the importance of cookies and managing them wisely enhances your online experience.

Firewalls for Protection and Function: A castle's moat is its watery shield, stopping many threats before they can even approach. In the world of computing, firewalls act as this protective barrier, warding off unwelcome intrusions and keeping the system safe.

Password Security: The keys to a castle are precious, ensuring access to its many rooms and treasures. Similarly, passwords are the keys to our digital treasures. Safeguarding them, choosing them wisely, and changing them periodically ensures that only the rightful owner can access the protected contents.

Safe Browsing Practices: Knowing where to tread and what places to avoid is critical in the digital realm. Just as guards patrol the castle walls and peer from watchtowers to oversee the kingdom's landscape, safe browsing practices are your vigilance in the digital domain. Being alert and informed, much like the ever-watchful eyes of a sentinel, ensures that you navigate the digital kingdom safely without being caught off guard.

In this vast digital kingdom, armed with knowledge and understanding, you're better equipped to defend your personal castle, ensuring it remains a sanctuary amid the ever-evolving landscape. With each threat understood and every defense in place, stride confidently, knowing you're well-prepared for any challenge the digital realm might present.

Cybercrime

Cybercrime is criminal activity committed with a computer.

Cybercrime

Cyber criminals use computer technology to access personal information, discover trade secrets, or use the internet for a variety of exploitative or malicious purposes.

The two common types of cybercrime are:

- Single event—Victims endure a single-event cybercrime while unknowingly downloading a **Trojan horse** virus or installing a keystroke logger; or from phishing, theft, or manipulation of data via hacking; or from viruses, identity theft, and e-commerce fraud.
- Ongoing series of events—This is more serious than the single event and includes cyberstalking, child predation, extortion, blackmail, and terrorist activities.

Cyberattacks

A cyberattack is a deliberate misuse of computers and networks via the internet. Cyberattacks are launched from one or multiple computers against a single computer, multiple computers, or computer networks. Cyberattacks are of the two following types:

- Attacks where the goal is to disable the target computer or disrupt network activity
- Attacks where the goal is to get unauthorized access to the target computer's data

Cybersecurity Threats

According to the National Institute of Standards and Technology (NIST), a cybersecurity threat is an event or condition that has the potential for causing asset loss and the undesirable consequences or impact from such loss. Assets include information, software, and hardware. The specific causes of asset loss arise from a variety of situations and events related to adversity, which are typically referred to as *disruptions*, *hazards*, or *threats*. Asset loss constitutes all forms of intentional, unintentional, accidental, incidental, misuse, abuse, error, weakness, defect, fault, and/or failure events and associated conditions.

Cybersecurity threats should be assessed and remediated to prevent ongoing or future impacts.

Cybersecurity Vulnerabilities

Cybersecurity vulnerabilities are weaknesses or flaws in system security procedures, design, implementation, and control. Individuals could potentially compromise these weaknesses accidentally or intentionally. System compromises can lead to security breaches, data loss, system outages, and violations of an organization's system security policy. Vulnerabilities also arise from security exposure in an operating system or application software. Many organizations maintain accessible databases of software vulnerabilities based on the software version. These accessible databases aid other organizations in preventing known vulnerabilities. Attackers actively seek out vulnerabilities to exploit them for launching attacks.

Cybersecurity Exploits

A cybersecurity exploit follows the identification of a system vulnerability. An exploit is the means through which a system vulnerability can be used by a hacker to execute a malicious activity on a system. Exploits include specific code, command sequences, and open-source exploit kits that are designed to take advantage of a software vulnerability or a flaw in system security. They allow an intruder to remotely access a network to gain unwarranted privileges, which allows the intruder to move deeper into the organization's network. A cybersecurity exploit comes after identifying a system vulnerability. An exploit is how a hacker can use a system vulnerability to cause harm to a system.

Cybersecurity Breaches

A cybersecurity breach occurs when a hacker gains unauthorized access to an organization's systems, data, and information. Hackers use malicious applications and other techniques to reach restricted areas of a system or network. In simplified terms, a security breach is like a break-in at your home. If someone breaks into your home in an unauthorized way, it is considered a security breach. Security breaches are early-stage intrusions that can lead to system damage, data loss, and network downtime. Security breaches can occur in a variety of ways, including viruses, spyware, impersonation, and distributed denial of service (DDoS) attacks.

Threat Mitigation

Cybersecurity threat mitigation refers to the rules and plans an organization uses to stop security problems such as data breaches and unauthorized access to its network. Mitigation also includes policies and procedures to mitigate damage when a security attack occurs.

There are three main components of threat mitigation:

- Threat prevention—the policies and procedures an organization has designed and implemented to protect systems and data
- Threat identification—the security tools and oversight designed to identify specific and active security threats
- Threat cure—the policies, tools, and strategies used to lessen the impact of active security threats

Cyberterrorists

A cyberterrorist is someone who misuses computers and the internet to harm or disrupt for political reasons. Think of it like someone trying to sabotage digital systems to make a point or send a message. The US FBI describes cyberterrorism as planned attacks on computers and data, which can lead to real-world harm. It's a serious issue in which technology can be used in harmful ways for larger goals.

Attacks include

- the disruption of e-commerce and government sites;
- targeting of the power grid and public utilities (In fact, 84% of Americans believe cyberterrorism now ranks as the top critical threat to US vital interests. This is the first time cyberterrorism has surpassed nuclear weapons and foreign terrorism as the most critical threat.);
- hack attempts to the Pentagon (an estimated 1,000 attempts per day).

Cyberbullying

Cyberbullying is harassing a person or group of people using information technology in a repeated and deliberate way. Cyberbullies use internet service and mobile technologies with the intention of harming another person.

Cyberstalking is a form of cyberbullying and often involves disturbed obsessions. Nearly all states have bullying laws in place, many with cyberbullying or electronic harassment provisions.

Most states now have laws against cyberbullying. According to the PEW Research Center, nearly half of US teens reported being bullied or harassed online. Older teen girls are more likely to report being targeted by online abuse due to their appearance.

What should you do if you experience cyberbullying?

- **Do not respond or reply.** It is important not to engage with a cyberbully.
- **Take screenshots.** Screenshots allow you to document the occurrences to serve as proof.
- **Report the abuse.** It is important to report the abuse to social media sites. Additionally, it may be appropriate to contact your local authorities.
- **Block the bully.** Make sure to block the bully from all social media sites you frequently use.

Hackers and Hacked Systems

Hacking

A hacker is a person who attempts to gain unauthorized access to networks via hacking. **Hacking** refers to activities that seek to compromise the security of digital devices such as laptop computers, smartphones, tablets, and entire networks.

Most hackers are involved in unlawful activities (cybercrimes) and are motivated by financial gain, protest, or information gathering (spying). Hackers may also be driven by the sheer challenge of hacking a system.

The term *hacker* first appeared in the 1960s and was used to describe individuals or computer programmers who could increase efficiency of computer code by removing or "hacking" extra machine code (instructions) from a computer program. In the early 1970s one of the first network hackers, John Draper (a.k.a. Cap'n Crunch) used a toy whistle to match the 2,600-Hertz tone used by AT&T's long distance switching system. The term *hacker* is now used to refer to an individual who most often illegally accesses networks, systems, and devices using programming techniques and malicious programs.

Under the US Computer Fraud and Abuse Act (CFAA), punishments for hacking crimes include the following:

- Accessing a computer or accessing information has a maximum sentence of one year in prison for the first violation and up to 10 years in prison for a second violation.
- Extortion has a maximum sentence of 5 years in prison for the first violation and a maximum 10-year sentence for a second violation.
- Obtaining national security information has a maximum of 10 years in prison for the first violation and a maximum of 20 years in prison for a second violation.

Individuals found guilty of hacking crimes may also have to pay restitution. Restitution is a payment for injury or loss that results from a crime.

Black Hat Hackers

Black hat hackers access computer systems with the intent of causing damage or stealing data. **Black hat hackers** are also known as hackers or "crackers." Generally, their main motivation is personal or financial gain. They can also be involved in cyber espionage or protest, or they may simply be addicted to the thrill of hacking devices. Richard Stallman invented the definition of a black hat hacker to express the maliciousness of a criminal hacker versus a white hat hacker, who performs hacking duties to identify places to repair.

System Safety Compromised

Copyright © McGraw Hill Rawpixel.com/Shutterstock

White Hat Hackers

White hat hackers are non-malicious computer security experts who choose to use their knowledge and skills for good rather than evil. They are also known as ethical hackers. Sometimes they are paid employees or contractors working for companies as security specialists who attempt to find issues or security breaches via hacking. **White hat hackers** are often very highly paid consultants. **White hat hackers** test the security measures of an organization's information systems to ensure that the organization is protected against malicious intrusions. **White hat hackers** employ the same methods of hacking as black hat hackers, but they seek permission from the owner of the system to do so; therefore, the process is completely legal.

Script Kiddies

Script kiddies are would-be hackers who attempt to gain unauthorized access to networks in order to steal and corrupt information and data. **Script kiddies** typically use programs written by others and do not have the skill or experience to write their own hacking programs. They often use tricks found on YouTube to hack into networks or social media accounts.

Hacktivism

Hacktivism is the act of hacking or breaking into a computer system for a politically or socially motivated purpose. The term is a combination of the words *hack* and *activism*. Individuals who take part in hacktivism are called *hacktivists*. **Hacktivists** use the same techniques as hackers but do so to disrupt services and to bring attention to a political or social cause. "Anonymous" is one hacktivism group that has received significant notoriety.

How to Diagnose a Hacked System

Digital device hacking has become so common that almost everyone must deal with it at least once in their lives. It is extremely difficult to find the culprits of most hacking attacks, but you can often tell when a system has been hacked.

Some signs of a system or device that has been hacked are:

- fake antivirus messages (If you notice these messages, power down your computer and reboot it in Safe mode. In Safe mode, run your antivirus program.);
- unwanted browser toolbars;
- frequent random pop-ups;
- fake email or social media messages from your account;
- programs automatically connecting to the internet;
- unusual activities such as passwords being reset;
- a device starting or shutting down on its own;
- apps you didn't download mysteriously appearing on your phone;
- your phone seems to be running more slowly than usual;
- random spikes in data usage;
- malicious software and processes running without your knowledge;
- changes in phone behavior such as apps switching on and off unexpectedly.

Malware: Types and Protection

Malicious Software

Malware is a contraction of the terms *malicious* and *software*. It refers to any software written with the intent to damage devices, steal data, or disrupt networks. Hackers often create these nuisances to make money. Hackers may spread the malware themselves or try selling it on the Dark Web.

Malware is designed for the following purposes:

- stealing information
- destroying data
- incapacitating a computer or network
- frustrating the user

Common types of malware include:

- viruses
- worms
- Trojan horses

Computer Viruses

A computer virus is software that infects computers using computer code. Viruses need to be "run" to attack and cause damage. They replicate themselves and can spread like a biological flu virus, often requiring proper programming within files or documents to propagate.

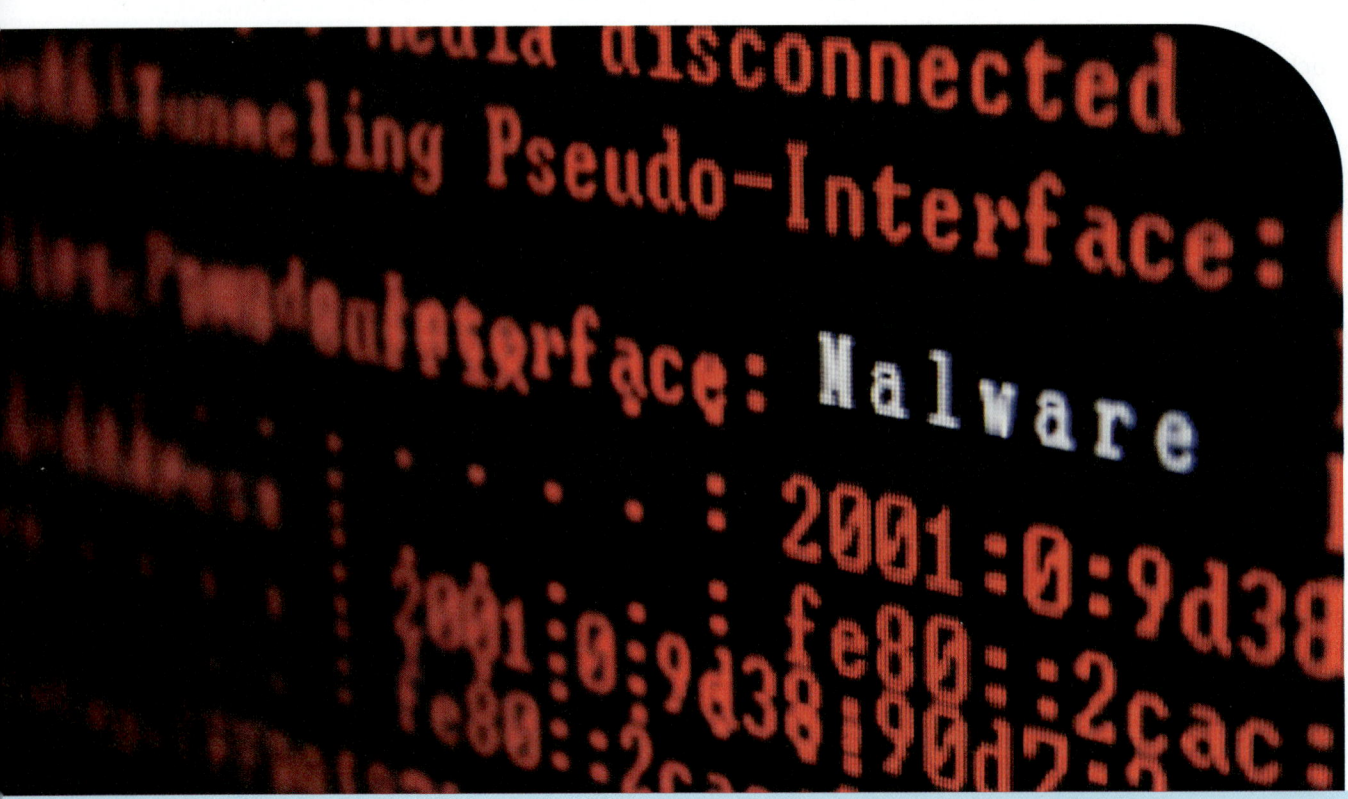

Copyright © McGraw Hill   Yuttanas/Shutterstock

How a Virus Infects Computers

Computer viruses attack digital devices by arriving via infected email attachments or downloads from the internet, sharing infected files, visiting contaminated websites, disguising themselves as images or media files, inserting infected discs or drives, and exploiting vulnerabilities in antivirus programs.

Viruses can infect a digital device in several ways, including

- by opening an infected email attachment;
- by downloading an infected file from the internet;
- by sharing an infected file;
- by visiting a contaminated website;
- by being disguised as images, audio, or video files;
- by inserting or connecting an infected disc or drive;
- by not running or maintaining your antivirus program;
- via spam email messages that are disguised as legitimate communications;
- by being bundled with other software.

Be cautious about downloading software that is not hosted on the vendor's site.

How Do Computer Viruses Work?

A computer virus attacks a digital device using a series of actions. These actions are repeated over and over, resulting in a full-blown virus attack.

- First, the virus arrives via email attachment, file download, or by visiting a website that has been infected.
- Next, an action such as running or opening a file activates the virus.
- Once activated, the virus copies itself into files and other locations on a computer.
- As files are shared, the infection spreads to other computers via infected email, files, or contact with infected websites.
- At any point after it is activated, the virus may execute malicious code in the infected devices.

Virus Symptoms

Viruses are very harmful to your computer. There are many symptoms a computer may exhibit when it has been infected with a virus. Some symptoms may include the following issues.

- The operating system (OS) may not launch properly, and the computer user may need to reboot and restart the computer frequently to ensure all programs are starting and working correctly.
- Critical files may get deleted automatically, which can happen periodically or all at once.

- Error messages will become prevalent.
- It may become difficult to save documents, and the computer may be running slower than usual. If a system or network is infected severely, it may even black out or not launch the startup process.

How to Avoid Viruses

To avoid computer viruses, you can:

- install updated antivirus software;
- use a firewall;
- visit trusted websites;
- download content safely;
- remove USB drives before booting;
- update your router regularly;
- adjust browser security settings;
- avoid bypassing browser security features;
- never open files or messages from unknown sources.

Macro Viruses

Macro viruses are embedded in documents or spreadsheets. When opened, they execute and trigger destructive events. They can be avoided by not downloading or opening unknown file attachments containing Microsoft Office files.

Macro viruses can spread very quickly. Additionally, they can be difficult to detect. Some macro viruses can cause issues within text documents such as changes to formatting, spelling, and grammar issues. **Macro viruses** can attack both Mac and PCs.

Trojans

In ancient Greek mythology, the city of Troy was protected by a massive wall that the Greek soldiers couldn't breach. After many failed attempts to break through, the Greeks devised a clever plan. They constructed a giant wooden horse, hollowed it out, hid a group of their best soldiers inside, and left it outside the city.

The Trojans believed the Greeks had admitted defeat and left them a gift. So they pulled the horse inside their city walls. That night, while the Trojans were asleep, the Greek soldiers emerged from the wooden horse. The Greek soldiers then opened the gates for the rest of the Greek army.

Drawing from this tale, in computer terminology, a "**Trojan horse**" or simply "Trojan" is software that appears harmless or even beneficial but hides a malicious intent. Like the wooden horse in the story, it tricks users into bringing it inside their system's "walls." Once inside, it can then perform harmful actions or give access to other malicious software. Trojans are often used by cyberthieves and hackers to gain unauthorized access, find passwords, destroy data, or bypass firewalls.

Email Viruses

An email virus is attached in a file and sent via email message. When that infected file is opened, the virus infects the computer. There are three common types:

- **Phishing**—this tricks users into revealing usernames and passwords by appearing legitimate
- **Ransomware**—when activated, this encrypts files on the user's hard drive and then sells a decryption program
- Keystroke logging Trojan—when activated, it saves all keystrokes for subsequent transmission to the virus creator

Here are some steps you can take to avoid facilitating an email virus attack:

- Be cautious when opening files attached to email messages or clicking on hyperlinks to outside sites contained in the message. Email viruses are often contained in attached files. Outbound links can draw a user to an external site designed to steal information and launch viruses.
- Verify the identity of the message sender before opening the message. Make sure to scrutinize the email address. You can often uncover any potential issues by reviewing the sender's email address.
- Keep your antivirus software up to date. This helps to ensure your device is optimally protected.

Worms

A worm is a destructive program that replicates itself throughout a single computer or across a network. Worms are designed to attack both wired and wireless networks. **Worm** replication exhausts network bandwidth or available storage.

Ransomware

Ransomware is malware that makes a computer's data inaccessible until a ransom is paid. **Ransomware** usually invades a computer with a **Trojan horse** in a legitimate-looking email or with a worm in a networked computer.

Ransomware typically encrypts the victim's data files. A message offers to decrypt the files if the victim makes a ransom payment to the perpetrator. Payment is made via a means that is difficult to trace such as Bitcoin or prepaid cards. Once payment is made, the perpetrator may or may not send a decrypting code that allows the victim to open their data files again. Another version of ransomware threatens to make the victim's personal files public unless the ransom is paid. The US Cybersecurity and Infrastructure Security Agency (CISA) hosts the StopRansomware. gov site. It provides a one-stop location to combat ransomware more effectively. The CISA recommends the following mitigations against ransomware:

- Update software and operating systems with the latest security patches, as outdated applications and operating systems are the target of most attacks.
- Do not click links in unsolicited email messages.
- Back up data on a regular basis.
- Store data on a separate, offline device.

Rootkits

A rootkit is a type of malware that hides in the operating system (OS) and is triggered each time you boot your computer.

A rootkit allows someone, either legitimately or maliciously, to gain and maintain command over the computer system without the computer system user knowing about it. This means that the owner of the rootkit can execute files, delete files, change system configurations, access log files, or monitor activity on the target machine.

Types of rootkits include

- firmware or hardware rootkit, which impacts hard drives, routers, and system BIOS and allows hackers to log keystrokes and monitor activity;
- bootloader rootkit, which conducts the loading of an operating system on a device and replaces the device's legitimate bootloader with a hacked one;
- memory rootkit, which hides in a device's RAM and uses its resources to conduct malicious activities;
- virtual rootkit, which is loaded underneath a device's operating system and hosts the target OS on a virtual machine, allowing the interception of hardware calls.

Zombies

A zombie is a computer that has been secretly taken over by an outsider, typically using a rootkit. The rootkit is programmed to perform a variety of damaging tasks. Millions of PCs around the world have been taken over by third parties to transmit messages without the user's knowledge. A group of compromised computers controlled by a hacker is called a botnet or a zombie army. Zombies are used to transmit spam messages from many different internet protocol (IP) addresses, thus avoiding detection. Zombies are widely used to relay spam so that messages come from thousands of different IP addresses, avoiding detection and expanding volume at the same time.

DoS and DDoS Attacks

A denial of service (DoS) attack on a network is designed to interrupt or stop network traffic by flooding it with too many requests. A distributed denial of service (DDoS) attack launches a virus on a network's computers. Once the computers are infected, they act as zombies and work together to send out legitimate messages, creating huge volumes of network traffic and resulting in a network crash.

Compare and Contrast DoS and DDoS Attacks

DoS Attack	DDoS Attack
Single system targets a victim's system	Multiple systems target a victim's system
Generally slower than a DDoS attack	Generally faster than a DoS attack
More easily blocked because only one system is used	Difficult to block, as multiple devices are sending packets from multiple locations
Relatively easy to trace	Often difficult to trace
Less traffic volume than with a DDoS attack	Massive traffic volume to a victim's network

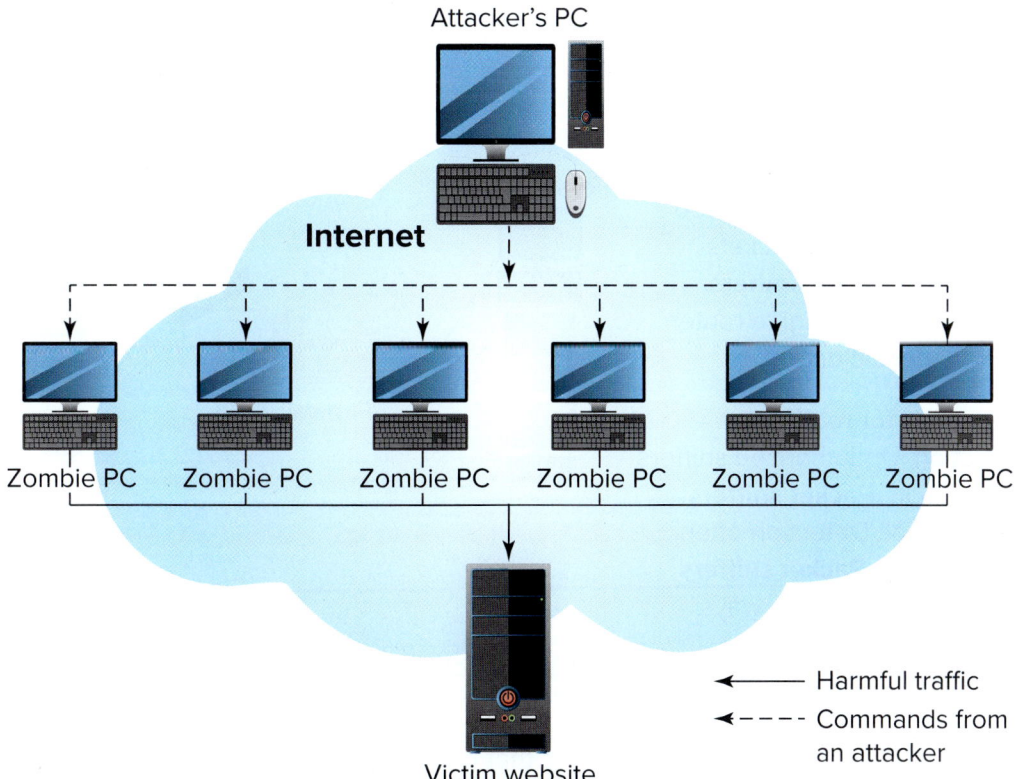

Attacker's PC

Internet

Zombie PC   Zombie PC   Zombie PC   Zombie PC   Zombie PC   Zombie PC

Victim website

◄——— Harmful traffic
◄– – – Commands from an attacker

Spyware

Spyware is software that collects information about your internet surfing habits and behaviors. The information collected by spyware includes keystrokes, passwords, account numbers, and other confidential information. Spyware is often installed via free downloads or by visiting certain illegitimate websites.

Keystroke Loggers

A keystroke logger, sometimes called a key logger or system monitor, is a form of spyware/surveillance technology that records all actions typed on a keyboard. A keystroke logger can be software, but it can also be hardware. Keystroke hardware can come from a USB keystroke logger that plugs into a USB device with or without a user's knowledge. Any USB device could contain a hidden keystroke logger, so be careful about what you plug into your computer.

Keystroke loggers

- can be used on smartphones, tablets, and laptop computers;
- are hardware devices and software applications;
- record passwords and confidential information;
- may be Trojans that are installed without the user's knowledge.

Key loggers can also be designed for legitimate purposes, including monitoring employee activity for detection of unauthorized use of company resources such as printers and networks.

Network Sniffers

Network sniffers (also referred to as "packet analyzers" or "packet sniffers") are specialized hardware or software that captures packets transmitted over a network. **Packet sniffers** examine each packet passing through a router gateway and search for specific data.

Legitimate sniffers are used for routine examination and detection. Unauthorized sniffers are used to steal information. **Packet sniffers** can be very difficult to detect. Detection often requires specialized software. **Packet sniffers** can be configured in the following two ways:

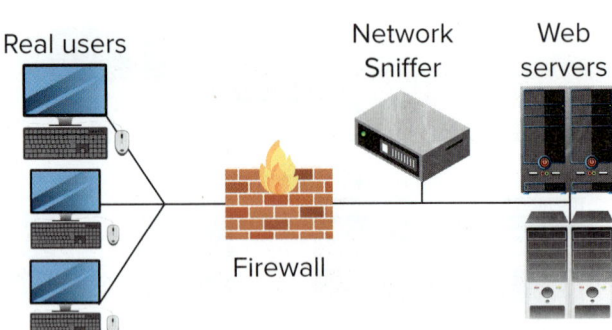

1. Unfiltered sniffers—These sniffers capture all packets possible and store them on a hard drive for future inspection.

2. Filtered packet sniffers—These sniffers use analyzers that capture only packets containing specific elements of the data being transmitted.

CIA Triad: Confidentiality, Integrity, and Data Availability

Confidentiality in the CIA Triad

In the confidentiality, integrity, and availability (CIA) triad, *confidentiality* is roughly equivalent to privacy and refers to avoiding the unauthorized disclosure of information. It involves the protection of data, providing access for those who are allowed to see it while disallowing others from learning anything about its content. It prevents essential information from reaching the wrong people while making sure that the right people can get it. Data encryption is a good example of ensuring confidentiality. The tools to ensure confidentiality include encryption, access control, authentication, authorization, and physical security.

Integrity in the CIA Triad

In the CIA triad, *integrity* refers to the methods used to ensure that data are real, accurate, and safeguarded from unauthorized user modification. It is important that information has not been altered in an unauthorized way and that the source of the information is genuine. Tools that can be used to ensure data integrity include data backup, CheckSums, data correction codes, and availability of information.

Data Availability in the CIA Triad

In the CIA triad, *availability* refers to the process of ensuring that data and systems are available when needed. This is best achieved by rigorously maintaining all hardware, performing necessary hardware repairs immediately, and maintaining a correctly functioning operating system environment that is free of software conflicts. It is also important to keep current with all necessary system upgrades. Providing adequate communication bandwidth and preventing the occurrence of bottlenecks are equally important. Redundancy, failover, and maintaining a redundant array of independent disks (RAID) can mitigate serious consequences when hardware issues do occur.

Online Nuisances and Protection

Annoying Software

Online nuisances are annoying software programs that can slow down the operations of a computer, clog email inboxes, and lead to theft of information and money.

Online nuisances can be downloaded from content found on the internet, linked to corrupt websites, and found on illegitimate websites.

Online nuisances include

- spam—unsolicited email messages;
- spyware—malware designed to steal information or monitor activity;
- phishing and pharming—attainment of sensitive information by impersonating a legitimate site;

- trolling and spoofing—posting provocative messages or content in online forums or social media sites to damage reputations or get reactions;
- clickbait—using sensitive or misleading information to attract clicks to sites or social media posts;
- pop-up ads—unwanted ads that randomly appear on the screen when you are browsing the web.

Spam

Spam is an unsolicited email message, typically received from an unknown sender. **Spam** messages are sometimes called *unsolicited bulk email* (UBE) and are used for a variety of marketing purposes, including selling products and services.

Messages may contain a variety of malicious programs. It is estimated that there are 293.6 billion emails sent every day. Of these, 50 percent are considered spam.

Here are some steps to reduce the amount and impact of spam messages:

- **Install and use spam filters.** Most email services (including Outlook and Gmail) offer spam filters that will identify spam messages and place them in a designated spam folder. Some legitimate messages may make their way to the spam folder, so it is good practice to check this folder frequently to ensure messages are not being missed.
- **Do not overshare your email address.** Oversharing your email address can lead to an overabundance of spam email messages.
- **Use a "burner" email account.** A burner email account is created as a secondary email account that is different from work or school accounts. Burner accounts are often used to receive email communications from marketers and can be used when an email is expected to create an account to access information. They are often temporary and can be closed when too much spam email is filling the inbox.
- **Unsubscribe from email lists.** Most businesses will remove your email address from distribution lists, which can slow down the flow of spam email messages into an inbox.

Spam Filters

A spam filter is computer software that is designed to prevent spam messages from entering a user's email inbox. **Spam** filters are also known as "spam blockers" or "anti-spam utilities." These filters search for spam by watching for keywords and by monitoring suspicious word patterns and frequencies.

Phishing

Phishing is the illegitimate use of an email message that appears to be from an established organization such as a bank, financial institution, or insurance company.

To appear legitimate, the message often contains the company's logo and identifying information. **Phishing** uses legitimate-looking email messages to con a user into giving up private information such as account numbers, Social Security numbers, and personal information. **Phishing** scams direct users to a fake website, where they are asked to enter or update personal information.

What to Do If You Get a Phishing Message

Banks and credit card companies will never ask you to provide personal information via email messages. Never give out personal information through email messages or over the phone. If you receive a suspicious message, contact the institution that the message was allegedly sent from. You should also contact the US Computer Emergency Readiness Team (US-CERT). You can contact US-CERT using the following preferred methods:

- In Outlook, you can create a new message then drag and drop the phishing email into the new message. Send the message to phishingreport@us-cert.gov.
- If you cannot forward the email message, at least send the URL of the phishing website to US-CERT. You can also call the number on the US-CERT website.

Pharming

Pharming is a type of phishing that seeks to obtain personal information through malicious software that is inserted on a victim's computer.

The malicious software redirects the user to a phony web page, even when the correct web address is used. When users go to the fake web page, they are encouraged to enter their username, password, and other sensitive personal information.

Pharming attacks are launched in one of two ways:

1. A hacker installs a virus or Trojan on a user's computer that changes the computer's Hosts file to direct traffic away from its intended target and toward a fake website instead.

2. The hacker may poison a DNS server, causing multiple users to inadvertently visit the fake site. The fake website can be used to install viruses or Trojans on the users' computers, or it may attempt to collect personal and financial information for use in identity theft.

Adware

Adware is software that collects a user's web browsing history. These programs are designed to display advertisements on your computer, redirect your search requests to advertising websites, and collect marketing-type data about you. **Adware** tracks the types of websites you visit and serves up customized ads based on your browsing behavior. Data are collected to create unsolicited targeted pop-up advertisements. **Adware** often accompanies downloaded programs.

Be sure to read before you click "Next," "Yes," or "I agree" when pop-up ads are present. **Adware** often is hidden in these pop-ups.

How to Diagnose a Computer Impacted by Online Nuisances

Online nuisances include spam, ads, pop-ups, online scams, and other activities.

Signs your computer may be affected by an online nuisance include

- receiving spam emails and bounced-back email;
- frequent pop-ups while searching the web or working in programs;
- pop-ups that appear after visiting unfamiliar sites;
- targeted pop-up ads based on recent internet searches.

If you start receiving pop-ups after visiting a website, you should delete your cookies and be sure to have antivirus software installed.

Cookies and Online Browsing

What Is a Cookie?

A cookie is a small text file of information created by a website that is stored by the web browser on the computer's hard disk. When you revisit a website that uses cookies, your browser will send the cookie to the web server, which uses this information to customize and optimize your experience.

First-party cookies are created by a website you visit. They keep track of your personal preferences and the current web browsing session.

Third-party cookies are created by a website other than the one you are currently visiting and are used to track your surfing habits. Third-party cookies are considered an invasion of privacy.

Session Cookies

Session cookies are small text files that are stored in temporary memory. **Session cookies** are lost when the web browser is closed.

The US Federal Trade Commission (FTC) states that session cookies (single-session cookies) are designed to

- help with navigation on the website;
- only record information temporarily and to be erased when the user quits the session or closes the browser;
- be enabled by default in order to provide the smoothest navigation experience possible.

Session cookies

- determine the start and end of a session;
- analyze and measure traffic on a web page;
- determine the web browser being used.

There is no need to delete session cookies. Just close your browser to remove them.

Persistent Cookies

Persistent cookies, which are small text files, are stored on the hard drive and are not lost when the web browser is closed. **Persistent cookies** are only lost if they are designed with an expiration date.

Persistent cookies collect the following information:

- user preferences
- password and username information
- internet protocol (IP) address
- data on web surfing behavior

Make sure to clear your web browsing history to remove stored persistent cookies.

What Are the Privacy Risks Associated with Cookies?

Cookies pose many potential privacy risks. It is important to understand the types of cookies used and what types of information they collect.

Websites that use cookies can collect information about surfing habits and sell that information to a variety of third parties.

Websites can use cookies to track your surfing behavior and use this information to create specific user profiles. Corporate and government entities can use cookies to monitor your surfing behavior.

How to Manage Cookies

Managing cookies is an important part of safe and efficient web browsing. Cookies can be managed by determining which cookie settings should be considered. Be aware that cookie settings must be adjusted in each browser you use.

Cookie settings in web browsers are as follows:

- delete cookies
- block/customize cookies
- allow cookies

Firewalls: Protection and Function

A firewall is hardware or software used to protect a computer from outside threats such as hackers and viruses. Firewalls allow or block internet traffic in and out of a network or computer.

The most ideal firewall configuration consists of both hardware and software. **Personal software firewalls** are typically included with the operating system and can be configured based on user preferences.

Hardware-based firewalls can be purchased as a stand-alone product but are often also included in broadband routers.

How Do Firewalls Work?

Typical firewall programs or hardware devices filter all information coming through the internet to your network or computer system. A variety of techniques can be used to minimize the risk of harmful intrusions into a computer or network, including

- **packet filters**, which inspect each packet leaving or entering a network and either accept or reject a packet based on a predetermined set of rules;
- **proxy servers**, which intercept all messages between clients and servers and help avert a hacker or other intruder from attacking a network.

Typical firewall programs or hardware devices filter all information coming through the internet to your network or computer system.

NAT

Network Address Translation (NAT) was developed by Cisco. It is used by firewalls, routers, and computers that are connected to the internet. It is used in firewalls to provide protection from outside network intrusions by hiding internet protocol (IP) addresses. NAT can be used to prevent many types of network attacks. To maximize security, it must be used in conjunction with the firewall built into the router or the firewall provided by the operating system (OS).

Firewall Options

A firewall application executes tasks different from those executed by antivirus or anti-malware apps. Firewalls help protect against worms and hackers; antivirus apps help protect against viruses; and anti-malware apps help protect against malware. You need all three applications to be fully protected. You need one firewall app on your personal computer (PC). Having more than one firewall app on your PC can cause conflicts and problems. Microsoft Windows and Mac OS X operating systems include firewall software.

Password Security

Passwords

A password is a secret code used to help prevent unauthorized access to data and user accounts. It can be used to secure computers, networks, software, personal accounts, and digital devices. Passwords identify only the authenticity of the password, not the user. Because the user cannot be verified through a password, other security measures may take place. These can include biometrics, which are used to verify the authenticity of a user. The terms *passcode* and *key code* are synonymous with *password*. Passwords and passcodes are often generated by the user. Keys are often generated by encryption software or by the vendor who produced the software.

Password Security

Here are some password tips and techniques:

- **Change your passwords frequently.** The longer you use a password, the higher the risk.
- **Use generic passwords.** Do not use persons, places, or things that can be identified with you.
- **Keep your password private.** Do not share your passwords with others.
- **Use a combination of letters, symbols, and numbers.** The more characters you use, the more secure your password will be.
- **Check your accounts.** Regularly check your accounts to ensure they have not been tampered with.

The estimated time it takes to crack a password increases dramatically as the password becomes longer and more complex.

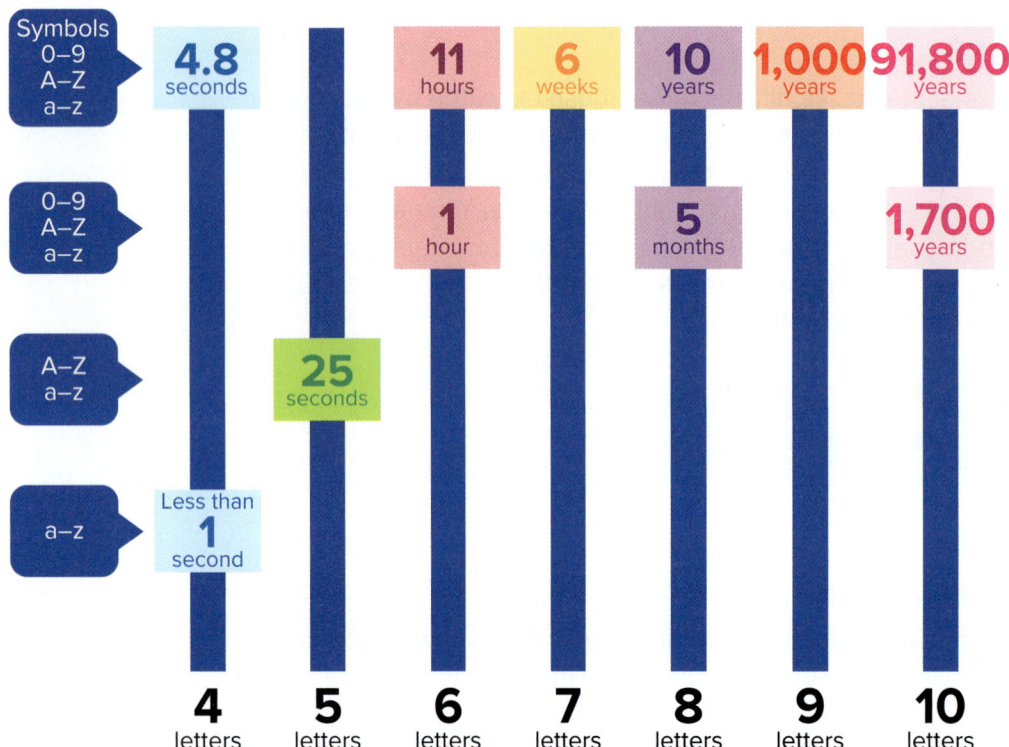

KEY
0–9 = Numbers
A–Z = Capital letters
a–z = Lowercase letters

	4 letters	5 letters	6 letters	7 letters	8 letters	9 letters	10 letters
Symbols 0–9 A–Z a–z	4.8 seconds		11 hours	6 weeks	10 years	1,000 years	91,800 years
0–9 A–Z a–z			1 hour		5 months		1,700 years
A–Z a–z		25 seconds					
a–z	Less than 1 second						

Password Management Options

Password security is one of the most important issues for today's computer users.

The best way to keep your internet login safe is to use a strong password and to never use the same password twice. This would be very difficult to do on your own, so experts suggest using password management software to keep your passwords safe.

A typical password manager installs as a browser plug-in to handle password seizure and repeat. When you log in to a secure site, the password management software offers to save your identification. When you return to that site, it will automatically complete the login using your saved information but will generate a new password for the next login.

Safe Browsing Practices

Internet Filters

An internet filter is firewall software used for blocking a user's access to specific internet content. Filters can be used by organizations, parents, and businesses to restrict access to file transfers, websites, or internet downloads. The software can reside on a router, a user's computer, or on a network. Internet and web filters are often called *parental controls*.

Safe Browsing Tactics

Some tactics that can help keep you safe on the web are as follows:

- **Go incognito.** The four most popular web browsers have a private browsing mode. When private browsing has been activated, your browser will not store cookies. Private browsing does not securely hide your identity beyond the device because your internet protocol (IP) address can still be tracked.
- **Be cautious when using social media.** Go to Facebook, click Settings, and then download a copy of your Facebook data. You will be surprised how much data Facebook has collected about your usage habits. Many other social media sites also collect data on usage patterns, so be aware!
- **Use a virtual private network (VPN) to protect your online identity.** A VPN essentially hides your IP address from outside websites by running your communications through a secure network.
- **Look for HTTPS.** Make sure the sites you are visiting have "https" at the beginning of the uniform resource locator (URL). HTTPS is the secure version of the HTTP protocol. You can force the sites you are visiting to use HTTPS by employing an extension such as the one offered by HTTPS Everywhere.

Antivirus Software: Role and Options

Identifying and Removing Malicious Software

Antivirus software is a computer program used to scan files to identify and remove computer viruses and other malicious programs.

Antivirus programs use a variety of techniques to identify and remove viruses and malware. Two of the most common techniques are signature-based detection and heuristic-based detection.

- Signature-based detection—Viruses are created with a specific set of data and instructions that constitute its signature. Antivirus programs look for these signatures to find and remove the virus.
- Heuristic-based detection—New malware is detected by examining files for suspicious characteristics without an exact signature match.

Other techniques include behavioral detection and cloud-based detection.

- **Body:** The body of the message should succinctly discuss the purpose of the email message.
- **Close:** The close should include any action you would like the reader to take and pleasantly end the message. Examples include the following: *Thank you*, *Look forward to hearing back from you soon*, *All the best, Kind regards*, and so on.

Note: Proper formatting adds professionalism to a business email.

Business Email Etiquette

The key to business email etiquette is professionalism. Remember that any communication represents your organization.

Before sending an email, first consider if it is appropriate. Potentially negative news is usually best given face to face or by telephone to minimize the potential for misinterpretation. Next, remain formal and always use the highest level of courtesy in your salutation, especially with initial contacts. Carefully format your message, avoiding jargon or text-style shortcuts, and always spell check. When sending attachments, make sure the files are small enough to go through any email server. Finally, always be courteous. Because emails don't readily convey voice inflections or physical mannerisms, it can be easy to unintentionally offend.

The Mechanics of Instant Messaging

Instant Messaging (IM) is a tool that lets you send short messages to someone over the internet in real-time. It's like texting on a phone, but you need an internet connection. Popular IM platforms include Facebook Messenger, Snapchat, and WhatsApp. For IM to work, both the sender and receiver need to be online and using the same app.

Video Conferencing

Video conferencing is a technology that allows multiple people to have a meeting using video and audio through the internet. Participants can see and hear each other, making it useful for discussions and presentations or just catching up. To participate, you need a device with a camera, microphone, and speakers. Some commonly used video conferencing apps are Zoom, Google Meet, and Microsoft Teams.

Best Practices for Video Conferences

- **Know the Agenda:** Plan what you'll discuss during the call.
- **Check Your Equipment:** Ensure your camera, microphone, and speakers work properly.
- **Stay Safe:** Use meetings with passwords and only invite trusted participants.
- **Set the Scene:** Use a quiet space with no distractions and proper lighting.
- **Dress Appropriately:** Consider the purpose of the call and dress accordingly.
- **Limit Background Noise:** Mute your microphone when not speaking.

Conference Calling

Conference calling lets a group of people have a shared phone call. Instead of seeing each other, participants only hear each other. It's useful for situations where video isn't necessary but group communication is needed.

Conference calls are useful when members of the call already know each other so that nonverbal or body language cues are less necessary.

Components of a Professional Voicemail Greeting

A voicemail greeting helps callers know what to do when you can't answer. Here's what you might include:

- **Your Name:** Provide your name to confirm callers reached the right person.
- **Reason for Absence:** Mention briefly why you can't answer.
- **Return Call Information:** Tell them when you might call back.
- **Alternate Contact:** Provide another way to reach you or someone else.
- **Guidance for Their Message:** Ask callers to leave their name, number, and reason for calling.

   Note: Avoid generic phrases like "Your call is important"; it's more useful to give clear instructions.

Data Security in Business

Businesses store vital data, such as financial records, employee details, and customer information, on their networks. Protecting this data is crucial as breaches can harm a company's reputation and financial health.

Basic Security Measures

- **Firewalls:** These act as barriers between untrusted external networks and trusted internal networks.
- **Passwords:** Strong, unique passwords are essential for safeguarding data.
- **Antivirus Software:** This software helps detect and remove malicious software.
- **Regular Software Updates:** Keeping software updated ensures known vulnerabilities are patched.
- **Physical Security:** Securing physical computers can reduce theft risks.
- **Data Backups:** Regular backups help in recovering data in case of loss or corruption.

Data Security Statistics

According to the US Federal Bureau of Investigation (FBI),

- 22 percent of data breaches were due to phishing;
- 45 percent of breaches involved cloud storage;
- hospitals accounted for 30 percent of large data breaches;
- 90 percent of health care institutions faced at least one security data breach;
- 77 percent of companies lacked a robust strategy against data breaches;
- the average cost of a significant data breach is $4.85 million.

IDS

Intrusion Detection System (IDS) tools identify and counteract network breaches. Their capabilities include

- detecting and alerting about potential intrusions;
- stopping ongoing attacks;
- blocking future intrusion attempts from the same source;
- gathering data about intrusion attempts;
- "honeypots," which are used to trap intruders by presenting vulnerable points.

Data Theft

Data theft is unauthorized extraction of company data. Information stolen can include emails, confidential documents, copyrighted content, or Personally Identifiable Information (PII). Often, insiders, such as employees, commit these thefts. One common method is "Thumb-sucking," where USB drives are used to steal data.

Wire–Wire Scams

These scams trick businesses into transferring money to fraudulent accounts. Scammers often pose as executives or suppliers. Once the money is transferred, especially to international accounts like those in China, recovery becomes challenging.

Data Corruption in Business

Data corruption introduces errors into information during various stages. Errors can be detected or go unnoticed (silent corruption). Methods to minimize corruption include error-correcting codes and data scrubbing.

Ethical Employee Behavior

Doing the right thing at work means being honest, fair, and treating everyone equally. If a company doesn't encourage good behavior, it can lead to problems like money issues, legal troubles, and unhappy workers.

Using computers and other devices at work should also be done the right way. Companies should have clear rules about how to use these tools and teach their workers about them. Just because someone has a computer at work doesn't mean they can use it like they do at home. Using a work computer in the wrong way can slow things down, cause security problems, or even get someone into a lot of trouble.

What is an ethics violation?

An ethics violation occurs when someone does something against a group's rules or values. This can happen with computers and networks too. There are two main types of ethics violations:

- **Illegal Activities:** This means doing things that break the law using computers or the internet, like stealing computer parts, downloading material without permission, or sending computer viruses.
- **Breaking Company Rules:** Every company has its own rules. Some common rules are about not sending personal emails, shopping, checking social media, or playing games on the computer.

How to Report an Ethics Violation

An **Acceptable Use Policy (AUP)** is like a rulebook. It tells you what you can and can't do when using a company's computer network. Before you can use a company's network, you usually need to agree to follow the AUP.

If you see someone breaking computer rules, you should tell your boss or the person in charge of the computer network. When you tell them, just be straight to the point and say what you saw. But make sure what you saw was really a rule being broken. Remember, your main job isn't watching what others are doing on their computers.

Employee-Monitoring Software

Employee-monitoring software lets a person in charge see what's happening on all the computers in one place. It can be useful but can also be misused. Big companies often use internet filters and remote desktop software instead. This software lets them see what's on a computer's screen from somewhere else.

Remote Desktop Software

Have you ever contacted a help desk and had them ask if they could "take over" your computer? If so, you've seen remote desktop software in action. **Remote desktop software** lets one computer control or see what's on another computer through the internet. It's useful for people who manage computer networks because they can help or check on any computer without physically being there.

Computer companies also use this when they're helping customers. **Remote desktop software** is becoming more common now, especially since many people work from their homes.

Information Systems and Management

In today's digital age, businesses heavily rely on advanced computer systems to operate, make decisions, and grow. These systems, often referred to as Business Information Systems (IS), provide tools to collect, process, and distribute vital data seamlessly across different departments.

Business Information System Roles

The rise of technology in the business world has paved the way for specialized roles that focus on harnessing the power of digital tools.

IS Program Manager

IS program managers are like the captain of a ship. They work with computer experts and bosses to make sure computer systems run smoothly in the company.

- **Degree:** Typically, a bachelor's in IT
- **Average 2023 Salary:** $113,209

IS Security Manager

IS security managers focus on safeguarding an organization's IS and Wi-Fi networks, collaborating with senior leadership on risk assessments and ensuring the overall cybersecurity of the organization.

- **Degree:** Typically, a bachelor's in IT plus experience in IS security
- **Average 2023 Salary:** $102,410

IS Technician

IS technicians are responsible for installing, troubleshooting, and maintaining computer equipment and software, occasionally offering training to users.

- **Degree:** Typically, an associate's degree in IT
- **Average 2023 Salary:** $47,802

Data Analyst

Data analysts gather and interpret data to create insightful reports for decision-makers within the organization.

- **Degree:** An associate's degree or higher in IT
- **Average 2023 Salary:** $64,777

Database Administrator (DBA)

DBAs are responsible for monitoring database performance, ensuring data security, and maintaining the overall health and efficiency of an organization's databases.

- **Degree:** A bachelor's in computer science or a related field plus experience as a data analyst
- **Average 2023 Salary:** $77,733

MIS: Management Information Systems

MIS software helps managers work better and make smarter choices. For the top bosses, MIS can give a full view of everything happening in the business. Think of it like this: It's like software that can track how much product a store sells or predicts if the store will need more of an item soon.

TPS: Transaction Processing Systems

TPS keeps track of things like sales and how much stock is left in a store. One of the first TPS was made for airlines. Back in 1960, American Airlines could handle 80,000 transactions in one day!

DSS: Decision Support Systems

DSS tools help managers make decisions by showing them data. A **DSS** combines the power of databases with decision-modeling software. It can

- analyze a lot of data;
- predict future sales;
- suggest decisions;
- help teams work together.

When DDS systems pull these abilities together, they can help in the following:

- **Route Planning: DSS** often assists GPS-guided software to determine the most efficient route. **DSS** can collect data on current traffic patterns and any points of congestion.
- **Agricultural Planning: DSS** helps assist with data-driven decisions on the optimal time to plant, fertilize, and harvest crops.
- **Health Care: DSS** assists physicians with a variety of clinical decisions including the data-driven interventions to efficiently get patients off ventilators.

SCMS

Supply chain management software (SCMS) combines the capability of TPS software with the forecasting of **DSS** software to help businesses manage inventories efficiently. Typically, SCMS includes purchasing software, inventory management software, shipping and receipt software, and supply management software.

SCMS focuses on inventory management efficiency and effectiveness, making processes such as just-in-time inventory practical.

SAP offers a suite of digital supply solutions that give it leading market share at 19.1 percent. Their solutions are designed for the variety of business areas including

- supply chain planning;
- supply chain logistics;
- manufacturing;
- product lifecycle management;
- enterprise asset management.

ESS

Executive support systems (ESS) or executive information systems (EIS) provide management information for senior managers, such as company presidents and chief executive officers. ESS have easy-to-use interfaces and menus to maximize executive efficiency. They offer drill-down capability so that executives can see all business operations at all levels. ESS allow executives to monitor and analyze trends throughout their businesses.

ERP: Enterprise Resource Planning

Think of Enterprise Resource Planning (**ERP**) as the big brain of a company's software. It sees everything, from making products to ordering pencils! One of the most popular ERPs is Microsoft Dynamics 365. It connects different parts of a business, for example sales and finance.

Information Management

Levels of Business Information Management

Tech has made businesses smarter and faster. And data has become super important!

- Information systems are a big deal in companies today.
- The top bosses use them to see how the company is doing and predict the future.
- Middle managers use them to check on and ensure everything is on track.
- The first-level managers, like team leaders, use information systems for daily tasks and monitoring their team's work.

Business Information Flow

Information is important to all parts of a business. Companies use modern business information systems to allow data to flow smoothly between departments. This helps different departments work together and make better decisions.

In marketing, using online sales, social media, and things like search engine tricks are super important. In accounting, computers are crucial for keeping track of money and making reports.

For finance, having the latest information is key for making smart money choices. In the day-to-day operations, software helps make the best decisions. And in research, strong computers are a must-have.

Blockchain and Cryptocurrency

Blockchain

Think of blockchain as a special kind of diary. Instead of writing daily entries, we add "blocks" of information to it. Every time we add a block of information, it's connected to the one before it, making a chain (hence the name "blockchain"). These blocks are like puzzle pieces because they fit together in a certain way that makes it difficult to change anything without everyone noticing.

Blockchain Process

Let's explore how blockchain works when you buy something like a bike.

1. **Initial Transaction:** Imagine you buy a mountain bike online from a store like REI.com. That's a transaction.

2. **Checking the Purchase:** Other computers make sure your bike purchase is legitimate. They look at details like when you bought it and for how much.

3. **Storing the Purchase Information:** Once everything checks out, the details about your bike purchase get put into a block.

4. **Adding a Unique Code:** Before this block joins the others in the blockchain, it gets a special code, kind of like a unique sticker.

Cryptocurrency

Cryptocurrency is like online money, but there's no bank involved. Instead, all the money transactions get recorded on our special diary—the blockchain.

1. **Sending and Receiving:** If you want to send or get this online money, you do it directly with the other person, using special digital wallets.

2. **Safety First:** All the information is locked up safely using codes, which is why it's called "crypto" (short for cryptography).

3. **No Bosses Here:** Unlike regular money that government or banks control, cryptocurrency is managed by its users and special math rules.

4. **Making New Cryptocurrency:** People can create new cryptocurrency by using their computers to help check and record transactions in the blockchain. This process is called "mining." Bitcoin is a well-known type of cryptocurrency, but there are many others too!

Business Information Beyond the Office Walls

Business in the IoT

Let's think about the many ways our world is connected. Once, our phones were just devices to call friends; now phones help us navigate through cities, find a nearby place to eat, or even control the lights in our houses. And it doesn't stop there. How about using a smartwatch to adjust the temperature of a room, respond to a meeting invitation, or alert someone if you're running late to an appointment? This network of interconnected devices, from our phones to our household appliances, is known as the Internet of Things (IoT). For businesses, the possibilities with IoT are endless. Imagine a retail store that can track its inventory in real-time through smart shelves, or a farm where sensors measure the moisture in the soil to automatically water crops. Companies can understand their customers better, improve efficiency, and even create new products and services that we haven't thought of yet.

More to Connect = More to Protect

While it's fantastic that we can access information anywhere, there's a hidden challenge: keeping our computer systems and data safe. As we connect more devices to the internet, we have more things to protect. Being connected everywhere means hackers have more doors to try and enter. Your personal computer, if not well protected, might just be the weak link that jeopardizes the entire business.

Working Remotely

Working remotely gives us flexibility, but it also demands a new kind of awareness. In this vast network of IoT, it's crucial to ensure the safety of our devices, not just for our sake, but for everyone relying on the same network. As we embrace the future of work and play, let's also arm ourselves with the knowledge and tools to make our interconnected world secure.

Protecting Your Computer

There are many ways to protect your computer:

1. **Antivirus Software:** Think of antivirus software as your computer's shield. It fights off harmful viruses. With cybercrime costing a lot of money each year, it's like having a guard dog for your computer.

2. **Webcams:** You know the camera on your computer? Hackers might peek through it. Yikes! Use a cover to block it when you're not using it. If you're on a video call, you can blur your background so people can't see everything behind you.

3. **Use of Work Devices:** Keep your work computer for work only. Think of it as your special workspace – you wouldn't want your little brother doodling in your work notebook, right?

4. **VPN:** Your virtual private network is like a secret tunnel for your data on the internet. It hides where you are and keeps your information secret. Some workplaces will ask you to use one.

5. **Passwords:** Passwords are like keys. Don't use the same key for everything, and make them tricky! Use long ones with a mix of letters, numbers, and symbols. You can use tools like NordPass or Keeper to remember them for you.

6. **Home Wi-Fi:** Your home Wi-Fi can be like your front door. You wouldn't leave it open for anyone to come in. Keep it safe! Get a good router, change its name and password, and turn on encryption. That's like putting extra locks on your door.

7. **Cloud Storage:** Instead of keeping all your important stuff on your computer, you can store it in the cloud. Imagine the cloud as a large, safe storage room on the internet. Services like Google Cloud or Amazon Web Services offer this.

   Note: When you're working from home, you want to make sure you're doing it safely. Just like you'd lock your doors at night, you want to protect your computer and data.

Business and Artificial Intelligence

Types of AI

Artificial Intelligence (AI) involves creating intelligent machines and computer programs that mimic human intelligence. It extends beyond merely emulating human ways of thinking, allowing for a broader scope of capabilities. In simpler terms, AI is all about using computer science and data to solve problems or automate tasks. This field includes subcategories like machine learning and deep learning, which focus on developing algorithms that enable computers to make predictions or decisions based on input data.

Machine Learning

Machine learning is like teaching computers to learn from experience, just as humans do. These systems improve their accuracy over time through continuous learning processes. They rely on human input to refine their knowledge and predictions. AI can learn from machine learning models by analyzing inputs and expected outputs, contributing to their ability to adapt and improve.

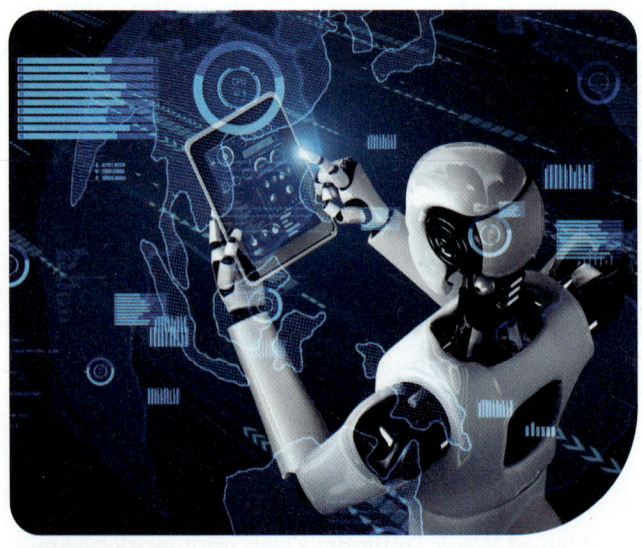

Deep Learning

Deep learning takes this a step further by automating parts of the learning process, making it even more scalable. This advancement reduces the need for constant human intervention. Lex Friedman, an AI researcher at MIT, aptly describes deep learning as "scalable machine learning." Today, AI finds application in various domains, such as speech recognition, customer service, image analysis, and automated stock trading.

Algorithm

An algorithm is a step-by-step set of instructions designed to solve problems or perform calculations. It's the backbone of AI, guiding computers through complex tasks by breaking them down into manageable steps. With AI's transformative capabilities, algorithms play a pivotal role in enabling machines to accomplish tasks that were once limited to human capabilities.

Potential Benefits of AI

There are many potential benefits of AI:

- **Improved Work Efficiency:** Artificial intelligence can make tasks faster and more efficient. A study showed that AI can boost work speed by 40 percent and increase profits by 38 percent. By using AI to make decisions based on a lot of data, industries, such as health care, can save up to $100 billion a year.

- **Safety in Dangerous Jobs:** Some jobs can be risky for people. AI can take over these tasks, especially in fields like manufacturing, mining, and chemicals. This means fewer people might get hurt.

- **Reduction in Errors:** People, even when they try their best, can make errors. Computers with AI can often avoid these mistakes. They gather data, analyze it, and then decide what to do.

- **Creating New Products and Services:** Computers with AI "think" in a different way than humans. Therefore, they can find new opportunities or ideas for products and services that might not have been seen before.

Potential Drawbacks of AI

- **Job Loss:** AI might replace some jobs, but it can also create new ones. It's predicted that AI will lead to 97 million new jobs by 2025, but also cause 85 million jobs to disappear. We need to study this more to understand the real effect on jobs.

- **Personal Data Concerns:** Protecting personal information, like names, addresses, and other private details, is important. Specific concerns with AI include the following:

  a. **Data Repurposing:** Using data for a different reason than first intended.

  b. **Data Spillover:** Gathering data from people who weren't originally meant to be included.

  c. **Data Security:** Making sure data are kept safe from harm or theft.

How Data Science Uses Technology

Data science is the work of making sense of vast amounts of data to make decisions.

Today's technology and information systems are all about creating smart sets of rules, called algorithms, that automatically collect, sort, and share data. This means decision-makers get the information they need quickly and can act on it right away.

By using data science, businesses can spot and fix issues in their processes faster. Think of it as a tool that helps companies run smoother and better. Data science is key for organizations that want to keep improving.

Chapter Review

1. What is an ethics violation, and what are the two main types of violations related to computer usage?

2. List the main elements of a properly formatted email message and describe the importance of each element in a business setting.

3. What are the primary reasons for businesses to emphasize the importance of data security?

4. Describe the primary functionalities of an IDS.

5. What is "Thumb-sucking" in the context of data theft?

6. Explain the primary objective of wire–wire scams.

7. What are the measures to minimize data corruption?

8. What are the primary components to ensure effective video conferencing?

9. List the main elements that should be included in a professional voicemail greeting.

10. Explain the concept of blockchain in simple terms.

11. How does cryptocurrency differ from traditional currency?

12. List some protective measures for working remotely to ensure computer system safety.

13. What is the main difference between machine learning and deep learning in the context of AI?

14. Describe the Internet of Things (IoT).

15. How does data science influence today's business decision-making processes?

Ethics Discussion

What are ethics? What constitutes an ethics violation? Conduct research on the internet about business ethics violations. Find at least two articles that discuss business ethics violations. What can businesses do to prevent ethics violations?

14 File Management

What To Expect

After completing this chapter, you will be able to

- explain folders and how to work with folders;
- identify how to view folders;
- explain files and how to save, open, and work with files;
- distinguish various file extensions;
- explain file compression and identify various file compression standards;
- explain cloud storage;
- distinguish between different cloud storage services.

Chapter Topics

- Context Menus
- Folders
- Tags
- Files
- The Recycle Bin
- Cloud Storage

Let's Explore File Management

Imagine a vast storage room in a natural history museum. This room represents your computer's root directory. Within this enormous space, there are countless filing cabinets, each labeled for different exhibits like "Dinosaur Era," "Ancient Civilizations," or "Underwater Life." These cabinets represent the primary folders or directories on a computer.

As a photo archivist, you're responsible for organizing countless images that come to the museum. You pull open a drawer in the Ancient Civilizations cabinet, and inside, there are individual folders separated by specific civilizations—Ancient Egypt, Mayan Empire, and Roman Empire. This further division is like the subfolders within your primary computer folders.

Within each of these folders, photographs are neatly organized, but some have sticky notes attached to them. The notes might read "Marketing Material" or "Africa Exhibit." These sticky notes can be seen as "tags" in the digital world. They offer an additional layer of organization, making it easier for the archivist to find and group images based on specific criteria without physically moving the photos around.

While this filing cabinet model helps us conceptually organize our data, it's essential to understand that computers don't really work like a physical storage room.

A computer stores all data as 1s and 0s on its hard drive, with no actual physical space dedicated to each file or folder. It doesn't keep the Ancient Egypt photos in one corner and the Mayan Empire photos in another. Instead, it relies on an indexing system, using the folder structure and tags we provide to find and display files almost instantly. So when we search for a tag, like "Africa Exhibit," the computer quickly references its index and displays every file with that tag, regardless of which "folder" it's stored in.

That's useful because you might organize your files by region, but you may want to pull together an exhibit about something else. Say you want a photo collage about ancient pyramids; you don't need to put them in a new place called "pyramids." If you've tagged "Mayan pyramids" and "Egyptian pyramids" with the label "pyramids," you can ask the computer to pull up everything with the tag "pyramids," no matter where it's located.

Understanding the mental model of cabinets and folders and the reality of digital storage gives us flexibility. By visualizing a storage room, we can organize our files in ways that make sense to us. But by leveraging the power of tags and understanding the computer's instant retrieval capability, we can access our files in more dynamic ways than a physical system could ever allow. As a digital archivist, you have the tools of both the past and the future at your disposal, ensuring that the museum's precious photos are always at your fingertips.

Context Menus

Context refers to the situation that something is in. In computing, context menus are important ways to manage files and folders, because all platforms use some type of context menu to help you identify and manage files. If you've played video games, you've probably used them as well. For example, in a video game, if you click on an object, a menu pops up showing you all the things you can do with it, like color it, throw it, put it in your inventory, or move it. This is a lot like the context menu on a computer!

The context menu is super helpful because it shows you options based on what you're looking at or selecting. So if you click on a picture, the menu might let you zoom in, print, or edit the picture. If you click on a music file, it might let you play the song or add it to a playlist. And if you click on a blank space on your desktop, it could let you change the wallpaper or add a new folder.

To see the context menu, you usually right-click (or tap and hold on touch screens). This menu is a computer's way of saying, "Hey, here are some things you can do with this!" It's a quick and handy tool that makes using a computer a whole lot easier.

Folders

What's a Folder?

A folder serves as a storage space or container for files and programs on a computer. It helps in categorizing and organizing different types of information for easy access. Inside a main folder, you can have smaller folders known as subfolders. This structure is useful for further organizing your files.

Root Folder

The root folder or root directory is like the base of a tree. Just as a tree's roots provide foundational support, the root folder offers a starting point for the organization of files and subfolders in a computer's storage system. It sets the groundwork from which all other directories branch out, like the way branches grow from a tree's main trunk.

On most computers, the root directory of the main hard drive is designated by the symbol "C:". This signifies the primary location where the operating system, applications, and most user data are stored. The entire hierarchy of folders and subfolders that originate from this root helps in keeping the data organized and easily accessible. As you navigate deeper into this system, you'll come across various subfolders that help categorize and store specific types of files, ensuring a streamlined and efficient data structure.

Paths

Each file saved on your computer has a unique address showing its location on the storage device. This address is called the file's path. The path tells you which folder (or subfolder) the file is in, starting from the root directory.

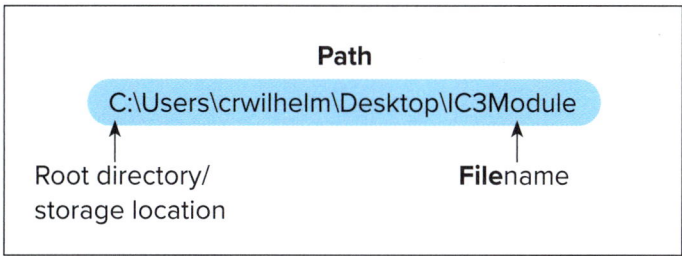

Path

C:\Users\crwilhelm\Desktop\IC3Module

Root directory/
storage location

Filename

Creating Folders

When it comes to file management, the creation of folders is a fundamental skill. While the exact methods to create folders might vary across devices and platforms, folders are used to organize and categorize our digital content.

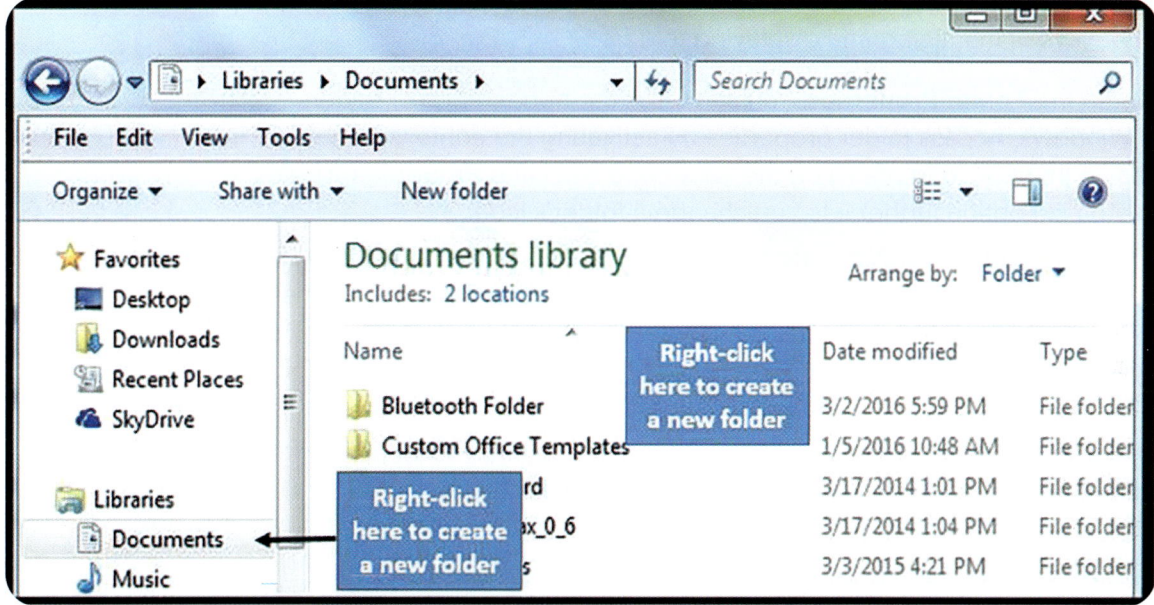

Why Create Folders?

- **Organization:** By categorizing files in folders, we can quickly locate and access them.
- **Efficiency:** Well-organized folders can save time, especially when dealing with large volumes of files.
- **Project Management:** For specific projects or tasks, creating dedicated folders can help keep all relevant files in one place.

Pros of Using Folders

- **Structured Approach:** Folders provide a systematic way to access files.
- **Minimized Clutter:** Folders prevent the accumulation of disorganized files, making searches faster.
- **Logical Grouping:** Folders allow for grouping by project, date, file type, or any other category that makes sense for the user.

Cons of Using Folders

- **Over-Complication:** Too many nested folders can make it challenging to find files.
- **Rigidity:** Once a folder structure is established, it might be time-consuming to reorganize.
- **Potential for Duplication:** Without careful management, the same file might be saved in multiple locations.

Folder Display and Properties in Microsoft Windows

Microsoft Windows provides extensive flexibility in terms of folder appearance and file views. Similarly, MacOS, popular cloud services, and other platforms offer their own unique ways of managing and displaying folders.

- **Modifying Folder Properties:** You can tailor the appearance of folders in Windows. Access folder properties by activating the context menu (e.g., right-click, long press) on the folder and selecting Properties. The Customize option further lets you change a folder's icon.
- **Changing File Views:** Navigate to **File** Explorer and click on the View tab to adjust file displays. Choices range from icon sizes (extra large to small) to detailed lists. **File** extensions, which identify the associated program of a file, can also be toggled on/off here. For instance, Word files use ".docx" while Notepad employs ".txt".
- **Quick Access to Views:** Activate the context menu in a vacant area of the Contents pane, then click View to swiftly change display settings.

Renaming Folders

You can rename a folder by accessing its properties through a file menu or a context menu. A context menu is a menu that changes based on the context. In simple terms, it's a menu that changes based on whatever you're selecting. So if you are selecting a folder, it will give you folder properties. If you are selecting a file, it will give you file properties. If you select your screen background, it will give you screen background properties.

Windows

Folder names can be extensive, with up to 255 characters. But remember, very long names might not always display fully in some views. To rename a folder in Windows:

1. Open **File** Explorer.

2. Activate the context menu on the desired folder (e.g., right-click, long press, etc.).

3. Choose Rename.

4. Once the folder name is highlighted, you're in edit mode—type away!

MacOS

Finder is the MacOS equivalent of Windows **File** Explorer. Within Finder, you can create, move, and modify folders. The context menu (e.g., secondary click, two-finger tap on touchpads) brings up options like those in Windows.

- **Cloud Services:** Platforms like **Google Drive** or Dropbox have web interfaces and apps. Here, folders can be created, shared, and organized using intuitive drag-and-drop or context menu actions.

- **File Systems and Extensions:** Different operating systems may use different file systems, impacting how data is stored and retrieved. This can lead to variations in how folders and files appear or are organized. Recognizing common file extensions, like ".docx" for Word or ".pdf" for PDF files, helps in identifying and using files across platforms.

Tags

Imagine you've captured an impressive photo of fireworks illuminating the beach sky on the fourth of July in 2023. Now, consider where you'd save it on your computer. Would it go in a folder called "Summer"? "Beach"? "Holidays"? "Vacation"? It seems fitting for any of those categories, right? And if you decide on one, where might another person search for it?

How about this? Instead of placing the photo in a single folder, you could tag it with several descriptors like "summer," "beach," "holidays," and "vacation". This way, regardless of which term someone thinks of, they'll be able to locate that photo. It's as though the photo exists in multiple spots at once, without creating numerous duplicates.

How to Tag

Tagging on a computer is straightforward. In most operating systems, when you access the Properties or Info menu, you can assign tags to files. Within this menu, there's often a Tags or Labels section where you can type in keywords related to that file. After entering your desired tags, you simply save or close the Properties window. From then on, when you use the computer's Search function and type in one of the tags, the file will appear in the search results, making it easier to locate. Different software applications, especially those designed for photos or documents, also have built-in tagging features that streamline the process further. Over time, as you consistently tag your files, you'll create a well-organized digital environment that's both intuitive and efficient.

Tags vs. Folders?

Computers are great at finding file—fast. If you know the tags or labels you've put on your stuff, the computer can show them to you right away without the need for folders. But if you're less familiar with the tags, or people don't use them in the same way, then folders can be better.

Think about a dresser. If you have a small drawer inside a bigger one, and then another one inside that, it gets confusing. That's how computer folders can seem if there are too many inside each other. So sometimes, it's easier to have a few main folders and use tags to find the exact thing you want. When it comes to keeping your computer organized, you have options. You might like folders, or maybe tags, or even both. It's all about finding what makes sense to you and your organization.

Files

A file is an item that contains information such as text or images or music.

On your desktop or in **File** Explorer, files are represented with icons; it is often easy to recognize a type of file by looking at its icon.

Common file types include:

- **Application Files:** Application files contain detailed instructions for the processor on what tasks to perform. Usually, these are stored in the Program Files folder on the hard drive.
- **Data Files:** Data files contain information you have entered and saved on a specific application on your computer. Common data files include Microsoft Excel spreadsheets and Microsoft Word documents. These files can be stored anywhere you designate.
- **System Files:** System files are included with the operating system. Many system files are hidden to prevent them from being altered or deleted.

Selecting Files and Folders

To select one file or folder, go to **File** Explorer, then point to that file or folder.

To select multiple files or folders in a contiguous list, point to the first file or folder in the list, press and hold the Shift key, then point to the last file or folder in the list.

To select multiple files or folders that are noncontiguous, point to the first file or folder to be selected and press and hold the Ctrl key as you select each file you want.

To turn off a selection of files or folders, simply click anywhere on the desktop.

You can also click Cortana, enter the filename, and Search.

Opening Files from the Documents Folder

The Documents folder contains any files you have saved to this location. However, it is a good idea to save files to folders you create so your files are organized and can be easy to locate.

Here are the steps to open a file that has been saved in the Documents folder:

1. Click the **File** Explorer button on the taskbar.
2. Click Documents. All files saved in Documents will be shown. Select the file you want to access.

How to Save Files

Knowing where your files are stored can save you endless frustration. It is important to be organized and to be able to save files in different areas on a device.

If, for example, you download an article for your economics class, it will probably be automatically stored in a folder called Downloads. If you want to find this article again, though, you may want to store it in a different folder.

To save a file to a different folder on your Windows 10 computer, first determine in which folder you would like to save the file. Next, while the file is open, check your ribbon to ensure that you're on the Home tab, and click **File**.

When the drop-down menu opens, click Save As.

When the Save As dialog box opens, navigate to the folder where you want to save the file (on the left side of the dialog box) and enter the file's name. You can also create a new folder to save the file in by clicking the New **Folder** button.

Once you select the correct site and file name, click Save.

How to Copy Files or Folders

When you copy a file or folder, the original remains in the source location, and a copy is placed in the destination location—the same information will be in both locations.

Use one of the following methods to copy a file or folder:

- Press Ctrl+C, then move to the new location where the file will be copied and press Ctrl+V.
- Right-click the selection and then select Copy. Then move to the new location where the file will be copied, right-click, and select Paste.
- To copy a file or folder from one device to another, select the file, hold down the left mouse button, and drag the selection to the new location.
- To copy files in the same drive, press Ctrl as you drag the selected file or folder to the new location.

If you have copied files with the same file name, you may get a message indicating there are duplicate files. Windows will provide a recommended new file name. It also allows you to replace the existing file.

How to Move Files and Folders

Moving a file or folder means it is deleted (cut) from its original storage location and copied (pasted) into the new storage location (the destination).

When you move a folder, all the contents of the folder are moved to the new storage location.

To move a file or folder in **File** Explorer, select the file or folder and then use one of the following methods:

- Press Ctrl+X and then move to the new location where the file will be copied and press Ctrl+V.
- Right-click the selection and select Cut; then move to the new location where the file will be copied, right-click, and select Paste.
- To move a file from one device to another, select the file and hold down the Shift key while you drag the file to the new location. If you don't hold down the Shift key, the file will be copied instead of moved.
- To move a file from one folder to another, drag the file from one folder to the other. You can also drag one folder into another to make it a subfolder.

Renaming Files and Folders

For organizational purposes, it is sometimes necessary to rename files or folders. There are several methods, but here are two easy ways. Most users simply right-click the file or folder and then select Rename from the shortcut menu. Another way is to select the file or folder by clicking it in **File** Explorer and using the Rename option on the Home ribbon. With either method, when you see the highlighted name, simply type in the new file or folder name, and then press Enter.

Searching for Files and Folders

Searching for a file or folder in Windows is quite simple. All you need to do is type your desired search criteria in the Search box. You can search for files or folders using name, file type, date created, date modified, and many other criteria. Any file or folder that closely matches the criteria you typed in the Search box will be displayed. Look through the results and select the file or folder you need.

How to View File Extensions

To always display the file extensions, open **File** Explorer and go to the View tab. Make sure the check box for **File** Name Extensions is checked.

Important file types such as system and data files will be hidden to prevent unintentional changes.

Audio File Extensions

Audio files can be played using specialized applications called players. Audio players that are freely available include Windows Media Player, RealNetworks RealPlayer, Winamp3, and XXMX (designed for UNIX systems).

Common audio file extensions include

- .mp3 and .m4a (Motion Pictures Experts Group)—requires a player such as iTunes, QuickTime, or Windows Media Player;
- .wav (Waveform Audio **File** Format)—Windows format for sound files;
- .au—audio format used on UNIX servers.

Common Video File Extensions

Video files, like audio files, can be played using specialized applications called players. Video players that are freely available include Windows Media Player, RealNetworks RealPlayer, and Apple QuickTime. When you play a video file, your designated player will automatically begin video playback.

Common video file extensions include

- .avi (Audio Video Interleave)—standard for Windows video files;
- .mov—video format for Apple QuickTime and macOS;
- .mpeg and .mp4 (Motion Picture Experts Group)—format for internet video files;
- .ram (Real Audio Metadata file)—RealNetworks video format.

Common Image File Extensions

Image files include photos, drawings, and charts that have been saved in a digital format. Many graphics files are supported by web browsers and operating systems due to built-in graphics viewers.

Graphics can be imported into a digital device from digital cameras or scanners, or they can be created using dedicated graphics apps such as Microsoft Paint and Adobe Illustrator. When you click an image, the image automatically will be displayed in a viewer or graphics editing program. If the file does not open, Windows will prompt you to select the appropriate app to open the file.

Common image file formats include

- .gif (Graphics Interchange Format)—used for drawings and illustrations;
- .jpg or .jpeg (Joint Photographic Experts Group)—used for photos and graphics;
- .tif or .tiff (Tagged Image **File** Format)—used in desktop publishing and medical imaging;
- .png (Portable Network Graphics)—commonly used on web pages.

Common Document File Extensions

Documents and other text files are created using Microsoft Office, OpenOffice, or Google Docs.

Some documents can be created to be cross-platform compatible, which means they can be viewed using most common operating systems. Common cross-platform file types include PDF and RTF.

Common document file types include the following:

File Extension	Description
.docx	Document format for Microsoft Word 2007 and later
.pdf	Can be viewed using the Adobe Reader plug-in (Portable Document Format)
.pptx	Presentation format for Microsoft PowerPoint 2007 and later
.rtf	Supported by most word processing programs and operating systems (Rich Text Format)
.xlsx	Spreadsheet format for Microsoft Excel 2007 and later
.one	Microsoft OneNote
.accdb	Database format for Microsoft Access 2007 and later

There are extensions for other types of documents. A document's type determines its file extension. Examples include:

- Microsoft PowerPoint ⟶ ".pptx"
- Prezi ⟶ ".pez"
- Creating a web page could use ⟶ ".html"
- Microsoft Excel ⟶ ".xlsx"
- Microsoft Access --> ".accdb"

Nearly every application has its own file extension.

Common Executable File Extensions

Executables are critical to the operations of your computer and are designed to launch programs and specific procedures. Be cautious when downloading or opening executable files you receive as email attachments or find on the internet. When you open an executable file, your computer may automatically run any operations that are included in the file, which may be detrimental to your digital device.

Common executable files include the following:

File Extension	Description
.cgi	**File** used to generate web content (Common Gateway Interface)
.exe	Compressed files that are self-extracting (Windows executable program)
.vbs	Created using the Visual Basic programming language. These files are known to carry viruses (Visual Basic script files)
.cmd	Windows command file
.msi	Windows installer file used to automate software installation

File Compression

File compression reduces the size of data files. This saves storage space as well as transmission capacity and can speed data transmission. Most file types can be compressed and decompressed using a file compression utility.

There are two types of file compression: lossy and lossless. Lossy compression removes less-valuable data. This means that when the file is expanded, some of the data in the original file are not included. Lossless compression eliminates redundancy but retains all the original data.

Common Compressed File Formats

Compression file formats reduce the amount of time needed to download a file and the amount of storage capacity a file takes up. Most file types can be compressed and decompressed using a file compression utility.

Common file compression extensions include the following:

File Extension	Description
.bz2	Compression type used by the Bunzip app
.rar	Compression type that is platform neutral. On Windows systems, the WinRAR app is used.
.zip	Compression type used by the WinZip app

Image File Compression Formats

Because a digital photo can be a large file, many users compress images before storing or sending them.

Popular image compression formats include JPEG, PNG, and GIF. JPEG is the most popular lossy image compression format. PNG is the most popular lossless image compression format.

Audio File Compression Formats

Audio file compression allows extremely small devices to store many music files. The most popular compression format is a lossy compression format called ".m4a". The most popular lossless compression format is ".wav".

What If I Cannot Open a File?

When you double-click a file, Windows should automatically open it. If it can't, Windows will prompt you to select the correct program to open the file. This often entails purchasing and downloading the necessary program from the internet.

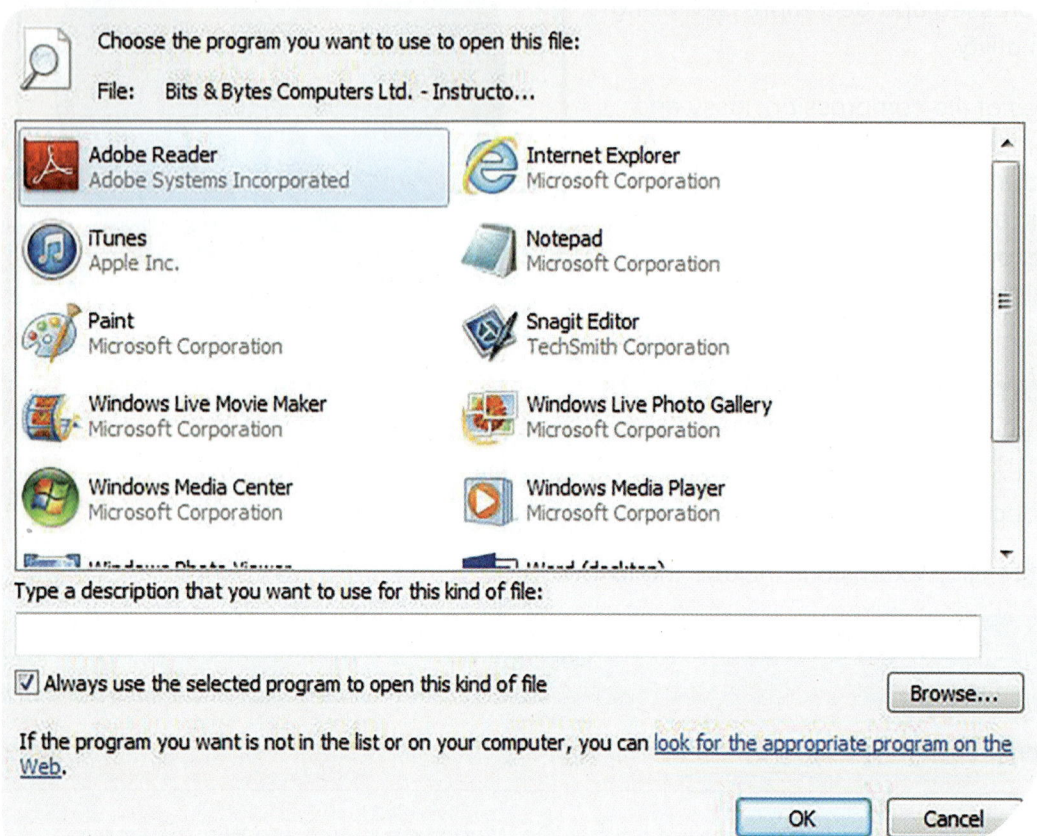

The Recycle Bin

The **Recycle Bin** is included in the Windows operating system (OS). The **Recycle Bin** is a storage area for files and folders you want to delete from a Windows computer's hard drive. It is like the Trash Can on Macintosh computers. Files deleted from other storage devices such as a memory card or flash drive are permanently deleted. Files sent to the **Recycle Bin** are not permanently deleted until you empty it. You can restore files in the **Recycle Bin** if it has not been emptied.

Empty   **With Deleted Items**

Copyright © McGraw Hill   Microsoft Corporation

To delete a file or folder, use one of the following methods:

- Select the file or folder and click Delete.
- Right-click the file or folder and click Delete.
- Select the file or folder, hold down the left mouse button, and drag the file to the **Recycle Bin** icon.
- To bypass the **Recycle Bin** and permanently delete a file or folder, select the file or folder to be deleted, hold down the Shift key, and then press the Delete key.

Restoring Items from the Recycle Bin

If you moved a file or folder to the **Recycle Bin** and want to restore the file to its original location, it is a relatively simple process. However, you can only restore files that were not deleted from the **Recycle Bin**.

Use one of the following methods to restore a file from the **Recycle Bin**:

- Select the file or folder you want to be restored and then click Restore This Item.
- If you want to restore all items in the **Recycle Bin**, select Restore All Items.
- Right-click the file or folder you want to restore and then select Restore.
- Select the items you want to restore and then select Restore the Selected Items.

Cloud Storage

Cloud storage is online data storage. When files are stored on the cloud, it means they are stored on servers connected to the internet instead of on your device.

If you have used your student email account to send or receive messages, you've used the cloud. All the emails in your inbox are stored on servers in the cloud.

Cloud storage offers many benefits, including greater accessibility, data protection, and protection against data loss.

Although there are many benefits to using cloud storage, there are some drawbacks. **Cloud storage** requires an internet connection and can be vulnerable to security breaches, although these are believed to be relatively rare.

Cloud Storage Options

Many sites allow users to store and share files on the cloud.

Cloud storage is increasing in popularity because of the ease of accessibility it offers, as well as the fact that it eliminates the need to house files on the hard drive of a device.

Here are some popular cloud storage options:

- **OneDrive: OneDrive** offers free cloud storage to anyone with a Microsoft account. You can upload and store files, videos, music, and photos.
- **Google Drive and Google Docs:** These allow users to upload and store documents, spreadsheets, and presentations to Google's data servers.
- **Dropbox:** Dropbox allows users to upload and share documents. It is popular with those sharing larger files.

Social media sites such as Facebook, Instagram, and LinkedIn allow you to save images and résumés on their servers.

Cloud Storage Terminology

Infrastructure-as-a-Service (IaaS)

IaaS describes an organization using servers in the cloud instead of buying and maintaining storage servers for their day-to-day business operations. This helps to reduce costs associated with hardware, software, and support employees. This is very beneficial for small organizations that do not have the expertise to run a complex information technology system.

Software-as-a-Service (SaaS)

SaaS refers to the distribution of computer applications over the internet. SaaS is commonly referred to as "web apps." Microsoft 365 is considered SaaS. SaaS has advantages over installing software on each device, including frequent updates and apps that are delivered on demand.

Platform-as-a-Service (PaaS)

PaaS describes a programming environment used to develop, test, and disseminate web apps via the internet. Businesses' use of PaaS software allows for greater collaboration and for less-specific programming language knowledge.

Cloud Storage Security

Two major concerns about cloud storage—security and reliability—are expressed by many users of cloud storage:

Security: Most cloud storage sites use encryption to protect the data housed on their servers. Encryption uses algorithms to encode information and keep it secure. Most cloud storage sites also require usernames and passwords as another layer of security.

Reliability: This is a major concern with many cloud storage users. What happens if I cannot access a cloud storage site? What if the cloud storage site is no longer in business?

These issues could arise, so it is best to back up your important data!

Microsoft OneDrive

OneDrive is a free online cloud storage service provided by Microsoft. **OneDrive** comes with your Microsoft email account. However, to access **OneDrive**, you must create a Microsoft **OneDrive** account.

OneDrive is like having an extra hard drive that is accessible from most internet-enabled devices. Using **OneDrive** removes the need to save files to a USB flash drive or to email files to yourself.

Creating a Microsoft Account

Creating a Microsoft account is the first step to saving and working with files in **OneDrive**. It is a good idea to create a new personal email account that you can use for business correspondence. When you create a new account, keep it professional. It is a good idea to use your first and last name.

Here are the steps to creating a new Microsoft account:

1. Type "onedrive.live.com" into your web browser.

2. Select Personal and then select Sign Up for Free.

3. Select Create a Microsoft Account (this is for new users).

4. Select Get a New Email Address and then create your new email address and password. Your new email address will have your chosen name@outlook.com. Make sure to save your email address password information where you can access it for future reference.

You may be asked by Microsoft to verify your account. You need to verify your account to get the full functionality of **OneDrive**.

Uploading Files to OneDrive

Uploading files to **OneDrive** frees up hard drive space and allows you to back up files.

Here are the steps to upload a file to **OneDrive**:

1. Browse to the location where you want to add the files.

2. Using your web browser, visit **OneDrive** and log in using the information from your Microsoft account.

3. Tap or click Upload.

4. Pick the files you want to upload and then tap or click Copy to **OneDrive**.

Sharing a File Using OneDrive

Here are the steps for sharing a file using **OneDrive**:

1. Open **OneDrive**.

2. Pick the file or folder you want to share by selecting the circle in the upper corner of the item. You can also pick multiple items to share them together.

3. Tap or click Share at the top of the page. In the Share box that appears, choose one of the sharing options: Get a Link or Email.

Choose the "Email" option if you want to send an email invitation to people or groups and keep track of whom you invited. This also lets you remove permission for specific individuals or groups later if necessary.

Choose the "Get a Link" option to share items with several people you might not even know personally. For example, you can use these links to post to Facebook, Twitter, or LinkedIn or to share in email or IM. Anyone who gets the link can view or edit the item, depending on the permission you set. Keep in mind that the link can also be forwarded, and sign-in is not required.

Google Drive

Google Drive is a free cloud storage site for Google account holders that also provides collaboration and sharing. To use **Google Drive**, you need to open a Google Gmail account. **Google Drive** encompasses Google Docs for word processing, a spreadsheet program called Google Sheets, and presentation software called Google Slides.

Creating a Google Drive Account

Creating a **Google Drive** account is the same process as signing up for a Gmail account.

You can enter the following in your browser window:

"https://accounts.google.com/signup" or simply type "Google signup" in a search engine.

Once you've reached the sign-up page, fill out the form and you'll get your Google account.

Create your Google Account

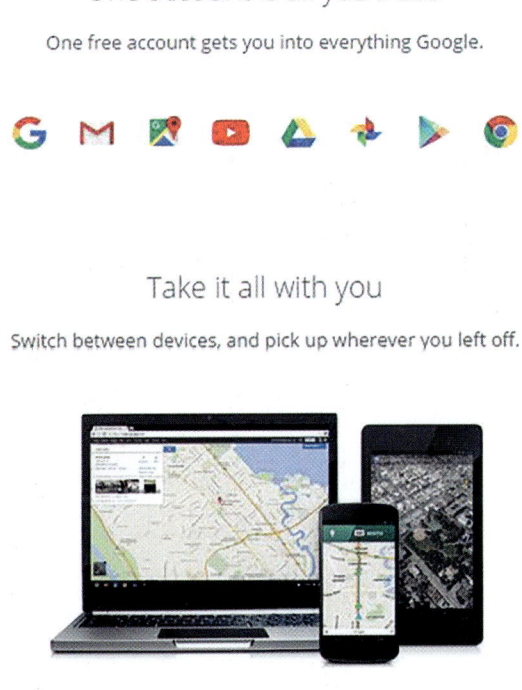

One account is all you need
One free account gets you into everything Google.

Take it all with you
Switch between devices, and pick up wherever you left off.

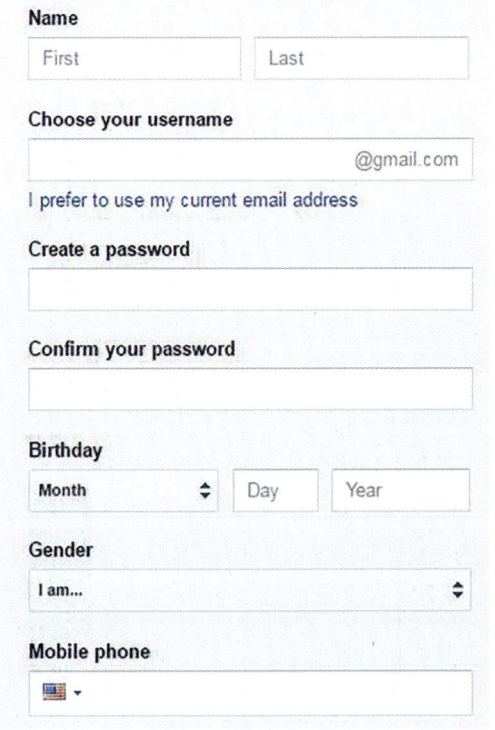

Uploading a File to Google Drive

Saving a file to **Google Drive** is quite easy, and each account you open provides 15 gigabytes of storage.

To save a file, go to your **Google Drive** account and sign in if necessary. Then click New to open the drop-down menu and select **File** Upload. Browse to the file and click Open. Now your file is on **Google Drive**!

Sharing Files Using Google Drive

You can share **Google Drive** files and folders and choose whether the people you share them with can view, edit, or make comments. You can share files with a link or email attachment.

Sharing files using **Google Drive** is easy. Open your **Google Drive** account and sign in as necessary. Right-click on the file or folder you want to share and click Share. Enter the email address of the person with whom you want to share the file.

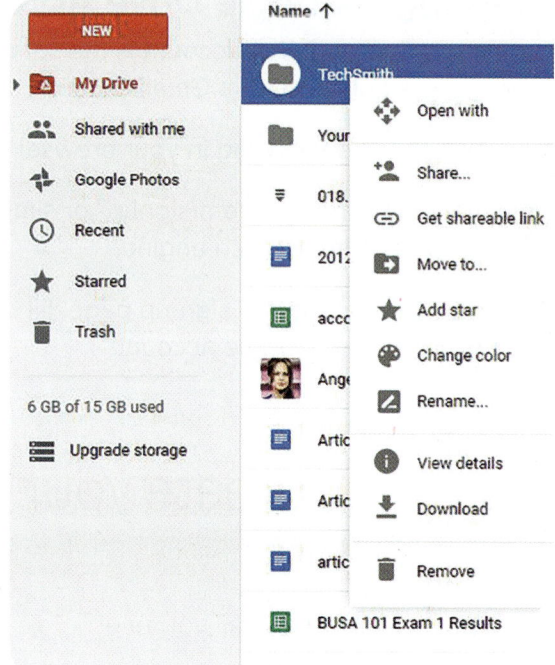

Chapter Review

1. What is a folder? How are folders used to store files? Why is it a good idea to use folders?

2. How do you create and rename a folder?

3. How do you change folder options and folder views?

4. Define a file. What types of files do you use and create?

5. What is file compression? What is the main benefit of compressing files? Discuss various audio, video, and graphics file compression standards.

6. Discuss common word processing file compression standards.

7. Discuss common executable file compression standards.

8. What are the steps used to view the extension of a file?

9. What is the **Recycle Bin**? How do you send files to the **Recycle Bin**? How do you remove or clear files from the **Recycle Bin**?

10. What is cloud storage? How does cloud storage work? Conduct some research on the internet about cloud storage. Identify and discuss at least three cloud storage services available on the web.

11. What is Microsoft **OneDrive**? What are the pros and cons of using Microsoft **OneDrive**? Explain the process for creating a **OneDrive** account.

12. What does it mean to share a file? Why would you share files? Discuss the process for sharing and uploading files to a **OneDrive** account.

13. What is **Google Drive**? What are the pros and cons of using **Google Drive**? Explain the process for creating a **Google Drive** account.

14. Discuss the process for sharing and uploading a file to your personal **Google Drive**.

Ethics Discussion

You discover a shared folder at school or work that mistakenly allows anyone to view and edit its contents. Inside, you find important documents and information that aren't meant for everyone. What do you do? Is it ethical to browse through the documents, share the folder with others, or inform the person responsible for the mistake?

15 Databases

What To Expect

After completing this chapter, you will be able to:

- describe databases, database management systems, types of databases, and the advantages of databases;
- describe the structure and components of a database and how these components are connected;
- discuss relationships between tables in databases;
- describe the ways to enter and access data in a database;
- describe what Big Data is through the four *V*s;
- explain what business intelligence (BI) is;
- describe data warehouses and data marts;
- describe and compare data mining and web mining.

Chapter Topics

- Database Basics
- Types of Databases
- Advantages of Using Databases
- Database Elements
- Making Use of Database
- Data Mining

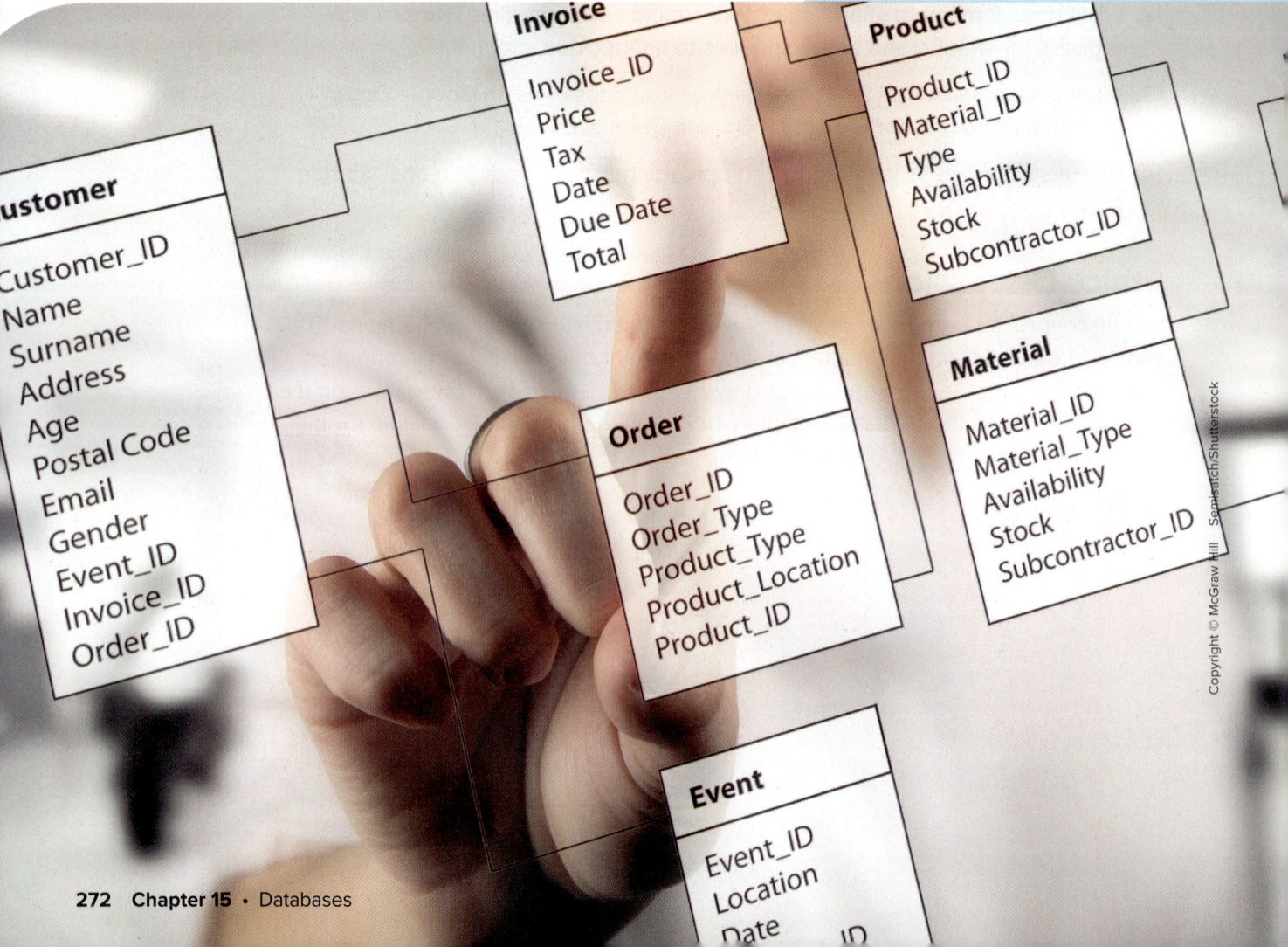

Copyright © McGraw Hill Sopisatch/Shutterstock

Let's Explore Databases

Meet Mia, a student with a flair for creativity. She's recently taken up the hobby of making crafts and selling them to friends, family, and even at local craft fairs. Initially, Mia started by keeping simple lists to manage her new venture.

Lists: Mia's first list was about the crafts she made—bracelets, earrings, paintings, and so forth. She noted how many of each she made, the materials she used, and how much they cost her. She also had a separate list for customer names and what they bought from her. As Mia's business grew, she realized she was flipping between lists frequently, trying to match which customer bought which craft and calculating profits.

Tables: To make things a bit more organized, Mia created tables. One table was dedicated to the crafts, with columns for craft type, materials, cost, and selling price. Another table listed her customers, their contact details, and their purchases. This made it easier for Mia to see all related information in one place. However, she still had to look up items across two different tables, and as her inventory and customer base grew, this became tedious.

Databases: Mia learned about databases in her computer class and realized they might be the solution to her growing organizational challenges. She set up a relational database to keep track of her crafts venture. Now, she had several tables:

- **Crafts Table:** Listing all her products, materials, and pricing
- **Customers Table:** Detailing her buyers and their contact info
- **Sales Table:** Recording every sale, linking to both the crafts and customers tables (This allowed her to see who bought what, when, and for how much, all in one place.)

But Mia's database did more than just store tables. She set up relationships between these tables. If she wanted to see all the crafts a particular customer bought, it was just a click away. Or she could click on the craft and view the customers who bought them. So she could easily pull up information about which items were her best sellers.

By connecting her tables in this relational database, Mia could quickly and efficiently manage her growing craft business, ensuring she always had a clear overview of her sales, stock, and customers. It also made it simpler to track her profits and understand her customers' preferences, helping her decide what to craft next.

Database Basics

A database is like a big digital cabinet, and inside this cabinet, you have multiple sections or compartments where specific pieces of information are stored. All this information, or data, is put together in a way that makes it easy to find, sort, and use. Every time you buy something online or take a test, there's data being made and saved. Guess where? Yep, in a database.

Tables

A table is a structured set of data within the database. It's made up of rows and columns.

Table Structure

In a database, a row in a table represents a unique record. A record could be information about a single student, product, or event. Columns represent the different types of data for each record. Think of them as categories of information.

This table has three columns (Column1, Column2, Column3) and three rows. Each row contains data specific to that row across the different columns.

Column1	Column2	Column3
Row1_Data1	Row1_Data2	Row1_Data3
Row2_Data1	Row2_Data2	Row2_Data3
Row3_Data1	Row3_Data2	Row3_Data3

Sample Table

Now let's look at an example with some information. In a "Students" table, a row might represent all the information for a single student (name, grade, favorite subject, etc.), while each column would represent one of those specific details (first name, last initial, grade level, etc.). In this table, each row represents a student, and the columns provide specific details about that student, such as their name, grade level, and favorite subject.

Student_ID	First_Name	Last_Initial	Grade_Level	Favorite_Subject
082011	Cordelia	E	6	English
012010	Silas	G	7	Culinary Arts
022007	Griffin	P	8	Social Studies
082006	Owen	D	7	Computer Science
091971	Dawn	L	6	Heath
011972	Jeff	M	7	Science

What's in a Database?

Now that we understand tables as organized collections of rows and columns holding specific information, let's explore the broader world of databases.

A database is like a digital library, housing various tables, much like a library contains books. But, just like a library has more than books—it has catalogs, sections, reading areas, and more—a database offers more than just tables:

- **Multiple Tables:** A single database can contain numerous tables. Each table can represent different sets of data. For instance, a school database might have one table for students, another for teachers, and yet another for subjects.

- **Procedures and Functions:** Databases can store procedures (sequences of instructions) and functions that help manage, calculate, and retrieve data in specific ways. Think of them as the "rules" or "methods" that tell the database how to handle certain tasks.

- **Management Tools:** Databases come with tools that assist in organizing and safeguarding the data. These tools can help in tasks like searching for specific data, sorting data in different orders, or even ensuring that only authorized people can access certain information.

- **Connections:** Databases can create relationships between tables. Let's go back to our school example. We might want to know which students are in which classes. By establishing relationships between the "Students" table and a "Classes" table, the database can help us connect the dots.

In summary, a database isn't just a collection of tables. It's an integrated system that stores, organizes, and retrieves data in an efficient and meaningful way. Just as every book contributes to a library's value, every table, procedure, and tool contributes to the richness and utility of a database.

Types of Databases

There are four types of databases:

- **Individual database**—A combined set of data files intended to be used by one person.

- **Company database**—Created for use by organizations. Users access the database via computers linked to local or wide area networks.

- **Distributed database**—The data in a database are stored in different physical locations and accessed via client/server networks.

- **Commercial database**—Large database that covers specific subjects. Commercial databases are also referred to as *information utilities* or *data banks*.

Database Hierarchy

Databases are structured in a hierarchy, with characters at the lowest point and the database at the highest point. The standard hierarchy starts with a character, then a field, then a record, then a table, then a file, and finally a database.

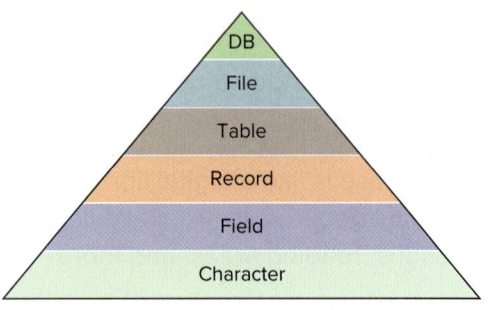

Example of a database hierarchy for student accounts at a university:

Database Property	Explanation
Character	Single letters, numbers, and symbols used to create fields
Field	Collection of related characters that convey meaning to a user (Field names come in various forms including account number, name, address, phone number, financial aid status, etc.)
Record	Collection of related fields (**Records** in the database at a university include individual information for students, faculty, staff, etc.)
Table	Collection of records with similar characteristics (In this example, the table would be "student accounts.")
File	Collection of related tables (In this example, the file would consist of related student tables including current, past, and future students.)
Database	Collection of files related to organizational processes (The university database would likely include files related to students, faculty, staff, facilities, etc.)

Relational Databases

Relational databases organize data into tables based on structured data groupings. **Relational databases** use links called *relationships* between tables. Information in tables is stored in rows (records) and columns (fields).

These relationships define how the data in the tables are related. A common field in both tables is used to create the relationship. **Structured Query Language (SQL)** is the most common programming language used with relational databases. SQL allows users to create and modify a database. Additionally, SQL statements can be used to create database queries.

Relational databases are designed to adhere to atomicity, consistency, isolation, and durability (ACID) properties. Adherence to these properties ensures that when a series of database operations is grouped together, the database will remain in a valid state, even if unexpected errors occur.

Advantages of Using Databases

Databases have many advantages when compared to other methods of managing and storing data.

Databases allow for efficient and secure storage of large amounts of data. Databases enable data sharing and help to ensure data integrity. They also allow for centralization, which is when data are maintained in only one file, and they provide the ability to logically structure data. This makes data analysis more efficient, as it increases data consistency and decreases data redundancy. In addition, databases enable data sharing via concurrent access, providing collaboration and permitting updates in real time.

Databases can integrate with other applications, including ERP and POS systems, making data exchange and analysis by other applications possible.

DBMS

A database management system (DBMS) maximizes the usefulness of a database. The DBMS allows the user to create, modify, and access a database. Common DBMSs include Microsoft Access, Oracle **Database**, and Microsoft SQL Server.

DBMS software is composed of five subsystems: DBMS engine, data definition, data manipulation, application administration, and data administration.

DBMS Subsystems

- **DBMS Engine:** The software that supports the database. Most DBMS engines include an application programming interface (API) in addition to a graphical user interface (GUI). The API allows designated users to interact with the backend of the database without having to access the GUI. DBMS engines are also referred to as *database servers*.

- **Data Definition:** Allows users to create a data dictionary. The data dictionary defines the fields in a database as well as the structure of files in a database.

- **Data Manipulation:** Enables users to edit data in a database including adding, changing, and deleting data. It is considered the primary interface between a user and a database.

- **Application Administration:** Environment for the management of the DBMS application including software updates and upgrades.

- **Data Administration:** Allows for the management of the database environment including security, data backup, and data recovery.

Data Integrity

Data integrity means the database is reliable, accurate, and aligned with the goals of the organization.

Data centralization is critical in increasing data integrity. When data is centralized, it is stored in only one place.

When multiple lists and data sources are maintained, information can become inconsistent, leading to decreased data integrity. Essentially, databases enable more efficient data maintenance.

5 Elements of Data Integrity

1. **Accuracy**—Data is entered in the correct form and accurately represents the intent of the data.

2. **Timeliness**—Data is up-to-date and reflects currency. Old or outdated data is often considered to be less useful than timely data.

3. **Completeness**—Critical data is not missing from the data set. Additionally, data is present and available to those who need it.

4. **Consistency**—Data is consistently formatted. This includes low-, medium-, and high-level data. Data across all levels should be formatted in a similar nature.

5. **Compliance**—Data is compliant with regulatory standards, including data privacy and data security.

Database Elements

What Is a Query?

A query is also known as a question. A query is a request for information from a table (or combination of tables) in a database. Information is generated using a specific query language. Structured query language (SQL) is one of the most popular query languages.

Queries allow you to find specific data quickly by filtering based on a specific criterion, to calculate/summarize data, and to automate data management tasks.

Tables, Fields, and Records

Tables: A table in a database is a collection of associated records. Databases often have more than one table.

Fields: A field is a group of related characters in a database table. A field is a column in a table that represents a characteristic of someone or something.

Records: A record is a collection of related fields in a data file. **Records** are a collection of characteristics that describe and identify an entity. A record is also referred to as a *row* in a table.

Primary and Foreign Keys and Relationships

Primary Key: A primary key is a special relational database field designated to uniquely identify all records in a table. A primary key must contain a unique value. Common primary keys include student ID numbers, Social Security numbers, and customer ID numbers.

Foreign Key: In a relational database, the foreign key is a common field between tables that is not the primary key.

Relationships: A relationship is a link between tables that defines how the data are related. A common field between the two tables is used to create the link.

Forms and Reports

Forms: Forms are used to control how data are entered into a database. **Forms** structure data input to ensure data integrity. Data are entered into the blank areas of the form. **Forms** turn data into information. **Forms** are created using database management system (DBMS) software.

Reports: Reports offer a way to view, format, and summarize the information in a database. **Reports** can be used to display or distribute a summary of data and archive snapshots of the data. **Reports** can also be used to provide details about individual records and to create labels.

Making Use of Database Information

Big Data

Big Data encompasses all the analysis tools and processes related to applying and managing large volumes of data. **Big Data** was conceived from the need for organizations to better understand the trends, patterns, and preferences that emerge from the interaction with different systems and databases. **Big Data** allows organizations to use analytics to help uncover a variety of predictive behaviors to help create new offerings. The four *V*s are the common characteristics of **Big Data**: volume, variety, veracity, and velocity.

The Four *V*s of Big Data

Volume refers to the scale of data. Enormous amounts of data are created every day. Most companies have over 120 terabytes of information stored (that's 120,000 gigabytes!). It is estimated that 2.3 trillion gigabytes of information are created each day.

Variety refers to the different forms of data. Data come from many structured and unstructured sources. These sources include social media platforms, email, photos, videos, and point-of-sale interactions. YouTube estimates that over 1 billion hours of video content is viewed each day. It is estimated that in 2020, over 2,400 exabytes of healthcare data will be generated and stored.

Veracity means truth. With all the data being generated and stored, it is important to ensure that data are meaningful, true, and useful. Poor-quality data are estimated to cost the US economy over $3.3 trillion per year. A recent survey found that one in three business leaders do not trust the information they use to make decisions.

Velocity is speed. With data streaming continuously from many sources, the pace is mind-blowing. **Big Data** must analyze data as they change! Every minute there are 510,000 comments posted on Facebook; 293,000 statuses updated; and 136,000 photos uploaded. Over 3.5 billion Google searches are conducted worldwide each minute of every day—that's 2 trillion searches per year!

Business Intelligence

Business intelligence (BI) includes the technologies, computer applications, and procedures for the collection, analysis, and presentation of business information to help support decision making. Fundamentally, **BI systems** are data-driven decision support systems (DSS) that aid businesses in making better strategic decisions.

BI systems provide a picture of historic, current, and future views of operations. BI systems use information stored in data warehouses, data marts, in-memory computing, and other analytic platforms to create information output.

Data Warehouses

A data warehouse is a repository of data and information that organizations analyze to make informed business and operational decisions. Data flow into a data warehouse from a variety of transactional systems such as point-of-sale and online transactions, from databases, and from other data-generating sources. Information flows into a data warehouse in regular intervals and is stored for later processing.

A variety of people within an organization have access to the data warehouse, including data scientists, key decision makers (KDMs), and data specialists. Data are analyzed using BI tools, SQL clients, and a variety of analytics applications designed to interpret the data. The output created from data warehouses includes reports, dashboards, and queries. Data and the analytics provided from the analysis of data allow organizations to create and maintain a competitive advantage.

Data Marts

Whereas data warehouses are considered multipurpose data and information storage facilities, a data mart is a subsection of a data warehouse that is designed and built specifically for individual departments or business functions. There are three types of data marts: dependent, independent, and hybrid.

A **dependent data mart** is constructed from existing data warehouses and utilizes a top-down approach where organizational data are stored in a centralized location. Specific data are extracted when analysis is needed.

An **independent data mart** is a stand-alone system that is separate from a data warehouse and focuses on specific organizational functions.

A **hybrid data mart** assimilates data from a data warehouse as well as from other data collection systems. It incorporates a top-down approach, end-user inputs, and enterprise-level integration.

Using Data Warehouses and Data Marts in Business

Data warehouses help to create a decision support system (DSS) environment, which allows businesses to gauge the performance of an enterprise over measurable periods of time. **Data warehouses** contain large amounts of historical data that represent data points at a specific time. This allows organizations the ability to compare different time periods to make more informed business decisions.

Data marts are designed to collect and measure data from specific operational areas of a business and are used by individual departments or groups. **Data marts** are used to track inventories, purchase transactions, and follow the supply chain. **Data marts** assist with the analysis of what data a user needs rather than focusing on existing data.

Data Mining

What Is Data Mining?

Imagine you're looking for treasure on the beach. People have left a lot of stuff there; some of it is trash, some of its treasure—and some trash might be useful if you get enough of it and can figure out what to do with it. A gold ring? Treasure for sure. A bottle cap? Trash; but if you get enough, you might be able to get some cash by recycling the metal. Snack wrappers? Trash, right? But wait. What if you had all the candy wrappers from all the beaches and categorized them by time of the year to form a picture of who is eating what while visiting the beach? That might be useful intelligence to a snack corporation or to a health sciences organization.

Data mining is like that, except we use computer tools to dig deep and find patterns or connections in lots of data. These tools use machine learning, statistics, and artificial intelligence to help us. We do this to

- automatically find patterns;
- guess what might happen next;
- make helpful decisions;
- work with lots of data at once.

How Data Mining Works

When we mine metal, we don't just dig a hole and find useful metal. We dig through a lot of rock and use special chemicals and physical sorting mechanisms to get the metal out of the rock and refine it. This yields the final outcome—a valuable metal. **Data mining** uses the same concept. The data are extracted from various sources and then refined down into precious information. **Data mining** is a way to analyze data that can help organizations find patterns and relationships within data sets. While data mining is a powerful tool, it does not replace the need to have an intimate knowledge of the organization, the data that is produced, and the analytic methods used to turn data into information. **Data mining** assists businesses in uncovering information that may be hidden in data sets, but it does not offer organization-specific reasons as to why this information may be valuable.

This step—interpreting why certain data may be related—is usually executed by managers and data analysts. Predictive information and relationships that are produced from data mining are not causal relationships. In other words, information yielded from data mining cannot be said, definitively, to cause specific outcomes. **Data mining** yields probabilities, not exact answers. For example, data mining might determine that females with incomes between $75,000 and $100,000 and who also subscribe to certain magazines may be more likely to purchase various products. But the analysts cannot assume that the population identified through data mining buys these products simply because those people belong to the identified population.

Web Mining

Web mining uses the principles of data mining to uncover and extract information from websites, social media sites, e-commerce platforms, and web services. Organizations use web mining to assist in gaining a better understanding of consumer behavior, website efficacy and usage patterns, and to analyze how web searches are used.

Three techniques are commonly used to gain information from the web, including web content mining (WCM), web structure mining (WSM), and web usage mining (WUM).

- **Content mining** includes the extraction of information from web pages and documents, including text, images, videos, and other interactives.
- **Web structure mining** includes the analysis of hyperlinks, nodes, and related web pages.
- **Web usage mining** (also called *log mining*) includes the analysis of web access logs or when, how, and with what frequency websites are accessed. Common web mining applications used by today's organizations include Google Analytics, Data Miner, and Tableau.

Visibility, Usability, and Accessibility

Businesses and organizations can improve a user's web experience and therefore improve outcomes using web mining. **Web mining** can lead to improved website visibility, usability, and accessibility.

- Site **visibility** includes how and when the site surfaces when queries are executed in search engines. Search engine optimization (SEO) can be enhanced through the information gained by web mining. This information can also assist marketers in online ad placement and search engine advertising.
- **Usability** refers to how easily website users interact with the site. Data gained from web mining can help web designers to optimize site navigation and the structure of website information.
- **Accessibility** includes the structure of websites and web pages to ensure access and scalability on multiple devices and platforms and to ensure that information is available to all types of users, including access for those with physical and cognitive limitations.

Databases and Web Technologies

Accessing the internal databases of a business or organization via web technologies is becoming increasingly important in today's competitive environment. Customers might access an online store's internal database to view current product offerings, compare products, and so on. Organizations might use the web to check the database of a supplier for product information or current inventory. Users access internal databases via the web on digital devices including smartphones, tablets, laptops, desktops, even wearables like Fitbit and Apple Watch (and whatever is coming next!).

Web browsers communicate with the web server via HTML, JavaScript, and a host of new technologies like node.js, REACT, and others. Using the web to access databases is an advantage for businesses because most users are comfortable operating web browsers and digital devices. Front-end web interfaces require little change to existing database structures, giving businesses and organizations an ever-increasing amount of global web connectivity and reach.

The Process of Data Analytics

Data analytics is like being a detective using numbers and information. It helps us figure out what's going on by looking at different kinds of data, like numbers, texts, or pictures. There are three main types of data analytics: descriptive, predictive, and prescriptive.

- **Descriptive Analytics:** This is about looking at past data to understand what happened. Think of it like reading a history book or checking last month's grades. It answers questions like "What were my sales last month?" or "How did people feel about my last social media post?" It gives you a clear picture of past events.
- **Predictive Analytics:** Here, we use data to guess what might happen in the future. Imagine trying to predict tomorrow's weather by looking at patterns from the past few days. We use tools like probability (chances of something happening), data digging, and even some cool techniques like machine learning to make these guesses. This type of analytics can be used in many ways, such online shopping sites predicting what you might like to buy next or doctors guessing where the flu might break out next.

- **Prescriptive Analytics:** This is the most advanced type. It not only tries to guess the future like predictive analytics, but it also gives advice on what steps to take next. For instance, airlines might use prescriptive analytics to watch fuel prices and guess whether they will go up or down.

In short, data analytics helps us understand the past, guess the future, and decide what actions to take next. It's like having a super-smart friend who gives you advice based on lots of information!

Responsibilities of a Data Analyst

A data analyst uses the knowledge of processing software, business strategy, and analytical skill to deliver data and reports that guide management and executives as they make well-informed decisions. Duties of a data analyst include working on teams to extract data from large data sets, creating reports that outline key findings, monitoring key performance indicators (KPIs) to identify success or failure, and analyzing data to identify trends. Data analysts are employed in a variety of fields including finance, e-commerce, healthcare, government, and science.

Additional responsibilities include

- collaboration with executives and other stakeholders to uncover areas for improvement;
- data visualization to aid in the interpretation of data;
- structuring large data sets to ensure data is accessible and usable;
- creating reports and presentations for management and executives that outline key findings and recommendations.

The US Bureau of Labor Statistics reports that the typical entry-level education for data analysts is a bachelor's degree; the median salary across different disciplines is $88,770; and the job outlook is 15 percent growth (much higher than average).

Key Skills of a Data Analyst

Data analysts must possess a variety of skills to be successful at their job. They are likely to have a bachelor's degree with emphasis in areas such as management information systems (MIS), computer science, mathematics, statistics, economics, or finance. Entry-level data analysts frequently start with no prior experience, but they go through extensive on-the-job training that teaches them how to apply information learned in college as well as training on organizational policies and procedures.

Specific skills data analysts possess include

- technical writing skills;
- experience with computer code including SQL, Python, and Oracle;
- strong analytical and problem-solving skills;
- experience with data visualization software including Tableau and Power BI;
- Microsoft Excel and spreadsheet experience;
- effective time management and the ability to multitask and to meet deadlines;
- oral communication and presentation software skills.

Data-Driven Decision Making

According to Tableau, data-driven decision making (DDDM) is the use of facts, metrics, and data to guide strategic business decisions that align with organizational goals, objectives, and initiatives. The analysis of facts is one key dimension of DDDM because facts lead to patterns that support effective decision-making. The amount of data and information that is collected continues to increase, as does the complexity of collection and analysis.

Chapter Review

1. What is a database? How are databases used in business?

2. Compare and contrast a list vs. a database. Why is using a database more efficient than using a list?

3. What is data integrity? How can you ensure your database has data integrity? Why is it important?

4. Define and discuss a DBMS engine.

5. Define and discuss DBMS data definition.

6. Define and discuss data manipulation.

7. Define and discuss application generation.

8. Define and discuss data administration.

9. Describe and contrast the four database types discussed in the module. Consider who would be the "owner" of each type of database. For instance, who would be the owner of an individual database?

10. Discuss relational databases. Conduct research on the internet about relational databases. Find an example of a commonly used relational database.

11. What is a database table? How are tables used to hold data and information?What is a database record? What information does a database record contain?

12. What is a primary key? What makes a primary key unique? Why do databases use primary keys? Give examples of four different potential primary keys.

13. What is a form? How and why are forms used?

14. What is a query? How are queries used? Why is the ability to query an important element of a database?

15. What is a report? Who uses reports? Why is it important to create and use reports generated from a DBMS?

Ethics Discussion

Two ethical areas of concern regarding database privacy are the security of the database and the implications of what can or should be stored in a database. Conduct research on the internet about database ethics. What are some of the ethical issues surrounding databases? Is there an intrinsic ethical duty placed on database security professionals to secure a database system? Are you concerned about how much of your information is available in databases? Why or why not?

16 Spreadsheets

What To Expect

After completing this chapter, you will be able to:

- identify the different spreadsheet programs;
- describe how to interact with spreadsheet programs;
- describe various spreadsheet terminology including *rows*, *columns*, *formulas*, *charts*, etc.;
- explain how to create formulas, tables, and charts;
- explain how to use color in spreadsheets;
- demonstrate knowledge about various math principles.

Chapter Topics

- What is a spreadsheet?
- Spreadsheet File Management
- Key Spreadsheet Tools
- Using Charts and Graphs
- Color Conventions in Spreadsheets
- Math In Spreadsheet Review

Copyright © McGraw Hill, chorma 123RF

Let's Explore Spreadsheets

Welcome to the world of spreadsheets! At first glance, spreadsheets might seem like a simple grid of boxes, but they're much more than that. A spreadsheet is a digital tool that allows you to organize, calculate, and analyze information in a structured manner. Think of it like a digital notebook that can automatically do math for you!

Why are spreadsheets useful? Let's look at some examples:

- **Organizing Information:** Imagine you're part of a school project where you need to track how many cans your class collects for a food drive each week. Instead of jotting numbers down on paper, you can neatly organize them in a spreadsheet. With columns labeled "Week 1," "Week 2," and so on, you can easily input and track the number of cans collected.

- **Automating Calculations:** Let's say you've saved up some money from your allowance, and you want to know how much you'll have after a few months. With spreadsheets, you can enter a formula to do the math automatically. If you add or change any amount, the spreadsheet updates the total without you having to recalculate.

- **Visualizing Data:** Have you ever wanted to see how your test scores improve over the school year? Enter your scores into a spreadsheet, and with a few clicks, you can create a graph or chart. This visual representation can give you a clear picture of your progress.

- **Making Informed Decisions:** Suppose you're planning a class party and need to stick to a budget. With a spreadsheet, you can list all potential expenses, adjust quantities, and instantly see if you're staying within the budget.

By understanding and using spreadsheets, you'll have a valuable skill that can help you in school projects, personal activities, and later, in various professional fields. So, let's dive in and discover the practical uses and benefits of spreadsheets!

What is a spreadsheet?

Spreadsheets are like digital versions of tables with computation and data management, and data visualization tools. On a computer, they help you organize and work with numbers and other data. Each box in the grid is called a cell, and you can enter information in it like text, numbers, or even dates. These cells are arranged in rows (going left to right) and columns (going up and down). One of the coolest things about spreadsheets is that they can do math for you. If you've ever wanted to add up a whole column of numbers, instead of doing it by hand, you can just tell the spreadsheet to do it, and it'll give you the answer instantly. Plus, if you have a bunch of data, like scores from different games or test grades, you can quickly turn that data into a chart or graph. This helps you see patterns or trends, like which game you scored the highest in or if your grades are getting better over time. So, in short, the spreadsheet is a powerful tool for organizing and understanding all kinds of information.

How Spreadsheets Are Used

Because spreadsheets help users organize numerical data, they are useful for anyone who needs to keep track of numbers. The most common uses for spreadsheets include the following:

- **Budgeting and Accounting: Spreadsheets** make keeping track of and displaying budgets easy. Programs such as QuickBooks incorporate many of the functions of spreadsheets.
- **Finance: Spreadsheets** allow users to track financial trends and convert the data into easily interpreted charts and graphs.
- **Grades:** Professors and teachers use spreadsheets in programs such as Canvas and Blackboard to calculate grades.
- **Sports: Spreadsheets** are used to track statistics, times, and scores in nearly every sporting event.

Comparing Spreadsheets

There are different types of spreadsheet programs. Let's explore a few.

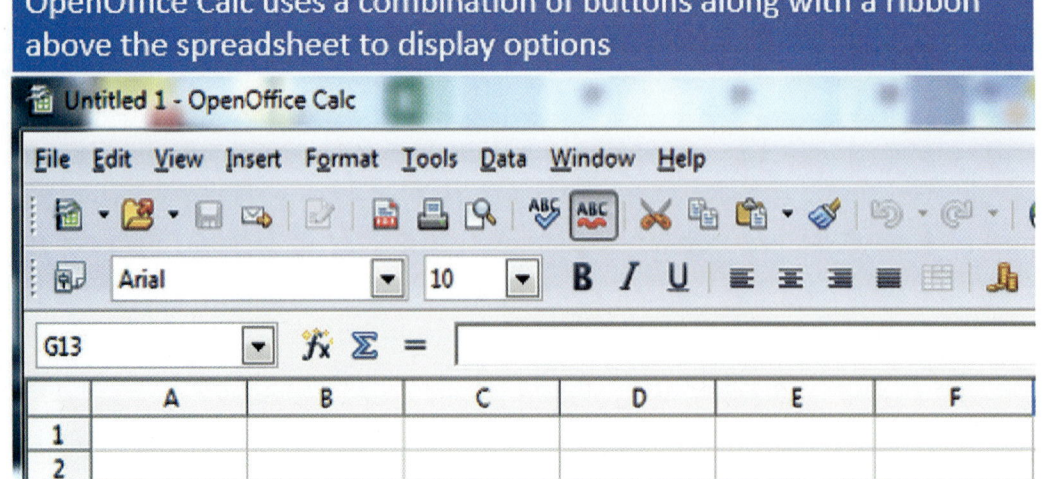

OpenOffice Calc uses a combination of buttons along with a ribbon above the spreadsheet to display options

Google Sheets

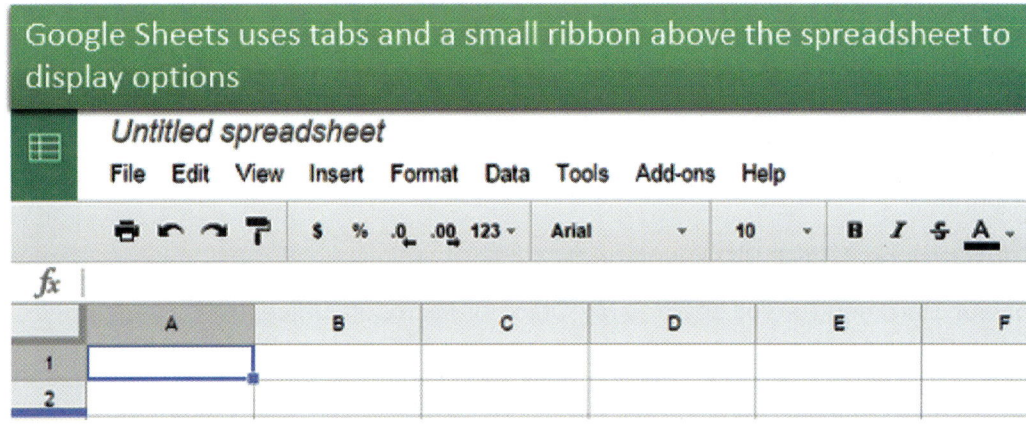

Apple Numbers

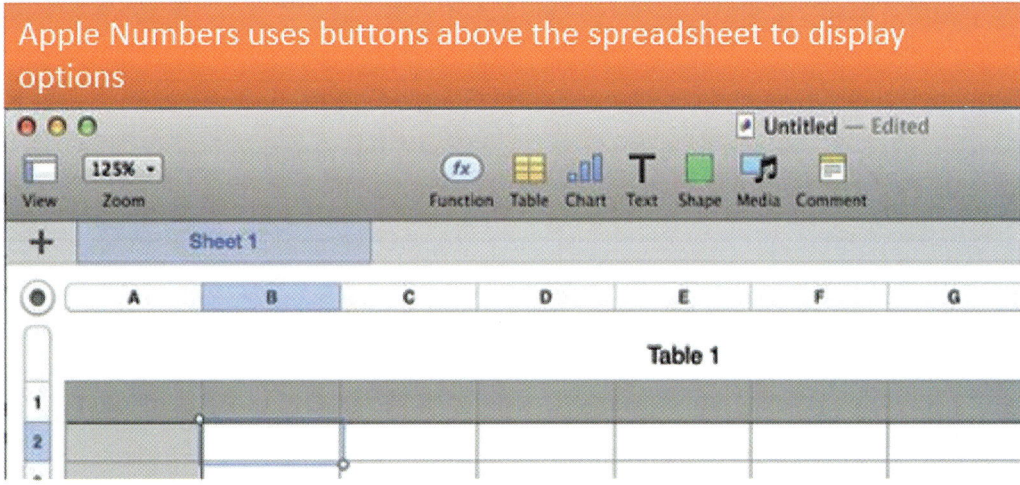

Excel

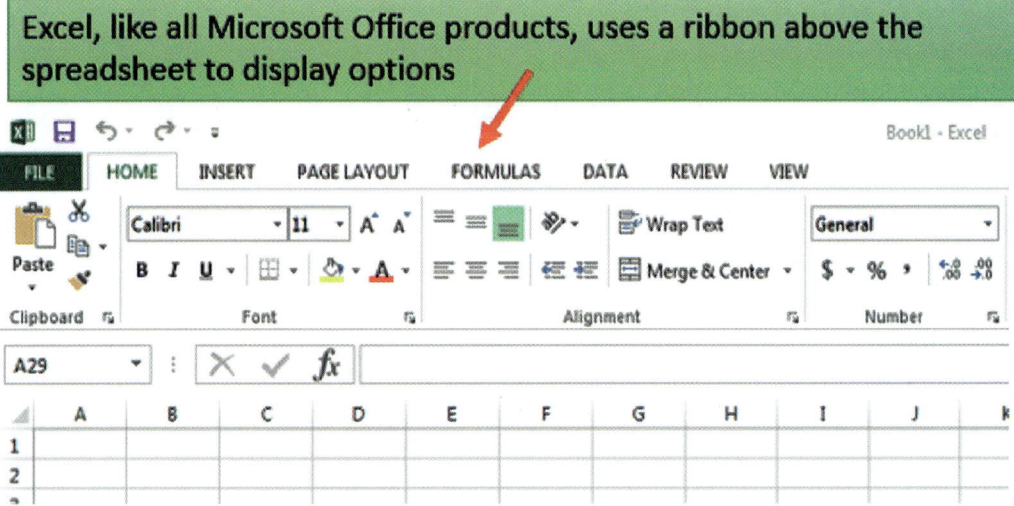

Spreadsheet File Management

Opening Excel

To open Microsoft Excel, follow these steps:

1. Left-click the Start button.

2. If you see Excel, just click on it. Click Blank workbook.

3. If Excel is not on your Start menu, scroll down to My Office or Microsoft Office, click Excel, and click Blank workbook. Alternatively, you can type "Excel" in the search window, select Excel, and, when it opens, click Blank workbook.

Opening Google Sheets

To open Google Sheets, follow these steps:

1. Log on to your Google account.

2. Click the Apps Launcher, click Drive, and click New. Select Google Sheets.

3. Click Blank spreadsheet. You may need to wait a few moments for the program to fully open. Alternatively, from the Apps Launcher, you can scroll down and click Docs.

4. Click the Main menu button.

5. Click Sheets and click Blank spreadsheet.

Saving a Spreadsheet

Saving a spreadsheet allows you to keep your file for future use. To save a spreadsheet, you should first name the file and determine where you wish to keep it. Saving a spreadsheet is a relatively simple process in any spreadsheet program. For example, to save a spreadsheet in Google Sheets, you simply name the sheet and then close the file. Any changes you made are automatically saved.

In Microsoft Excel or OpenOffice Calc, you simply name the file and then press Ctrl + S or open the File tab or menu and select Save.

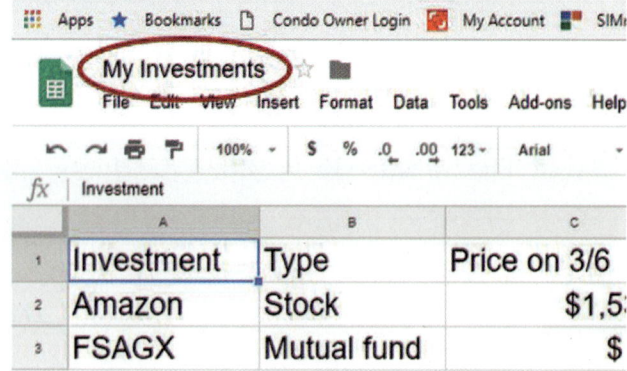

Comma-Separated Values

The comma-separated values, or CSV format, allows spreadsheet users to share data with other spreadsheet users regardless of which spreadsheet program is being used. CSV files can be opened with Microsoft Excel, Google Sheets, iWork Numbers, iCloud Numbers, OpenOffice Calc, and virtually every spreadsheet or database program.

The CSV format saves only the cell values. It does not save any styles or formulas. Business use the CSV format to send order, customer, or product data between stores or offices. It is commonly used by those who are sending data but do not know what spreadsheet program the recipient uses.

File name: Book1

Save as type: Excel Workbook

Excel Workbook
Excel Macro-Enabled Workbook
Excel Binary Workbook
Excel 97-2003 Workbook
XML Data
Single File Web Page
Web Page
Excel Template
Excel Macro-Enabled Template
Excel 97-2003 Template
Text (Tab delimited)
Unicode Text
XML Spreadsheet 2003
Microsoft Excel 5.0/95 Workbook
CSV (Comma delimited)
Formatted Text (Space delimited)
Text (Macintosh)
Text (MS-DOS)
CSV (Macintosh)
CSV (MS-DOS)

Saving a Google Sheets File as an Excel File

To save a Google Sheets file as an Excel file, be sure you have already named your file. Then follow these steps:

1. Open the File menu.

2. Use the Download as option and look at your choices.

3. Click Microsoft Excel. Many spreadsheet users prefer the .xlsx format.

4. Place the Excel file in the desired folder and click Save.

Using Save As in OpenOffice Calc

The Save As feature in OpenOffice Calc is essentially the same as in Excel. Use the Save As feature if you wish to send the file to someone in a different format as, for example, a .csv file.

1. First, open the File menu and select Save As.

2. Then, open the Save As type menu and select the format. Notice that you have a wide selection to choose from, including Excel, CSV format, and many others. Calc's Save As feature notes the file extension.

3. The Save As feature in Excel works the same way. Open the File menu and click Save As. Select the folder and name the file.

4. Then open the Save As Type menu. Notice the many formats to choose from. One difference is that Excel does not list the file extension.

Converting an iCloud Numbers File to an Excel File

You can convert an iCloud Numbers file to a different format. Because Numbers is a cloud-based spreadsheet, it is convenient when borrowing a computer or if you need to access a spreadsheet when you are not at your own computer. To save an iCloud file in a different format, use the Tools button. Select Download a copy and choose a format. You can choose from Numbers, PDF, Excel, or CSV. You may have to wait while iCloud downloads and converts your file. Then name your file and save it to a desired folder.

If you are using a public computer, such as those found in libraries or hotels, it may be best to save the file to the desktop and delete it later. Once the file has been downloaded, you can email it to yourself or save it on a thumb drive.

Key Spreadsheet Tools

Cells

A cell is the basic unit of any spreadsheet. It is a box where data can be entered. A cell is identified by the cell address, which is a letter and number combination where the letter is the column, and the number is the row in which the cell is located. For example, A3 would identify the cell found in column A and row 3 of a spreadsheet. This letter-number combination is called the cell address, cell reference, or cell identifier.

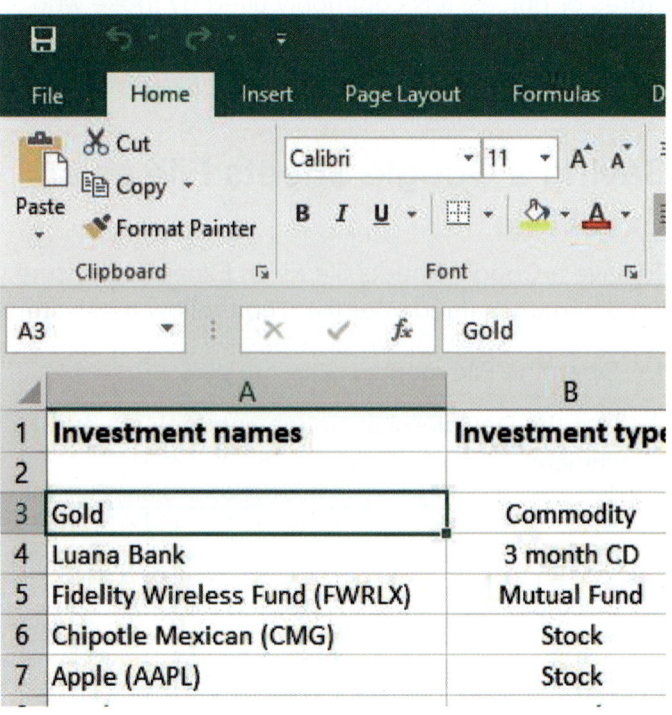

Active Cells

The active cell in a spreadsheet is the cell that a user has selected. By making a cell active, a user can enter data into the cell. A common way to make a cell active is to simply click on the cell.

An active cell can be identified by the heavy border around the cell. Another way to identify the active cell is to see the highlighted column and row headings. Some spreadsheet programs, such as OpenOffice Calc and Microsoft Excel, include a name box that shows the active cell.

Rows and Columns in Spreadsheets

In a spreadsheet, a row is a horizontal line of cells. **Rows** are identified by numbers. Clicking on a number selects the entire row.

In a spreadsheet, a column is a vertical line of cells. **Columns** are identified by letters. Clicking on a letter selects the entire column.

Entering Labels

A label helps the user identify data in a spreadsheet. Labels do not interact with other cells in the way that numbers or other values can, for example, by being added to or multiplied by values in other cells.

If you wish to enter a number but have it act as a label, type an apostrophe before the number. Labels remain on the left side of the cell when the label is entered into the cell by pressing the Enter key, Tab key, an arrow key, or by selecting the Enter button.

Values

A value in a spreadsheet refers to entries that the program recognizes as being able to interact with other cells. For example, values may be used in a calculation. Values include numbers, dates, formulas, and so on. Values jump to the right side of the cell when the value is entered into the cell by pressing the Enter key, Tab key, an arrow key, or by selecting the Enter button.

Equal (=) Sign

The equal sign (=) is the most important symbol in a spreadsheet. The equal sign indicates that the cell holds a formula or function, and the program performs a calculation. In the cell, the equal sign is entered first, and the formula afterward.

For example, instead of typing, **2 + 3 = __** in a spreadsheet, the formula is typed **= 2 + 3**. The program would then place the number **5** in the cell after you press the Enter key.

Pointers

Spreadsheet programs use several different pointers. Although different spreadsheet programs use different pointers, all are similar, and dragging your cursor across any worksheet quickly shows which pointer types the spreadsheet uses.

For example, Apple Numbers, OpenOffice Calc, and Google Sheets all use a normal cursor pointer to select or activate cells. Microsoft Excel uses a thick cross. The different pointers include a selection cross or cursor, a fill-handle cross, a heading expansion cross, and an insertion pointer.

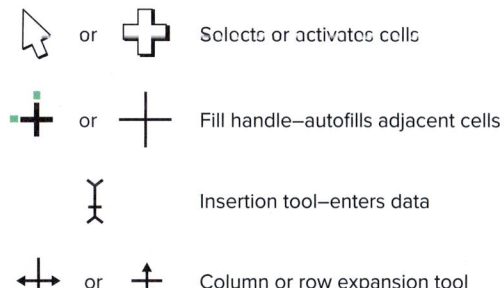

↖ or ✛		Selects or activates cells
✛ or ┼		Fill handle–autofills adjacent cells
⌶		Insertion tool–enters data
↔ or ↕		Column or row expansion tool

Adjusting Column Width and Row Height

To adjust a column's width, move your cursor to the side of the column until you see the adjustment cross. Then left-click and drag the column to the desired width. To adjust a row's height, use the same technique.

Common Error Values

Most spreadsheet programs share error values that indicate a problem with the formula being entered into a cell. One of the most commonly observed error values is #VALUE!. #VALUE! Means that the operation or function that is entered in the cell is not possible. For example, if you type a value, such as the number 20 into cell A1 and type a label, such as panda, into cell A2 and try to enter = A1 + A2 into another cell, the error #VALUE! Will appear.

Other error values include #DIV/0! Which is seen when attempting to divide a value by zero, and #REF!, if an invalid cell reference is used in a formula. Note that a series of hashtag symbols (####...) is not an error value; this simply indicates that the cell needs to be expanded to display the entire value.

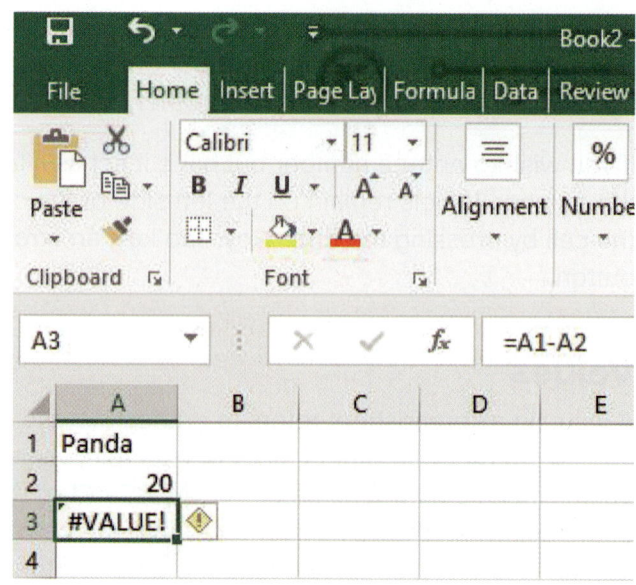

Moving Data

Moving data from one cell to another is easy. This section uses Google Sheets and Microsoft Excel as examples. In both programs, you can right-click on the cell. Then cut and paste the data into the new cell.

You can also select the cell and then move your cursor to the cell's edge. In Sheets the hand lets you drag the data while in Excel, a four-arrow cross lets you drag the data.

Inserting Rows and Columns

This section shows how to add a row or a column into a spreadsheet with Microsoft Excel and OpenOffice Calc. Both programs use basically the same methods.

One way is to right-click on the row directly beneath where you want the new row to be inserted and click **Insert**. In OpenOffice Calc you click **Insert Rows**.

Some users prefer to select a row and then select **Insert Cells** from the ribbon. In Calc you can select **Insert** from the top menu and then click **Row.** You can use the same methods to insert columns.

Deleting Rows and Columns

This section shows how to add a row or a column into a spreadsheet with Microsoft Excel and Apple Numbers.

In Excel you can select the row and then click the **Delete** button located in the **Cells** group on the **Home** tab. Alternatively, you can select the row, right-click, and then select **Delete**. In Numbers, just right-click on the row and select **Delete Row.**

You can use the same methods to delete columns. Select the column in Excel and then click the **Delete** button. Notice there is a drop-down menu where you can select **Delete Cells,** but it does not save any time. It is easier to simply click the **Delete** button. Most users simply use the right-click method.

With Apple Numbers, use the same methodology or double-tap your Mac's track pad and select **Delete Column.**

Using the Fill Handle

The **fill handle** makes completing a series easy. This section shows how to use the fill handle with Apple Numbers.

Click on the first cell of the series and drag down to the second cell. Then move the cursor to grab the fill handle in the lower-right corner and drag down as far as you want. You can also auto-fill to the right to fill in a series in a row.

You can auto-fill other patterns of numbers and even auto-fill things such as days of the week, months, or years.

Selecting Cells

This section uses Google Sheets to discuss how to select cells, but all spreadsheet programs use the same or very similar methods.

To select connected cells, assume you want to sum up the contents of cells **G4, G5, G6,** and **G7** and put the results in cell **G8.** Click on cell **G8,** select **SUM** from the drop-down menu when you click on the **Sigma** button (which looks like a Greek **E** on the top far right). Next, select cell **G4,** drag down to **G7,** and release. The sum of these numbers will appear in cell **G8.**

Another way to do this is to select the cell you want the sum to go into, select the **SUM** function, and then, holding down the **Shift** key, press on each of the cells you want to be part of your sum. Press **Enter** to complete the sum.

If the cells are not adjoining, use the **Ctrl** key (or the **Command** key on a Mac) to select the cells you want. You can use various functions such as **SUM, AVERAGE, MIN, MAX,** and others.

Freezing Panes

This section uses Google Sheets and Microsoft Excel to demonstrate how to freeze panes. In both cases, select the row or column you wish to freeze. Click the View tab and select Freeze.

The panes you freeze do not move as you scroll through the spreadsheet.

Columns are frozen in the same way. Select the column, click the View tab, and select Freeze.

Excel works almost the same way. Select the column, click the View tab, select Freeze Panes, and make your selection.

Cell Referencing

There are two ways to reference a cell—absolute referencing and relative referencing.

When you copy a calculation using the fill handle, the spreadsheet program copies the cell that is in the same relative position. Relative referencing is the default in spreadsheets.

For example, if you multiply the value in cell B4 by 3, put the result in cell C4, and then drag that formula down column C, the result in cell C5 will be the value in cell B5 multiplied by 3, the result in cell C6 will be the value in cell B6 multiplied by 3, and so on.

If you don't want the reference to change, it is possible to define an absolute reference. An absolute reference tells the program to refer to a specific cell when the formula is copied. For example, if you stored a value in cell G9 and wanted everything in column C to be multiplied by the specific value in cell G9, you would make the reference to cell G9 an absolute reference. To make a cell an absolute reference, you put a dollar sign ($) in front of each cell identifier. The reference to cell G9 then would be G9. A common Excel shortcut for this is to use the F4 key.

AutoSUM Function

The Sigma button at the top right of an Excel spreadsheet means summation. It works when the added values are in adjoining cells. Simply select the cell beneath the column of values that you want to sum up, press Sigma, press Enter, and you're done.

It works the same for a row. Simply select the row following the values you want to sum, click Sigma, and click Enter.

Average Function

Here is how to take the average of a set of values using Excel. Sheets works the same way.

The **Average** function can be found in various locations. Click on the Sigma button to find it on the drop-down menu. You can also find it in the Editing group on the Home tab, in the Formulas tab, and in the Function options.

It works when the cells are adjoining. Select the cell at the end of the column (or row, if you want to average values in a single row), select **Average**, and click Enter.

COUNT Function

Both Excel and Sheets include the COUNT function in the Sigma drop-down list. Google Sheets, however, does not auto-select cells as do Excel and Calc.

The COUNT function shows how many values are in a selected range and doesn't count empty cells. Select the cell beneath the range of values you want, select COUNT, then drag to select the cells you want to include, and click Enter. Counting the values in a row works the same way. You can also use the Shift or Ctrl keys to select a range.

Hiding and Unhiding Cells

This section will discuss how to hide or unhide cells using Microsoft Excel.

First, select the cells you wish to hide. Most users then just right-click and select Hide. To unhide, go to where the cells are hidden, right-click, and select Unhide.

The right-click method is also used in Google Sheets as well as iCloud Numbers. With a Mac, use Command Click.

In Excel, another method is to select the panes and then, in the Home tab, Cells group, click Format. Under Visibility, select Hide & Unhide and then make your selection. This works to both hide and unhide cells.

Max and Min Functions

The **Max and Min functions** display the maximum and minimum values in a range of cells in a spreadsheet. For example, if a teacher has a spreadsheet of student grades and wants to quickly see what the highest score was on an exam, the Max function would display the highest score in just two clicks. The Min function would display the lowest score in the same manner.

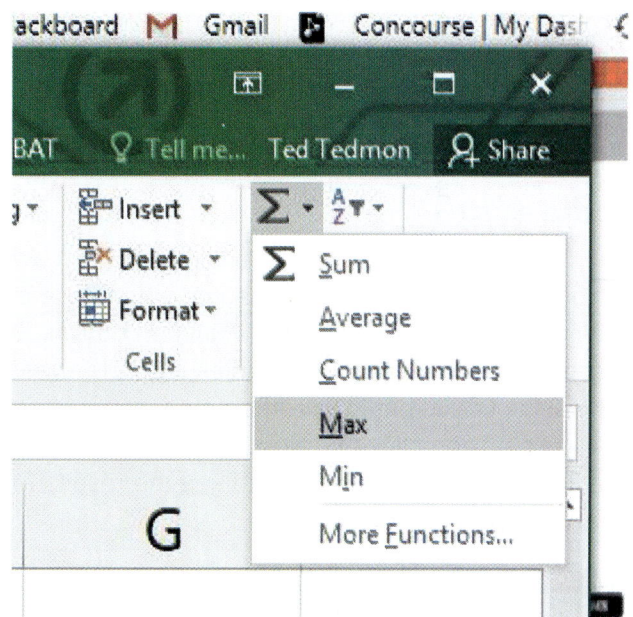

To use these functions, a user can first select the cell in which they want the Max or Min displayed. Then select the Max or Min button, located in several sites in Excel, but found most quickly on the Home ribbon under the Summation (Sigma) drop-down menu. After selecting the Max or Min function, the user can select the range and press Enter.

Payment Function

Let's say you need to buy a car and want to use a spreadsheet to help you find out what your monthly payment might be. You can use the **Payment function**. It will quickly show what a monthly payment would be for a loan such as a mortgage or a car loan.

To use the **Payment function**, you need three things:

- interest rate
- number of monthly payments
- loan amount

The interest rate is the amount of money the bank or credit union charges for the loan. This is advertised as an annual percentage rate (APR) or, simply, the Rate. Be sure to shop around to find the lowest APR. In this example, we will use 5 percent.

The rate must be divided by 12 because there are 12 payments per year. The number of payments depends on how long you are borrowing the money. For a car loan, this is often 4 years or 48 payments but can also be 3 years (36 payments) or 5 years (60 payments) or longer. In financial terms, every payment is referred to as a period or PER. In this example, we'll use 48 periods. The amount of the loan is how much money you are borrowing. In financial terms, this is referred to as the present value of the loan, or PV. In this example we'll say the loan amount is $20,000 and assume the loan will be for 4 years.

In OpenOffice Calc, you use the Function Wizard to find the payment function—PMT. Then select your arguments. Divide the rate by 12 because you are making 12 payments per year. Enter NPER (48) and PV (20,000) and then press Enter to see how much your car payment would be. You can see how the payment amount changes when you change the arguments. For example, change the loan amount (PV) to $15,000 and notice how the payment decreases. Increase the interest rate to 10 percent and see how it Increases. The payment function works the same in Microsoft Excel. Just select the **Payment function**, click the cells to select your arguments, and click Enter.

F Function

The IF function is used very often. Imagine you want to show what would happen if students passed an exam. Assume you have a column of students' grades. Click the first cell and click the Insert Function button. Select the IF function. The Logical test makes a numerical statement such as the value in cell B2 is greater than 64 (=B2 > 64). If that is true, then make the cell = Pass. If false, make the next cell = Fail. Then click OK. Now you can see if the student passed or failed. Copy the function down the column to see the results for all students.

Creating a Named Range

This section will demonstrate how to name a range of cells in a spreadsheet. When using tools such as vLookup, it is handy to name a range of cells. These instructions use Google Sheets. Click on the first cell in the range and drag down to select the rest of the range. Right-click to open the menu. If necessary, scroll down and select Define named range, name your range, and click Done. An alternate way to do this is to select the range by clicking on the first cell in the range, then hold the Shift key and click on the last cell in the range. As before, right-click on the range to open the menu, select Define named range, name the range, and click Done.

Pivot Tables

A pivot table lets you view the data in your spreadsheet more easily. Imagine that a group is raising money by selling cookies. To keep track of sales, the group shares a Google Sheet. Each time a sale occurs, the seller enters the information. This may include the seller's name, type of cookie sold, number of boxes sold, and the neighborhood where the sale took place. After a month of sales, the information needs to be organized. A pivot table can help.

First, clean up the data by removing any blank lines or columns. Then, highlight the data, open the Data menu, and select **Pivot table**. Choose the data you wish to organize. For example, in **Rows**, add the Seller. In **Columns**, add Cookie Type, and in Values, add the amount.

Creating a vLookup

In this section, we'll use Microsoft Excel to define the vLookup function and explain why you might want to use it. It works virtually the same in every spreadsheet. Continuing with our cookie example, let's imagine we have a long list of people who have prepaid for boxes of cookies.

Column A is populated with the purchasers' names, column B has the number of boxes purchased, and column C shows how much each person paid. The vLookup allows you to type in a person's name and find that person's name and how much that person paid so you can do that quickly when someone comes to pick up their cookies.

To make a vLookup, first highlight the range to search. Right-click on the highlighted range and name the range. In this example, the range will be named Paid. Click the Insert Function button and type vLookup in the search window. Click OK. Note that a vLookup (vertical lookup) searches columns. There is also an hLookup (horizontal lookup) that searches rows.

Now enter the arguments. In the Lookup_value argument, select the cell with the customer's name. In the Table_array argument, select the range that, in this case, is Paid. In the Col_index_num argument, select the column that, in this case, is 3 because the amount paid is in column C, the third column. In the Range_lookup argument, type FALSE.

Using Charts and Graphs

Inserting a Chart

This section shows how to insert a chart into a spreadsheet, using Microsoft Excel. Most spreadsheet programs work the same way.

First, highlight the desired range. Open the Insert ribbon and select the type of chart you wish to use. A pie chart works well when displaying a portfolio because it allows you to quickly see the proportion of each investment.

Notice that creating or selecting a chart opens two new tabs: Design and Format. Expanding the Chart Styles group provides more options. Hover your cursor over the options to view each one. Then click on the view that presents the data in the way you choose.

Types of Charts

Spreadsheet programs make it easy to display data using various chart types. Commonly used charts are bar graphs, line graphs, and pie charts.

- **Bar graphs** are often used to compare multiple items over a period of time or when comparing different items.
- **Line graphs** are used to compare changes that have occurred over time.
- **Pie charts** are used to compare different parts of a whole, for example, to show budget expenditures.

Elements of a Chart

When you create a chart using a spreadsheet, there are several elements to help explain the displayed data.

- A title can give context to the entire chart. Axes identifiers show what is displayed on the X-axis and the Y-axis.
- Data labels display the actual entry for each data point.
- A legend shows which items are being identified by the different colors, lines, bars, or pie segments.

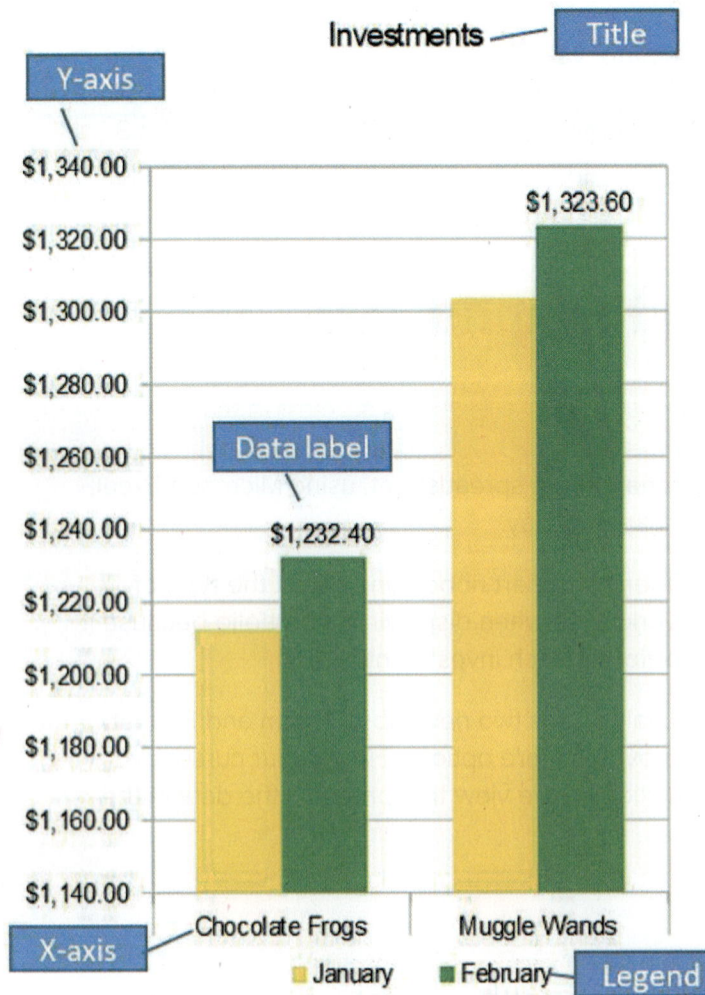

Adding Chart Elements

Adding elements can improve understanding and relevancy for the viewer of a chart. All spreadsheet programs allow users to add elements to the charts created in the spreadsheet.

The first step to adding elements is to select the chart. In most programs, this opens chart editing menus. In Excel selecting a chart by clicking on it adds the Chart Tools tabs to the ribbon. These tabs include the Design tab and the Format tab. The Add Chart Element menu is located in the Chart Layouts group in the Design tab. In Calc, selecting the menu changes the menu items under the tabs. For example, to add chart elements in Calc, select the chart and then open the Insert menu to add and edit chart elements.

Color Conventions in Spreadsheets

Colors can make spreadsheet data more readable, but they should be used thoughtfully. Too many colors can lead to confusion, especially for those with visual impairments.

Color Guidelines

It's important to understand the impact of color. Colors can communicate feelings or emphasize data trends. However, remember that colors like bright yellow, dark red, and medium green may be more noticeable due to their depth, but they also carry cultural meanings. For instance, while red might indicate danger in one culture, it can signify prosperity in another.

Basic Color Theory for Spreadsheets

Colors convey messages. The color wheel, rooted in color theory, distinguishes different colors and can guide you in choosing which ones to use. Warm colors like reds, yellows, and browns tend to pop out, while cool colors like greens, violets, and grays can recede in a viewer's perception.

Using the Color Wheel Wisely

Avoid using similar (analogous) colors next to each other; they might blur together. For contrast, use complementary colors, which are opposite each other on the color wheel. This can help highlight differences effectively.

General Color Guidelines for Spreadsheets

Financial data

- Red usually indicates negative values like losses.
- Black denotes positive ones like profits.

Data about Safety or Risk

- Consider using traffic signal colors.
- Green suggests safety.
- Yellow indicates caution.
- Red means danger or high risk.

Math In Spreadsheet Review

Order of Operations

The order of operations for a spreadsheet describes which calculation is performed first. **Spreadsheets** follow the mathematical protocol of Parenthesis, Exponents, Multiplication and Division (whichever comes first, left to right), Addition and Subtraction (whichever comes first, left to right).

The order of operations is important because it determines the result of the calculation. For example, if you enter $= 3 \times 2 - 1$ into a cell, the result will be 5. If you enter $= 3 \times (2 - 1)$ into a cell, the result will be 3.

Calculating Percentages

To calculate a percentage of a number, follow these steps:

1. Change the percentage to a decimal.

2. Multiply this decimal by the original number.

For example, let's say the restaurant bill is $26, and you decide to leave an 18 percent tip. How much should the tip be?

Note: The computer uses an asterisk (*) to denote multiplication.

The Average

An average is the sum of a set of values divided by the number of values. It is also called the mean. Other types of average include

- median—midpoint between highest and lowest values;
- mode—most frequently occurring value.

The average (mean) of 6, 8, 9, 7, and 4:

$$= (6 + 8 + 9 + 7 + 4)/5$$

$$= 6.8$$

Chapter Review

1. What is a spreadsheet?

2. Enter data into a spreadsheet in one of the spreadsheet apps and then save the data in the default file format of that app. This will require that you name the file. What is the default file format for that app? Now open a second spreadsheet app, create a spreadsheet, and save it, noting the default file format of that app. Do they each have their own file format?

3. What are the advantages of saving a spreadsheet file in a comma-separated format (.csv)? What are the disadvantages?

4. How are columns and rows used in a spreadsheet? Describe the identifiers for columns and rows and how they are used to locate data. Define the term *active cell*.

5. OpenOffice Calc and Microsoft Excel use something called a name box. Explain the purpose of a name box.

6. Describe the use of each of the following in a spreadsheet:
    a. label
    b. value
    c. equal (=) sign
    d. pointers

7. Why is it important to adjust column width and row height in a spreadsheet? Explain how you would widen a column or change a row height.

8. Explain what type of error is detected for each error value listed:
    a. #VALUE!
    b. #DIV/0!
    c. #REF!

9. While doing the previous tasks, did you find more than one method for a single task? If so, did you prefer one method? Describe your experience.

10. Describe the differences between average and mode.

Ethics Discussion

Spreadsheets often have important secrets from a company or private information from customers. Many people work from home now, but their computer and internet security might not be strong enough to protect this data. Even if you put a password on your spreadsheet, some people can figure it out and see the secret information. What are some things you can do at home to protect this information? Should you be allowed to have secret information at home?

17 Programming

What To Expect

After completing this chapter, you will be able to:

- define programming;
- discuss the reasons to study programming;
- explain the program development life cycle;
- define the major types of programing languages and their uses.

Chapter Topics

- What is Programming?
- Program Development Life Cycle
- Defining the Problem
- Planning Your Program
- Developing Code
- Major Programming Languages

Let's Explore Computer Programming

Imagine you're the captain of a cargo ship, and you need to decide where you want your ship to go. That decision, that destination you choose, is much like what we call the "problem statement" in programming. It's the main question or task you want to tackle.

Now that you have your destination in mind, how do you get there? Just like how a ship needs a map, in the world of programming, you'd create a clear set of instructions to guide you to your goal. This set of instructions is what we often refer to as "code." It's like giving your ship precise orders to ensure it sails smoothly toward its destination.

After plotting your course and giving out your orders, it's time for the fun part—setting sail! In our programming journey, this is when we run our code. It's an exciting moment, watching as the computer reads and follows each step you've laid out, taking you closer to your goal.

But let's be realistic—every adventure has its challenges. Maybe your ship might face a big wave, or perhaps an unexpected obstacle is on your path. Similarly, in programming, things don't always go perfectly. Sometimes you might run into "bugs." And no, we're not talking about creepy crawlies! In the tech world, bugs are those unexpected problems or hiccups that pop up in your instructions. But don't fret! Just like how you'd adjust your ship's route when something goes amiss, in programming, we have a process to tackle these issues. We call it "debugging." It's our way of ensuring we get back on track and continue our journey toward our destination.

What is Programming?

Let's define *programming* and talk about why you should care about it. **Programming** is giving a digital device, like a computer, to-do tasks for you. You take a problem, break it down into tiny steps, and then write those steps in a language the computer can understand. For example, if you have a remote-control vacuum cleaner, you could write a simple program to move around a room to clean it up.

So, programming is the method of converting a task into a sequence of commands that a digital device can understand to do the task.

Think of programming like solving a puzzle. The first thing you do is look at the big picture. What are you trying to solve? This thing you are trying to solve is called a problem statement.

For example, let's say you want to know how many people go to your school's football games. You start by asking, "How many tickets are sold in each football game?" That's your problem statement. Once you know the problem, you can start programming to find the answer.

Why Study Programming

Find a Career

Let's talk about the future. Every digital device or software application needs a programmer. People who learn advanced programming are in high demand and can work just about anywhere. People who learn the basics of programming are in demand because they can do things like customize software and plan as well as design and review programs.

Make Your Computer Your Sidekick

You know how you can buy ready-made programs that help you do stuff like write essays, edit photos, or play games? Well, guess what? If you know programming, you can create your own software that does exactly what you want.

In the programming world, you can create your own shortcuts called macros. These are like mini programs that you can add to existing software to make it work better for you. For example, instead of clicking five buttons to do something, you can program a macro to do it all with just one click!

Program Development Life Cycle

Steps in the Program Development Process

When you want to tell a computer what to do, you're basically creating a program. And just like baking a cake or building a treehouse, there are steps to follow. Here's how the program development life cycle (PDLC) works.

Step 1: Define the Problem

Before anything else, you've got to know what problem you're solving. Maybe you want a program that helps with math homework, or one that organizes your favorite songs. Whatever it is, this step is like figuring out what type of treehouse you want. Will it be high in the sky? Or maybe closer to the ground?

Step 2: Create a Plan

Once you know what you want, it's time to plan out how to get there. This is where you decide on all the smaller steps your program needs to take. If you're building a treehouse, this is the part where you sketch out how it'll look, how many windows it will have, and where the door will go.

Step 3: Develop the Code

Now comes the exciting part! You start writing out the instructions for the computer in a language it understands. Think of it as following a recipe. Each step is important and must be written clearly so the computer knows exactly what to do.

Step 4: Identify Bugs

When you're building something, sometimes things don't fit quite right. In programming, these little mistakes are called bugs. Every programmer, even the pros, runs into them. So, in this step, you play detective, finding and fixing those bugs.

Step 5: Testing, Testing, and More Testing

Your program is looking pretty good, and it's time to test it out. First, you and your pals give it a try. This is called Alpha testing. If you were building a treehouse, it's like asking your family to check it out first. After that, other people who haven't seen it yet (like neighbors or friends from school) give it a go. This is the Beta testing part. They might catch things you missed!

Step 6: Improve Your Program

After all that testing, you might find a few more bugs or get some cool ideas on how to make your program even better. So, you go back, make those changes, and then . . . voilà! You've got yourself a program ready to help with homework, play songs, or whatever you dreamed up!

Remember, every big project is just a bunch of smaller steps put together. So, take it one step at a time, and soon you'll be a pro at creating awesome programs!

Defining the Problem

Problem Statement

Let's say you need to figure out how much it costs for a group to go on vacation when there's a discount for 15 or more people in a group. Here's what we know.

- The cost per person is $500.
- If 15 or more people are in a group, the cost drops to $350.

Okay, let's lay down what we know. First things first, our main goal is to make a program that figures out how much a vacation will cost for groups. If a group is booking for fewer than 15 people, each person will need to pay $500. But here's the cool part: if the group is 15 people or more, each person only pays $350! The program will ask how many people are in the group. Then, it'll do the math and show how much the vacation will cost in total. Oh, and to make sure everything goes smoothly, the program will double-check that the number entered is a whole number and more than zero. So, no entering half a person or zero people!

Organize Your Information

Let's state the problem in the form of a table that will define what we want the program to do.

Goal	Compute the group cost for the vacation.
Input	Number of people in the group
Output	Total cost for the group
Process	Each ticket is $500 for groups of 14 or fewer. If the group is 15 or more, each ticket is $350.
Error handling	The input (number of people) must be a positive integer. Invalid inputs will require reentry.

Planning Your Program

Now, let's make a plan based on our expected input, output, and known information. Our problem statement makes it clear we need the number of people in a group and want the total cost of the group.

Testing Plan

First, figure out some test values for your plan, starting with your input and what you want for your output.

Input	Expected Output	Notes
5	5*$500 = $2500	Test for groups smaller than 15
14	14*$500 = $7000	Test the upper limit before the discount
15	15*$350 = $5250	Test the lower limit with the discount
20	20*$350 = $7000	Test for larger groups with the discount
0	Error message	Check for zero input
−5	Error message	Handling negative input
"Five"	Error message	Check for non-numeric input

Define the Variables

Now, let's define those as variables we'll need in our program:

- Num_People = Number of people
- Cost = cost for the group

Define the Calculations

Next, define the important calculations you'll use:

- If num_people < 15, then calculate the **cost = num_attendees * 500**
- If num_people > 15, then calculate the **cost = num_attendees * 350**

Define the Input

What kind of data can our program handle? We're calculating the number of people, so we need whole, positive numbers. We can't have a number like −2.5 people!

Write the Plan

What do we want the program to do? Let's write it out:

1. Get user input: number of people, stored in a variable named num_people

2. Check that num_people is a whole positive number

3. If num_people < 15, calculate the trip cost with the regular price:
   cost = num_people* 500

4. Else, calculate the trip cost using the discount: cost = num_people * 350·

5. Display result (total vacation cost)

Developing Code

When we say "code," we're talking about writing in a programing language, but developing the code refers to planning out how the code will work. Let's look at some examples of how to develop code.

Algorithms

An algorithm is like a set of directions telling a computer what to do step by step. You can use algorithms not just in computer stuff, but also in everyday decisions.

Programmers make pictures called flowcharts or write in plain English to show these steps. Using a flowchart makes it easier to turn the algorithm into computer code.

Flowcharts

A flowchart represents a graphical depiction of a sequence of events, a process, or an organizational structure.

People utilize flowcharts as a reference point when they engage with a process. Common flowchart elements include a diamond, which represents a decision point; a square, which represents a process; a rhombus or parallelogram, which represents an input or output point; an oval, which represents the beginning or end of a process; and an arrow, which describes the direction of flow for the chart.

Let's look at a sample flow chart for a fun, everyday decision.

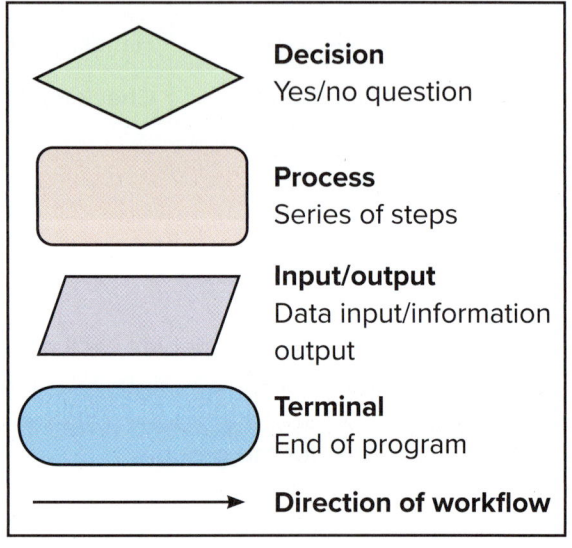

Decision
Yes/no question

Process
Series of steps

Input/output
Data input/information output

Terminal
End of program

Direction of workflow

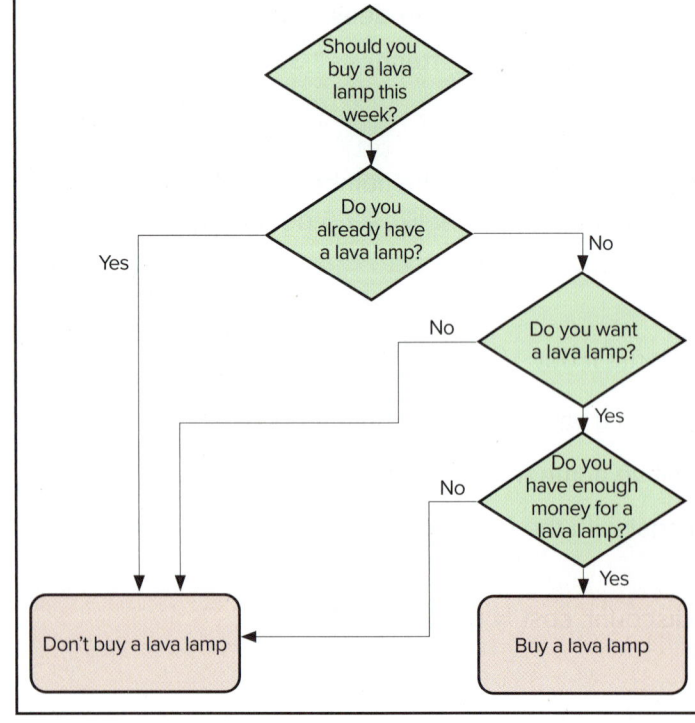

Copyright © McGraw Hill Incredible_movements/Shutterstock

Pseudocode

Pseudocode is like a set of instructions that tells a computer what to do. It's like the planning step before writing an essay. Just as you might use headings and bullet points for an essay plan, pseudocode uses steps to organize the plan for a computer program.

Pseudocode doesn't use a specific programming language. Programmers use everyday words and some special words that are like commands. These special words help when programmers start to write the real computer code.

Sample pseudocode for the program that calculates a group's vacation cost could be written as follows:

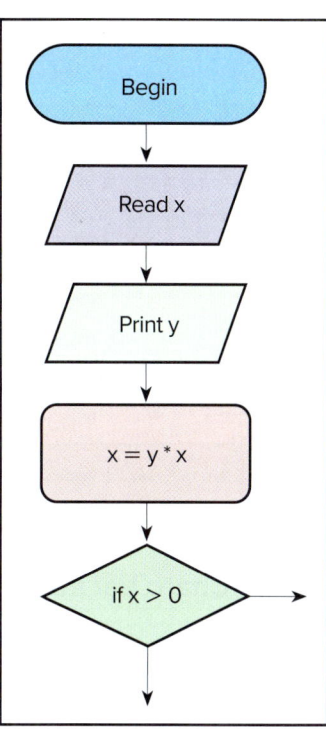

- Input the number of people in a group who want to go on vacation
- Check to ensure input is a whole positive number; if not, request new input
- If number of people < 15, cost = number of people * 500
- Else cost = number of people * 350
- Output tuition amount

Binary Decisions

Binary decisions act like a fork in the road. **Binary decisions** can be answered in only two ways: yes (the statement is true) or no (the statement is false).

For example, the answer to the question "Are more than 15 people in the group?" is either a yes or no response. If the answer is yes, the program follows a certain sequence of steps. If the answer is no, the program follows a different sequence of steps.

A computer can only understand yes/no or true/false questions, so all decisions must be phrased with binary results. However, using logical operators allows the programmer to create complex conditions.

Binary Chart: Vacation Problem

Looking back at our vacation problem, let's say you need to figure out how much it costs for a group to go on vacation when there's a discount for 15 or more people in a group. Here's what we know-

- The cost per person is $500.
- If 15 or more people are in a group, the cost drops to $350.

We will use this problem to create a binary chart.

Goal	Compute the group cost for the vacation
Input	Number of people in the group
Output	Total cost for the group
Process	Each ticket is $500 if group is less than 14. If the group is 15 or more, each ticket is $350.
Error handling	The input (number of people) must be a positive integer. Invalid inputs will require reentry.

Binary Chart: Vacation Problem

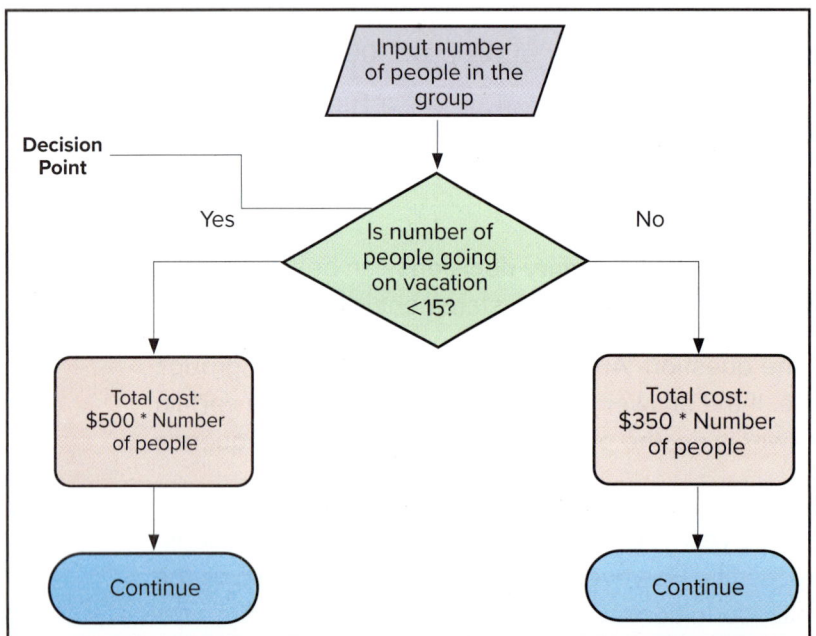

Loops

In a loop, the program poses a question. If the answer to the question is yes, it performs a sequence of actions. At some point, the program must change the condition being tested within the loop. Once the program completes that sequence of actions, it poses the question again, thereby creating a loop.

While the answer to the question remains yes, the algorithm continues to repeat the same set of actions. When the answer is no longer yes, the loop concludes, and the program proceeds to the instruction that follows the end of the loop.

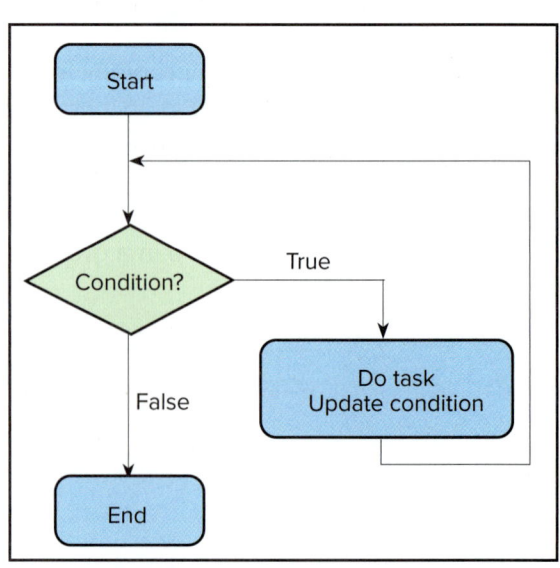

All loops must include an initial value, a test condition, a method for altering the condition, and one or more statements to execute in the loop body. If the initial value does not satisfy the test condition, the loop body never executes.

If the initial value remains unchanged within the loop body, preventing the test condition from ever returning no (or false), the loop never terminates. This situation is referred to as an infinite loop and should be avoided at all costs!

Top-Down Programming

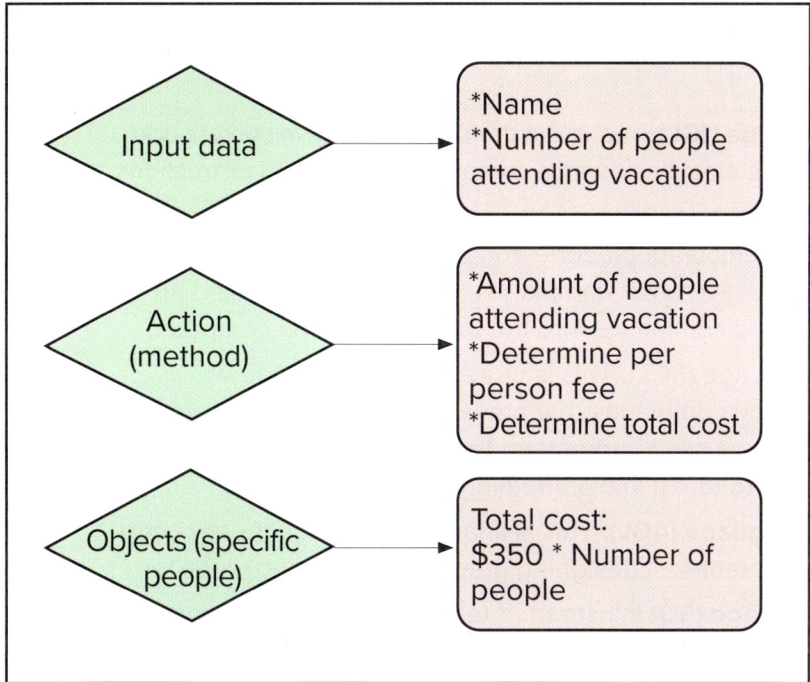

Object-Oriented Analysis

When object-oriented analysis is used, programmers first identify all the categories of inputs that are needed to solve the problem. The input categories are called classes, and objects are examples of the input items used in each class. Programmers using object-oriented analysis spend a lot of time identifying classes and establishing the relationships between each class.

One benefit of using this approach is the ability to reuse the objects that are created to produce new computer code. This is referred to as reusability.

Object-oriented programming is often referred to as OOP. It is useful for many types of programming, especially with a graphical user interface, known as a GUI (pronounced gooey). In OOP, programmers focus on the inputs to the program. **Classes** are used to define types of inputs. Each example of a class is known as an object of that class. For example, a program might have a class named VEHICLE. Some objects of the VEHICLE class might be TRUCKS, BUSES, SEDANS, and MOTORCYCLES.

Major Programming Languages

A programming language is a type of language, or code, for the set of instructions and commands the central processing unit (CPU) must perform. **Programming** languages use a collection of symbols, letters, and words that instruct the computer to perform specific operations. This is known as the language's syntax. Some examples of popular programming languages include **Java**, **JavaScript**, **C++**, **C#**, and **Python**.

Programming Language Generations

Imagine programming languages as different ways we talk to computers. Here's a simple breakdown:

- **First-generation Language (1GL):** This is like the computer's own secret code made of 0s and 1s. It's the most basic way a computer understands instructions.

- **Second-generation Language (2GL):** Think of this as a simpler language, kind of like a mix between computer code and human language. Programmers use this to give basic commands, and then a tool called an assembler changes these commands into the 0s and 1s for the computer.

- **Third-generation Language (3GL):** This is more like the language we speak. It has symbols and commands that are way easier for humans to understand. But computers can't understand the language directly, so we use tools like compilers to turn it into something the computer gets.

- **Fourth-generation Language (4GL):** This is a fancier language mostly used for things like making databases, designing graphics, or building websites.

- **Fifth-generation Language (5GL):** Instead of telling the computer how to do something, we just tell it what we want, and it figures out how to do it. It's often used in high-tech areas like artificial intelligence (AI).

So, from basic computer secret code to super advanced languages, that's how we chat with our computers!

Major Programing Languages and Their Applications

In this section, we provide simplified explanations of the major programming languages and the ways in which programmers use them.

Python and R

- **Python** is a language that's easy to read and learn. People use it for making websites, handling data, and many other things.

- **R** is mainly used for working with statistics and making charts. Both **Python** and **R** are popular because they help with analyzing data.

C and C++

- **C** is a very famous language that's been around for a long time. It was used to make many computer systems we use today.

- **C++** is like an upgraded version of C. Many computer games are made using C or **C++**.

Java and C#

- **Java** is used to make apps, especially for Android phones. It's special because you write it once, and it can run on many different devices.
- **C#** (pronounced "C sharp") was made by Microsoft. It's used to make programs for Windows computers and phones.

Objective-C and Swift

- **Objective-C** was used to make programs for Apple computers and iPhones.
- **Swift** is a newer Apple language that's easier to use. It's used to make apps for iPhones, iPads, and Apple computers.

HTML and HTML5

- **HTML** is what makes websites look the way they do. It uses "tags" to organize and show content like pictures and text on web pages.
- **HTML5** is the latest version of HTML. It has more features and works better with different web browsers.

JavaScript and VBScript

- **JavaScript** helps make websites interactive. It lets you click on things, fill out forms, and do other fun stuff on the web.
- **VBScript** is another language that makes websites dynamic and is also used for Windows apps.

XML

- **XML** is used to organize and share data on the internet. It wraps data with custom tags, making the language easy to read and use.

Ajax

- **Ajax** isn't a language, but a toolset. It lets web pages update without having to reload the whole page, making things feel faster.

JSON

- **JSON** is a way to write data that's easy for both humans and computers to understand. It's becoming more popular than XML for sharing data.

Interpreters

An interpreter is a program that converts a small portion or even a single line of code into machine language. When a programmer has compiled a program and needs to make a small change, they often use an interpreter to test a new line of code rather than compile the entire program again.

Interpreters are also used by some languages without compilers, such as **JavaScript**. **Python** uses both an interpreter and a compiler.

Interpreters allow programmers to see if these small changes will work when converted to machine language.

Compilers

An intermediary program, called a compiler, bridges the gap between the syntax of a programming language and the structure of the CPU and its machine language.

The compiler reads source code, containing programming instructions written in a high-level computer language, and then it translates this source code into the specific machine language. Most programming languages incorporate their own built-in compilers. However, if a programming language lacks a compiler, it typically resorts to using an interpreter. In this case, the interpreter translates source code line by line and executes each command as it is translated.

Chapter Review

1. Define the term *programming* and explain what programming involves.

2. Discuss at least three of the reasons stated for studying programming. Then pick one reason and discuss why it may be the most compelling reason for many people who study programming. Which reason appeals to you the most?

3. Briefly describe the four steps of the software development process.

4. One of the areas of discussion is the program development life cycle (PDLC). Research why this is described as a cycle and describe your findings, giving a real-life example to support the use of the word *cycle* in this case.

5. Why is an algorithm important to creating a plan for program development?

6. What is pseudocode, and why it is useful to use in a program flowchart?

7. Describe how a binary decision and a loop can be used together in a flowchart.

8. Research how top-down program design is used and create or find an example not included in this chapter. Describe the example in the following space provided.

Ethics Discussion

Conduct research on the internet about the ethics of programming. Consider sites such as softwareethics.org, freecodecamp.org, and the community blog at scrum.org. You may need to search a site to find discussions of ethics. Select one story that you found compelling. Should a programmer be held accountable for how a client company uses the code the programmer created? Discuss your findings and opinions on this topic.

18 Web Design and Development

What To Expect

After completing this chapter, you will be able to:

- describe emerging technologies in medicine.
- Discuss emerging technologies in commerce and business.
- Explain emerging technologies in society.
- Identify emerging technologies in the environment.
- Describe emerging technologies in science and computing.

Chapter Topics

- Why Websites Matter
- Web Navigation Essentials
- Website Creation
- HTML
- HTML5
- Cascading Style Sheet (CSS)
- Multimedia
- Anatomy of a Website
- Responsive Web Design (Mobile-Friendly Sites)
- Designing a Website
- Jobs of the Web Developer

Copyright © McGraw Hill · Roman Samborskyi/Shutterstock

Let's Explore Web Design and Development

Let's Explore the World of Web Design and Development!

Imagine a space where you can craft experiences, share stories, and bring ideas to life. This space exists, and it's called the web! Every time you browse your favorite website, watch videos, or chat with friends online, you're interacting with the intricate designs and codes crafted by web developers and designers.

Why is this important? Websites are more than just digital pages; they are platforms that businesses, artists, educators, and countless others use to reach out to the world. Every single tab you open is someone's creation designed to communicate, inform, entertain, or inspire.

Now, if you've ever wondered about the magic behind the scenes, you're in for an adventure! Dive deep into the foundation of web design with HTML and its advanced counterpart, HTML5. Discover the art of styling with CSS, ensuring every pixel is perfectly in place. Understand the crucial role multimedia plays in capturing users' attention and how responsive design ensures a seamless experience on any device.

Beyond just the technicalities, learn about the broader landscape. From the blueprint to the final design, witness the journey of a website coming to life. And if you're intrigued by the idea of shaping online experiences, get a glimpse into the dynamic roles and responsibilities of a web developer.

So, why embark on this journey? Because the web is an ever-evolving universe, and by diving into web design and development, you're not just learning a skill—you're opening doors to countless opportunities.

Ready to embark on this exploration? Let's get started!

Why Websites Matter

Power of Websites

While some say "everyone needs a website," it's not a must for everyone. Sure, you might be on Facebook or another social media platform, but having your own website? That's a game-changer. Today, from companies to schools, websites are the digital doorways to what they offer. They're key tools in today's digital world. But remember, it's all about what works best for you!

Reasons to Make a Website

Let's look at some ways a person or organization would use a website.

- **Launching Your Business.** When you're ready to introduce your business to the world, a website acts as your digital launch pad. This is where potential customers first learn about what you offer, who you are, and why you started. It's your chance to make a memorable first impression.

- **Supporting a Business Startup.** Maybe you're not at the very beginning, but you're still in the early stages. A website can be a hub of resources, tools, and community. It can connect you with mentors, potential partners, or even investors. It's where your business idea grows wings.

- **Amplifying Your Business.** If you've moved past the starting line and are looking to expand your reach, a website serves as a powerful booster. Share success stories, introduce new products or services, and engage with a wider audience. Amplify your message and reach the stars.

- **Project Management and Collaboration.** Behind every great launch is a lot of planning and teamwork. Websites can be platforms where teams collaborate, share updates, and track progress. It keeps everyone aligned and focused, ensuring your project is ready for liftoff.

No matter where you are in your journey, from the initial idea to ongoing growth, a website provides the platform to showcase, support, and propel your goals. Ready to launch? Your website awaits.

Businesses and Organizations Need a Web Presence

Big businesses need websites. That's how they sell products, communicate with employees, hire prospective employees, and much more. Small businesses simply can't compete in their markets without a significant web presence. Creating a website has become big business in itself. And ensuring visibility on the web is so important that SEO (search engine optimization) is an essential part of any website. Large and small organizations such as national, state, and local government agencies including educational institutions and nonprofit organizations also need websites. Without access to your school's website, you probably could not attend class.

Do You Need a Website?

Business experts agree that, if you're job-hunting, a personal website is a must-have. Why? Employers are going to Google you! Sites like LinkedIn let you shape your online image, but with your own website, you're the boss. It's your digital space that can grow with you, setting you apart from others. Thinking of building one? Plenty of places offer free websites with just a few clicks. Perfect for folks who want something straightforward. But if tech is your dream job, dive deeper into the world of web design. Even if tech isn't your thing, knowing the basics of websites is a cool skill to have!

Web Navigation Essentials

What Is a Browser?

A browser isn't just "how you reach the Internet." Think of it as a special app or program on your computer or phone. It's like a door that lets you step into the world of the Internet. Some computers already have a browser waiting for you—like Edge on Windows or Safari on Macs. you can, and should, have more than one browser! If you're designing a website, having a few different browsers helps you test things out to see how they'll look to different people.

Some big-name browsers include:

- Edge, by Microsoft
- Chrome, by Google
- Firefox, by Mozilla
- Safari, by Apple—which primarily works on Apple devices,

What Does a Browser Do?

A browser has two main jobs: to communicate with your ISP (Internet Service Provider) and to display web pages. When you type information into your browser's address bar, the browser sends that information out to the Internet through your ISP to find the website you want. Then it interprets the code from the pages that are returned to display your requested website. Before the page can be displayed, however, the browser reads and interprets all the HTML, CSS, and other code on the page to turn it into what you see on your screen.

URL: Uniform Resource Locator

URL stands for **Uniform Resource Locator.** It is a way to identify the location of any file on the Internet. Every web page on the Internet is a file, and each file has its own unique URL. This is why, if you type in a URL with a spelling error or add a space where none is required, you may get a File Not Found error. Browsers are smart but not smart enough to correct all human errors.

Today's browsers, however, are far better and smarter than they were a decade ago. If you want to visit the Amazon home page, all you need to type into the browser is "amazon" and, if you've ever visited the Amazon site before, you'll get to the site. The browser is translating your request into the actual address of www.amazon.com. The URL for Amazon's home page is https://www.amazon.com.

Internet Protocols

An **Internet protocol** is a set of rules that governs the format of data sent over the Internet or any other network. IP (Internet protocol) ensures that all machines on the Internet speak the same language; in other words, the IP standardizes how devices retrieve and interpret documents.

There are many Internet protocols, but the ones that are most often used are HTTP and HTTPS. Some of the other protocols that a web developer needs to know are FTP, SSL, TELNET, SMTP, POP3, and IMAP4.

HTTP, HTTPS, FTP, and SSL are used mainly for transferring files when web pages are under development from a client to a server and vice versa. TELNET is mainly used by businesses that may want employees to "remote in" to a server. SMTP, POP3, and IMAP4 are various protocols that deal with e-mail.

Domain Names and IP Addresses

The **domain name** is one of the pieces of a URL. In simplest terms, a domain is the name of a website. An **IP address** is a set of numerical instructions that tells the computer where to find a website. You can think of the **IP address** as the actual code that points to the website, and the domain name is a way for humans to easily point to that code.

Every domain name is unique. A person or a business can purchase a domain name from various sites, but once a name has been purchased, it can never be used by anyone else. Therefore, the key to success if you want to start your own business is to purchase a domain name that is associated with your business and is easy for people to remember. By now millions of domain names have been bought, so you might have to be creative when finding an appropriate one.

There are two parts to a domain name: the name itself and the **suffix**. The name expresses what the site is about, for example, a domain like petcare. The suffix tells you what type of site petcare is. In this example, petcare.com might be a business selling pet care items, petcare.org might be a nonprofit organization helping people care for their pets, and petcare.net might be a personal website with blogs about pet care.

The suffixes are also called **extensions**, and there are currently over 800 domain name extensions with more coming online every day.

What's in a URL

In reality, a website's address has a lot of parts, like a protocol (https) and a domain (amazon.com). But you only need to remember the simple part, like "amazon." If you type something common like "volcano," your browser will show you a list of websites related to volcanoes because it's not sure which one you want.

So, when you type an address, the browser adds some parts to it to make the full website address. This full address can have different parts like a protocol, a host, a domain, and sometimes other bits

- https://www.mheducation.com/highered/home-guest.html: https is the protocol.
- www means the site is on the World Wide Web.
- mheducation.com is the domain name for McGraw Hill, and the .com indicates that this is a business.
- *preK-12* is a folder name
- home-guest.html is the filename for the web page you see when you go to the URL.

Let's look at a few examples:

URL	Protocol	Subdomain	Domain	Folder(s)	Filename
https://www.nasa.gov/kidsclub/index.html	https	www	nasa.gov	kidsclub	index.html
https://www.stanford.edu/research/	https	www	stanford.edu	research	default page
https://www.bbc.co.uk/nature/animals	https	www	bbc.co.uk	nature/animals	default page

What is a Server?

People are often confused by the term *server* because it has two meanings. A **physical server** is a large computer that stores web pages that can be accessed over the internet. However, a **server** also refers to the computer software that manages and processes server requests and delivers data to other computers.

A company that hosts web pages will have one or more physical servers, all running server software. An individual can also turn any computer into a server by using server software.

There are several different types of servers that provide various services for clients. **Web servers** display web pages and run apps through browsers, while **email servers** facilitate sending and receiving email.

In the **client/server model**, the server program waits for and fulfills requests from client programs. When you enter a URL into your browser, your computer is the **client**. Your browser sends a request to a server. The server sends all necessary files or data back to your browser, which then interprets the information and shows you, the user, the requested web page.

What Is a Host?

A **web host**, or **web hosting service provider**, is a business that provides the technologies and services needed for a website or web page to be viewed on the internet. A web hosting company will store all the files and folders for a website on a server. Once a website is stored on a server, it can only be viewed on the internet through its URL.

Many hosting companies will help clients purchase a domain name, but a person can buy a domain name separately and then contract with a hosting company to provide space on a server for the website. Once you have purchased a domain name, you own it no matter where you keep your site. But when you contract with a hosting company, you normally must renew the contract once the initial period is over. Hosting is usually contracted with monthly fees for a year at a time. Some companies offer free web hosting, and some will even provide free domain names.

However, as with almost everything, you get what you pay for. Free hosting is usually bare bones. Services that keep your site secure, provide technical support, or help your site gain visibility are generally provided at extra cost.

Website Creation

Websites consist of many web pages but also contain other types of files such as style sheets, images and media files, **JavaScript** files, and databases. The websites of most businesses are created by teams of developers in specialized fields, including programmers, graphic designers, network administrators, and database administrators. While it takes a college degree to become any one of these specialists, it is still possible for a person with no specialized knowledge in any of these fields to create a website.

What Skills Are Needed?

Web programmers write the code to create the pages users will see, including code for interactivity on a site. This allows users to move from page to page, to communicate as necessary with the business, to ensure that the site is visible on all devices (such as smartphones or tablets and through various browsers), and to connect the front-end site (what the user sees) with the back-end (the server where the site is hosted, where the company's databases are stored, and so on).

But many other people are involved in creating a website. **Graphic designers** deal with the design of the pages. **Database** administrators work on the site's databases. Network administrators ensure that the back-end and front-end work together smoothly.

Other considerations that may require a team effort are ADA (Americans with Disabilities) compliance, security, and SEO, which attempts to increase a site's visibility during web searches.

Anatomy of a Web Page

Websites consist of linked web pages; therefore, to create a website you must understand how to create a web page. A web page has two main parts: the head section and the body.

The **head section** contains code that is visible only to the browser. It includes links to supporting files, **API calls**, documentation, meta tags, the page title, and the character set to be used for the display.

- **Supporting Files:** These include style sheets, scripts, images, and multimedia files.
- **API Calls:** *API* stands for "application program interface." **API calls** are links to code that can invoke various tasks the site may need. For example, you could use an API call to Google's geo-location application if you wanted a map to appear on your site.
- **Documentation:** This consists of information about the author of the page and any other information that may be necessary to other developers who might work on the site.
- **Meta Tags:** There are many types of meta tags, but some of the most common ones used in web pages are *description* and *keywords* that are used by search engines.
- **Page Title:** The page title tags in the head section tell the browser what to display in the browser tab at the top of the screen when a page is displayed to a user.
- **Character Set:** This identifies which character encoding to use when displaying the page.

The **body** of a web page includes the HTML tags that tell the browser what the content is and how to display it, working together with a style sheet. The body may also include links to images, audio and video files, or scripts that allow the user to interact with the page.

HTML

What Is HTML?

HTML stands for Hypertext Markup Language. It's the standard language used to create and design websites and web applications. Think of HTML as the skeleton of a web page: it provides the structure, while Cascading Style Sheets (CSS) and **JavaScript** give it style and interactivity. Each element on a web page, from paragraphs to images, is defined using HTML.

Understanding Tags

In HTML, the content is structured using "tags." **Tags** are used to define and categorize content, making it both readable by browsers and accessible to users. **Tags** usually come in pairs: an opening tag and a closing tag. The content resides between these tags. For example, the <p> tag is used to define a paragraph, and the content of the paragraph is placed between <p> (opening tag) and </p> (closing tag).

Common HTML Tags

Tag	Description
<html>	Defines the start and end of an HTML document
<head>	Contains meta information about the document and links to style sheets and scripts
<body>	Contains the content of the document
<h1>, <h2>, ... <h6>	Orders headers, with <h1> being the largest and most important, and <h6> the smallest
<p>	Defines a paragraph
<a href="URL">	Creates a hyperlink (Replace URL with the link address.)
<img src="path/to/image. jpg" alt="description">	Displays an image (src specifies the image path, and alt provides a text alternative for screen readers.)
<ul>	Defines an unordered list (Used with <li>, which defines list items.)
<ol>	Defines an ordered list (Also used with <li>.)
<table>	Creates a table (Inside, you use <tr> for rows, <td> for data cells, and <th> for header cells.)
** **	Inserts a line break
<div>	Defines a block-level container for content or other **HTML elements**
<span>	Defines an inline container for content or other **HTML elements**

Remember, this is just a basic overview. There are many more HTML tags available, each with its own specific purpose and attributes. When creating a web page, you'll often use a combination of these tags to structure your content effectively.

Header and Paragraph Tags

All **HTML elements** have a default display. The following shows how the basic elements would be displayed, including text size, font face, and color (unless the developer changes these defaults with other code).

This is how content within <h1></h1> tags are displayed by default.

This is how content within <h2></h2> tags are displayed by default.

This is how content within <h3></h3> tags are displayed by default.

This is how content within <h4></h4> tags are displayed by default.

This is how content within <h5></h5> tags are displayed by default.

This is how content within <h6></h6> tags are displayed by default.

This is how content within <p></p> tags are displayed by default.

Ordered and Unordered Lists

Lists also have a default display. The following shows how the ordered and unordered lists would be displayed, including text size, font face, color, spacing, and bullets or numbering (unless the developer changes these defaults with other code).

- The <ul> tag opens an unordered (bulleted) list.
- The <li></li> tag pair displays elements of the list.
- The </ul> tag closes an unordered (bulleted) list.

1. The <ol> tag opens an ordered (numbered) list.
2. The <li></li> tag pair displays elements of the list.
3. The </ol> tag closes an ordered (numbered) list.

Images

Images are not part of a web page. They must be saved as separate files and linked from the body of a web page to that file. When a web page is stored on a server, if it includes any images, the image files must also be included on the server. A common beginner's error is to create a page on a computer with images and forget to upload the images to the server or upload the images into an incorrect location. Then, when displayed on the internet, the images appear to be missing.

IMAGE

An image tag has no closing tag. A sample image tag looks like this:

<img src="path to image file" alt="text description of image">

Img, as you may have guessed, stands for "image," and *src* stands for "source." The source is the path to the file and must include exact directions about how to find the file. If an image is inside a folder named "Images" and the actual file is named "duckling.jpg," then the path would be images/duckling.jpg.

A Basic HTML Web Page

Sample Page

Here's a sample of an HTML web page named "index.html." The name "index" is commonly used for the main or default page of a website, serving as its starting point or home page.

<table>
<tr><th>Index.html</th></tr>
</table>

```
<!DOCTYPE html>
<html>
<!-- author: Cordelia L date created: May 2023 -->
<head>
<meta charset=utf-8>
<title>My First Web Page</title>
<meta name="description" content="creating a web page with HTML5">
<meta name="keywords" content="HTML,CSS,beginner,web,developer">
<meta name="viewport" content="width=device-width,initial-scale=1.0">
<link rel="stylesheet" href="styles/main_presentation.css">
<script src="js/swaps.js"></script>
</head>
<body>
<header>
<h2>
<img src="images/working_on_webpage.jpg" alt="creating a web page"> My First Web Page!</h2>
<h3>My favorite things</h3>
</header>
<main>
<section>
<p>I always enjoy working on the computer. This is where I will create my first web page:</p>
<img src="images/laptop.jpg" alt="a laptop computer">
</section>
<aside>
<p>But I also need to have my page display on my tablet and my phone.</p>
<img src="images/tablet.jpg" alt="a tablet">
<img src="images/smartphone.jpg" alt="a smartphone">
</aside>
</main>
<footer>
<p>&copy; 2022, My first web page</p>
</footer>
</body>
</html>
```

How It Works:

The HTML page has elements and tags that define the layout and content. Here's a breakdown of the key components:

DOCTYPE Declaration

Before any HTML tags, pages start with a DOCTYPE declaration. This tells the web browser which version of HTML the page is using, ensuring proper rendering and behavior.

HTML Element (<html> ... </html>)

This is the root element that wraps all content on the entire web page. Everything that defines the web page is nested inside the <html> tags.

Head Element (<head> ... </head>)

Nested within the HTML element, the head contains meta-information about the document, such as its title, linked files like style sheets, and other setup information. This section doesn't display content directly to the user but prepares the browser to correctly render the page.

Body Element (<body> ... </body>)

This is where the content that's displayed to the user goes. Everything inside the body tags—from text to images to videos—will appear in the browser window. It can further be broken down into various structural and content elements.

Header Element (<header> ... </header>)

Often used at the top of a web page or at the beginning of sections, the header typically contains a logo, the website's title, and the main site navigation.

Main Element (<main> ... </main>)

As the name suggests, this tag holds the main content of the web page. It's used to encompass content that's directly related to or expands upon the central topic of a document or the central functionality of an application.

Section Element (<section> ... </section>)

This tag represents standalone sections of content or functionality, like a tabbed content box or a set of related content grouped together.

Aside Element (<aside> ... </aside>)

This tag is used for content that is indirectly related to the main content. Typically, this might include sidebars, inserts, or explanatory boxes.

Footer Element (<footer> ... </footer>)

Situated at the bottom, the footer often provides additional information about the page or the site, such as authorship, copyright information, and related links.

Comments (<!-- ... -->)

Not displayed in the browser, comments allow developers to leave notes or explanations in the code, useful for both personal reference and collaboration with others.

By understanding these primary sections of an HTML document, one can grasp how content is organized, laid out, and presented to users. This structured approach ensures web pages are both accessible to users and interpretable by web browsers.

Line-by-Line Explanation

Line #	HTML	Explanation
1	<!DOCTYPE html>	Declares the document type and version of HTML
2	<html>	Opens the HTML document
3	<!-- author: Cordelia L date created: May 2023 -->	Comment stating the author and date (Comments are not displayed in the browser.)
4	<head>	Begins the head section, which contains meta information about the document
5	<meta charset="utf-8">	Specifies the character set for the document
6	<title>My First Web Page</title>	Sets the title of the document, which appears in the browser's title bar or tab
7	<meta name="description" content="creating a web page with **HTML5**">	Sets a description for the page, which may be used by search engines
8	<meta name="keywords" content="HTML,CSS,beginner, web,developer">	Sets keywords for search engines
9	<meta name="viewport" content="width=device-width,initial-scale=1.0">	Helps with responsive design by setting the viewport width and initial scale
10	<link rel="stylesheet" href="styles/main_presentation.css">	Links to an external CSS file for styling
11	<script src="js/swaps.js"></script>	Links to an external **JavaScript** file
12	</head>	Closes the head section
13	<body>	Begins the body section, which contains the content to be displayed in the browser
14	<header>	Designates a container for introductory content or a set of navigational links

Line #	HTML	Explanation
15	<h2>	Opens a level 2 heading
16	<img src="images/working_on_ webpage.jpg" alt="creating a web page"> My First Web Page!</h2>	An image with its source and alt text followed by text content, and then closes the h2 tag
17	<h3>My favorite things</h3>	A level 3 heading
18	</header>	Closes the header section
19	<main>	The main content of a document
20	<section>	Represents a standalone section of functionality contained within an HTML document, typically with a heading
21	<p>I always enjoy working on the computer. This is where I will create my first web page:</p>	Paragraph describing the author's enjoyment of working on the computer
22	<img src="images/laptop.jpg" alt="a laptop computer">	Image of a laptop with alt text for accessibility
23	</section>	Closes the section
24	<aside>	Designates content that is tangentially related to the content around it and that can be considered separate from the main content
25	<p>But I also need to have my page display on my tablet and my phone.</p>	Paragraph about the author's intention to make the page responsive
26	<img src="images/tablet.jpg" alt="a tablet">	Image of a tablet with alt text for accessibility
27	<img src="images/smartphone.jpg" alt="a smartphone">	Image of a smartphone with alt text for accessibility
28	</aside>	Closes the aside section
29	</main>	Closes the main content of the document
30	<footer>	Designates a container for the footer, which can contain information about the author, copyright information, etc.
31	<p>© 2022, My first web page</p>	Copyright notice for the web page
32	</footer>	Closes the footer section
33	</body>	Closes the body section of the HTML document
34	</html>	Closes the HTML document

The Basic Attributes of Elements

Elements also have attributes that can affect how the content is displayed. Here are some of the most common attributes:

- An **id** denotes a unique identifier for an element in a page and is used to locate elements through links or scripting languages, like **JavaScript**.
- A **class** refers to a CSS class of an element and is one of the tools used to change the appearance of the content of an element.
- A **style** describes the CSS styles that will be applied to an element.

The tags used in this sample HTML page are the most basic tags. The page bears no resemblance to most of the web pages you see on the internet today. There is a great deal more to learn about web development, but it all starts with these building blocks.

Creating a Basic Web Page

Step 1: Set Up Your Workspace

1. Open a text editor on your computer. This can be Notepad (Windows), TextEdit (Mac), or any other simple text editor.

2. Save the file immediately as "all_about_me.html." Make sure it's saved with the .html extension.

Step 2: Start Your HTML Document

Write the basic structure of an HTML document.

```
<!DOCTYPE html>
<html>
<head>
<title>All About Me</title>
</head>
<body>
</body>
</html>
```

Step 3: Add Metadata

Inside the <head> section, add metadata for character encoding. This ensures your web page displays characters correctly.

```
<meta charset="utf-8">
```

Step 4: Add a Header

Inside the <body> section, start with a header.

```
<header>
<h1>All About Me</h1>
</header>
```

Step 5: Introduce Yourself

Below the header, add a section for your introduction. Replace [Your Name] with your name.

```
<section>
<h2>Introduction</h2>
<p>Hi, there! I'm [Your Name]. I'm learning to build my own website!</p>
</section>
```

Step 6: Share Your Hobbies

Customize the content to match your hobbies.

```
<section>
<h2>My Hobbies</h2>
<p>I love playing soccer, reading science fiction books, and coding!</p>
</section>
```

Step 7: Conclusion and Contact

Conclude your page and provide a way for friends to contact you (maybe via email).

```
<section>
<h2>Contact Me</h2>
<p>If you'd like to chat or work on a project together, reach out at:
<a href="mailto:youremail@example.com">youremail@example.com</a></p>
</section>
```

Step 9: Add a Footer

At the bottom, add a footer with the current year.

```
<footer>
<p>&copy; 2023, All About [Your Name]</p>
</footer>
```

Step 10: Save and View

Save your all_about_me.html file. Double-click on the file to open it in your web browser.

Congratulations! You've just created your "All About Me" web page.

HTML5

HTML5 is the latest version of HTML. **HTML5** really represents two different concepts. It is a new version of the scripting language, HTML, which provides the structure of a web page. But it is also a larger set of technologies that allow for building more diverse and powerful websites and applications.

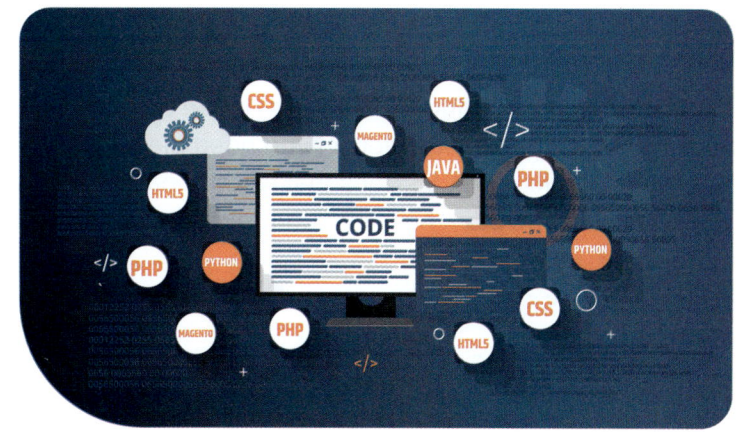

Older versions of HTML had elements that determined the structure of a web page. **HTML5** adds many new elements that allow developers to describe content more precisely. **HTML5** also supports multimedia on mobile devices and changes the way users interact with web pages. **HTML5** pages load faster, use less bandwidth, and reduce the need for many plugins and addons. All these things make web browsing a far better, faster, and smoother experience.

HTML5 Tags

HTML5 has expanded the list of basic elements discussed earlier to new tags that describe content explicitly. Here are some tags that define new elements:

- **Semantic Elements:** <header>, <footer>, <article>, <aside>, and <section>
- **Graphic Elements:** <svg> and <canvas>
- **Multimedia Elements:** <audio> and <video>

HTML5 pages are divided into sections that are defined by these new tags. This makes it easier for a developer to define presentational aspects of various sections of the page. Some are self-evident, such as <header></header>, <footer></footer>, and <aside></aside>. Others are not as clear.

SVG stands for Scalable Vector **Graphics**, and the <svg> element is a container for SVG graphics. The <canvas> tag is used to draw graphics on the fly via a scripting language such as **JavaScript**. The script is written inside the <canvas></canvas> element.

The <audio></audio> element allows the developer to include audio files and specify alternative formats of such a file because not all browsers support the same types of audio files. The concept for the <video></video> element is similar. These elements also allow the developer to specify whether audio or video controls should be visible to the user, the size of the display, and so on.

Cascading Style Sheet (CSS)

What are CSS and Styles?

CSS (short for Cascading Style Sheets) is like the design blueprint for a webpage. Just like when designing a house, you need plans to decide where everything goes and what it looks like; CSS does that for web pages. It decides things like text color, size, position, and even how a webpage looks on different devices.

How Does CSS Work?

- **Where Styles Are Set:** Styles can be set in three spots:
- Right inside the page's content.
- At the beginning of the page.
- In a separate file.
- Think of this like choosing the color for a room in your house. You could paint a wall with a unique color (inside the content), paint the whole room a single color (at the beginning of the page), or use a general color scheme for the entire house (a separate file).
- **How Styles Get Picked:** CSS uses a system where the style set closest to something gets used. So, if we paint a wall blue, then red, and then green, it will end up being green because that's the last color we chose.
- **Consistency with CSS:** A lot of websites use separate files for styles. It's like using the same wallpaper or paint in every room. It makes sure everything looks consistent.
- **Adapting to Different Screens:** Websites today need to look good on different devices. Using CSS, designers can set different designs for different screens. It's like having a house layout that can change based on who's visiting.

Changing the Look of Text with CSS

By default, text on a webpage has a certain look. But with CSS, we can change how it appears. Here's an example of how this is done:

```
/* Changing the font of the page */
body { font-family: Arial, Helvetica, sans-serif; }

/* Making the title big, green, and a bit slanted */
header h2 { font-size: 250%; color: green; font-style: italic; }
```

Arranging Things on the Page

With CSS, we can move things around and decide where they go on the page. It's like deciding where to put furniture in a room. Here's a quick look:

- If we have two parts on our page, one with text and one with pictures, we can decide how they appear:

-
```
<main>
<section>
<p>This is where I write stuff.</p>
</section>

<aside>
<p>And here are my pictures.</p>
</aside>
</main>
```

- And then, with CSS, we decide how each part looks and where it goes:

```
/* Setting up the text area */
section { width: 300px; }

/* Setting up the picture area */
aside { width: 300px; }
```

This way, our text and pictures can be side by side or one above the other, depending on how we want it.

Conclusion

CSS helps designers decide how a webpage looks and behaves. It's a set of rules that the browser follows to display the page just the way the designer wanted. And just like in designing a house, it's crucial to have a good plan!

Multimedia

Graphics and Images

Today you would be hard-pressed to find a website that does not include images. This makes understanding how images are included in a web page extremely important. Each image is not part of the web page but is, rather, a separate file that must be linked to the page. The link should be placed in the part of the page where the image will appear, and the link must include a path to the file so the browser can find that file. A link to an image file uses the <img src> tag. It does not require a closing tag.

Most websites include a folder named "Images," and all the images for the site are stored in that folder. If you want to include an image named "kitten.jpg" on your index.html page, you must include the exact path to where that kitten.jpg file is. The path would look like this:

<img src = "images/kitten.jpg" alt = "my kitten"

My kitten enjoys working on his tablet:

However, if the image is not in the correct folder or if the path is incorrect, a box will appear where an image should be. If you include the alt property (text that describes the image), then that text will appear but no image. People who are new to website creation often forget to upload the image files to the correct folders, or they write the wrong path to the images.

My kitten enjoys working on his tablet:

🖼️my kitten

Images may not always be pictures on a web page. The tiny image you may see in the browser tab of a site is called a favicon.

Using Images for Text

Not all browsers or computers have access to every font, so you cannot always be sure a page will display exactly as you want when people look at it on the internet. If a user's computer does not have the font you used when you created your site, the computer will substitute something else. If you really want certain text to use a specific font, you might create an image of that text with the font you want. Then, instead of typing in the text, you can insert it on your page as an image. That ensures it will look exactly as you want all the time, regardless of the user's computer or browser.

Audio, Video, and More

Multimedia files include audio files, video files, and slide shows. Using multimedia is a good way to enhance your site and keep viewers interested. It also is a good way to include instructions about how to do something; people prefer short videos to long text instructions.

It is relatively easy to include multimedia on a website in a manner similar to embedding an image, but when you do, you need to make certain choices. You will need to use different commands if you want the file to play automatically when the page loads or if you want the user to click on the file to start play. You also will need to decide whether you want the user to have access to controls, such as volume, start-pause-stop, and so on. If you make your own video files, you may want to include those files as part of your website or create YouTube videos and link to those instead.

The following code, embedded in a web page, brings the user to a YouTube video. This video, however, has been made private so it cannot be viewed at this time.

Example:

```
<a href=https://www.youtube.com/watch?v=
ZDDMDO_fRW"> Kitten playing
a computer game</a>
```

Interactivity

There are many ways to make a website that allows the user to interact with the site, although these require more advanced skills.

JavaScript

JavaScript is a programming language that allows the **client-side** (the browser) to interact with the user, alter the document content, communicate asynchronously, and more. For example, you can write **JavaScript** code to allow the user to change the color of a web page on the user's screen. Or you can create simple games, such as Tic-Tac-Toe, on your site.

jQuery

jQuery is a **JavaScript** library designed to ease the client-side scripting of HTML. A **library**, in programming, is a collection of programs that you can bring into your website, customize if necessary, and use without much knowledge of programming logic or syntax. In fact, if you want to do something interactive on your site that would require **JavaScript**, you can probably find a **jQuery** library that has already been created to do just that.

PHP

PHP stands for Hypertext Preprocessor. It is a server-side scripting language. This makes it different from everything we have discussed so far because **HTML5**, CSS, **JavaScript**, and **jQuery** are all client-side. **PHP** scripts run on the server, not on the user's computer. **PHP** allows the web page that the user sees to communicate with the server. Using **PHP** is probably best left to a professional, but it is a valuable tool.

A few of its functions are listed here, but it can do much more:

- **PHP** performs system functions. It can create, open, read, write, and close files.
- **PHP** can handle forms. It can gather data from files, save data to a file, send data by email to the website owner, or return data to the user. **PHP** allows the website owner to add, delete, and modify elements within a database.
- **PHP** allows a website access to cookies and to set cookies. **PHP** can encrypt data.

Anatomy of a Website

A website consists of many web pages. Each page includes various items such as style sheets, images, multimedia files, and **JavaScript** files. It's necessary, therefore, to keep everything organized, and this organization usually consists of folders that each hold similar items. For example, most websites have a folder named "Images" that includes all the image files used on the whole site. By organizing a site's files into folders, files can be easily used on several pages.

Files and Folders

A typical website might include the following folders as well as other folders specific to that site:

- The **Images** folder will hold all the images used on the site.
- The **Documents** folder might hold relevant documents such as PDF files.
- The **JS** folder is for **JavaScript** files. **JavaScript** code allows pages to be interactive between the browser and the user.
- The **Multimedia** folder might hold audio and video files that the site owner wants included.
- The **Styles** folder will hold various style sheets. Often a site needs different style sheets for different browsers or for different device sizes.

documents
images
js
multimedia
styles

By keeping all the files needed for a site in folders, it is easy to use relevant files, such as the site logo, on every web page.

Responsive Web Design (Mobile-Friendly Sites)

Less than a decade ago, all websites were designed to be viewed on a laptop or desktop computer. The biggest worry a developer had back then was to make sure a site did not look awkward on a very large monitor or on a computer where the user had an unusual screen resolution. But today it is hard to imagine creating a website that can't be viewed on a smartphone or tablet as well as on a large screen.

Designing sites for all devices is known as **Responsive Web Design (RWD)**. RWD requires a lot more coding than simply changing font sizes for smaller screens. For example, the navigation (links to various pages on the site) on a laptop screen normally consists of a list of links spread horizontally across the top of the page or vertically along the right or left side of the page. But most smartphone websites use the "hamburger" (the three dashes) to indicate a navigation drop-down menu.

Often elements on a page need to be repositioned for a small screen. To display three columns of content that appear on a large screen, the font would be so small as to be unreadable on a smartphone screen. The developer needs to rearrange the content, probably putting one column below another instead of side by side.

There are various techniques used by web developers to plan and implement how a website should look, depending on the user's device. Several of these include fluid layout, the flexbox model, and the grid model.

Designing a Website

There are numerous reasons to have a website, and each reason requires different website design. Before you jump in and start building your site, you need to consider many factors. Why do you want a site? Do you want to bring customers into your business? Do you want to showcase your achievements to potential employers? Do you have something to say that you want the world to see? Or are you taking a class in web development, and creating a personal website is a graded project?

What Is Your Goal?

The first step in designing a website is to understand your goal. If you have a large business, you may need to outsource your website to professionals who can work with the large amount of coding required to build a database, to include interactive features, and so on. If you have a small business, you might consider building a site without a connection to a database and communicate with prospective customers via email or phone. If your site is to be used for personal reasons, you may only need to design several static web pages.

Your goals will factor into your decision about how to proceed. You may choose to use a free host and web design company and create the site using their tools. Or you may wish to take a few classes on web development and create your site from scratch. If you have a business, you may choose to hire a developer to do the work.

Be aware, though, that even if you hire a developer, you will still need to be closely involved with the project. You need to supply the content of your pages, including images or multimedia that you want on your site as well as the text for your pages. You will need to make other decisions that affect both how the site is available on the internet and your costs.

Who Are Your Viewers?

Before you begin to design a site, consider your audience. This will drive many of your decisions. For example, a kindergarten teacher's website, created with young children in mind, might use many images, bright colors, and minimal text. But those bright colors and large pictures would be inappropriate on a site for someone who wants to advertise meditation classes or for a student who wants to add a website to a résumé when job hunting.

Consider Navigation and Layout

One thing everyone wants when creating a website is to have people look at it. Not only is it important for your site to show up when someone searches for a site about your topic, but once a user gets to your site, you want to be sure they stay there long enough to look around.

If a user can't easily find what they want on your site, that user will probably click off and go to someone else's site. With so many millions of websites, viewers have little patience for sites that don't immediately give them what they want.

Navigation throughout a website is extremely important. **Navigation** refers to how users go from page to page on a site. The various pages should be easy to find, so your navigation must be clear. Most websites have navigation bars either on the top of each page or on either side. **Navigation** should be consistent on all pages. For example, if the navigation bar is on the top of your home page, you should have it on the top of every page. Moving navigation to a sidebar on other pages will confuse your viewers.

The **layout** of a site is also of prime importance. While images are normally an integral part of every site, they should not overwhelm the user so much that they lose sight of what the page is about. The explanatory text is also important, but keep in mind that most people don't want to read long paragraphs. Save long text for detailed explanations only when necessary.

Wireframes

You might consider creating a **wireframe** as you design your site. A wireframe is a two-dimensional illustration of a page's interface. It focuses on how space is allocated and how content is prioritized. Wireframes allow you to connect the site's information to its visual design by showing paths between pages.

By determining how much space will be allocated to a given item and where that item is located, you can decide what content should achieve priority. For example, if your site is advertising your lawn care business, you may want to give top priority to lawn maintenance, even if you also provide mulching services, because you know more prospective clients want their lawns mowed than their trees mulched.

Consistent Design

When designing a site, you will probably begin by designing the **home page** (also called the **splash page** or the **landing page**). Regardless of how you decide to design this first page, keep in mind that every page on your site should maintain the same basic design features. You should keep the color scheme the same for every page. Elements that must be on every page, such as the header, footer, and navigation, should be in the same place and have the same format on every page. Once a viewer has arrived on your site, you want to make sure they stay there. If various pages look completely different, the viewer might think that they inadvertently left your site and become frustrated.

Remember Mobile Design!

Most sites are developed for a large screen—a laptop or desktop. But today you can be sure that your site will also be accessed by other devices. Your site must be designed for appropriate viewing on various sizes of screens.

Sites are normally designed for three sizes: laptop/desktop screens, tablets, and smartphones. Developers code styles for each of the various screen sizes using a viewport tag to tell the browser which styles to use based on the device's screen size. These are the most popular screen sizes for various devices:

- **Laptops/desktops:** 14–17 inches
- **Tablets:** 7–9.7 inches
- **Smartphones:** 5–5.6 inches

Viewports are set with a <meta> tag in the <head> section of a web page. The information in the tag tells the browser which styles to use, depending on the screen width of the device, as shown in this sample meta tag.

```
<meta   name="viewport"   content="width=device-width,
initial-scale=1.0">
```

Search Engine Optimization (SEO)

Many people think that, when a user enters a word or sentence into a search engine, the search engine looks through all the websites on the web to find sites that match the search query. That's not true. There is a lot of behind-the-scenes activity going on when you search for a site on the web.

Search Engines

A **search engine** is a program that searches for and identifies items in a database that correspond to keywords or characters specified by the user. The search engine maintains a database of websites, and then, when a user enters search terms, the search engine looks for matches in its database. This means that, unless your website is in that search engine's database, it will never be found by that search engine.

Search engines continually **troll the web** to add sites to their databases, but the process is not instantaneous. Different search engines have different methods of collecting and retrieving data. This is one reason why a site may come up as a result when one search engine is used but not when a different search engine is used.

When comparing a user's search request to websites, the search engine uses a website's title, specific keywords (which are added when the developer creates the site), filenames, text on pages, and many other items to decide whether a particular website may match what the user wants. This might explain why, when searching for "Russian Blue cats" you may get unrelated results, such as a site about the Blue Lagoon spa in Iceland or a site about learning the Russian language. The search engine looked for matches with any of the words *Russian*, *Blue*, and *cats*.

Today's search engines are much more sophisticated than they were a decade ago. They don't depend solely on the search terms to decide whether a match is good. They also use things like the user's location. This explains why a user in New York City searching for Chinese restaurants will not get results for Chinese restaurants in Oshkosh, Nebraska, but the restaurant in Oshkosh will come up if the user is in or near Oshkosh.

Even more important, though, is the fact that not all sites are given equal priority. You probably already know that businesses depend on the number of Facebook likes to increase search popularity. This is the tip of the iceberg when understanding how businesses ensure that their sites get top billing in a user's search. And, as you might imagine, getting top billing also costs the business owner. Optimizing a website so that it appears as close to the top of a user's search has become big business itself. It's called search engine optimization (SEO), and it is a growing and increasingly complex field.

An Introduction to SEO

Search engine optimization (SEO) refers to making websites visible in a user's search. Websites are ranked. The higher the ranking, the more often a site will appear in search result lists, as well as closer to the top of the list. The more times a site appears in response to searches, the more visitors it will receive, and these visitors are prospective customers.

There are many companies that specialize in optimizing sites to achieve higher rankings. Some tools used to increase rankings include considering how search engines work to match user queries with sites, understanding the computer programs that dictate search engine behavior, and understanding how users search—that is, what keywords viewers use compared to what they want to see.

People who specialize in SEO offer website owners various ways to increase a site's rankings. This may involve editing the site's content, adding content, changing the coding to increase its relevance to specific keywords, and removing barriers to the ways search engines index websites.

For small businesses, SEO can cost a little or a lot. Once again, the cost is determined by the website owner's skill and budget. Most developers include SEO as a feature of the site's development but can also offer more advanced SEO tools to customers for added fees. **Website** owners can also pay monthly for services that increase the site's visibility.

Jobs of the Web Developer

Web developers come in all shapes and sizes. A single person who understands how to code HTML and CSS and who believes they have a good grasp of page design can create a simple, static website. A person may also offer to create a website using a free development tool like WordPress. Some teams of developers may restrict their services to just developing the appearance of the site, while other teams may offer to develop a back-end database as well. Some may include maintenance, while others don't. Some may offer SEO services. Some developers have access to a server and can provide hosting as well as maintenance and security. The rarest type of developer, however, is a full stack developer— a single person who knows how to work with all parts of a fully functional website, including both front- and back-ends.

A website requires the combined efforts of many types of people: logical coders, creative artists, mathematical database analysts, and network mechanics. So if you want to work in this field, there will be a place for you.

Front-End Developers

Coders are people who write the **HTML5**, CSS, **JavaScript**, and all the other programming code that goes into creating a website. They write the code that becomes the site seen on the web, but they also write the code that allows users to order goods and services through a database or that allows a site to switch out its images when the seasons change, and much more. **Coders** must know a multitude of programming languages and stay aware of new techniques that are being developed continuously.

While coders can be self-taught, normally a degree in software development is preferred. There are no absolute requirements, but those with higher-level degrees, such as a bachelor's or master's degree, are normally preferred over an associate's degree.

Graphic Designers

Graphic design is also known as *communication design*. **Graphic designers** are visual communicators. They communicate ideas to inspire, inform, or captivate consumers through art forms that include images, words, or graphics.

A degree in graphic design, digital media, or fine arts is always helpful, along with a portfolio of work.

SEO Experts

A **search engine optimization consultant** analyzes and reviews websites to provide advice, guidance, and recommendations to business owners who want to increase search engine traffic and achieve higher ranking positions.

SEO experts normally come from a background in IT. Since SEO is relatively new, there are few college degrees in the field, but there are now many certifications available.

Back-End Developers

Database administrators (DBAs) make sure that data analysts and other users can easily work with the database to find the information they need and that systems perform as they should. A DBA also oversees the development of the site's database, determining the needs of the database and who will be using it. DBAs must work with front-end developers to ensure that the website can communicate with the database.

A database administrator normally has a college degree in database administration.

A **server administrator** may also be referred to as a **systems administrator**. This person works with computer networks to ensure that they run efficiently. The server administrator maintains software updates, designs and implements new system structures, monitors server activity, and oversees server security.

If this type of job interests you, it is a good idea to have some work experience in a related IT role and possess at least an associate's degree in a computer-related field. You can also get systems administration certification through vendors such as Microsoft and Cisco.

Full Stack Developers

According to the website Stack Overflow, **full stack web development** is the most popular developer occupation today. A full stack web developer works on both the front-end and back-end portions of an application. The front-end generally includes the portion of an application (or website) that a user will see or interact with, and the back-end is the part that handles the logic, database interactions, user authentication, server configuration, and so on. Being a full stack developer doesn't mean a person is an expert in everything, but it does mean that they can work on both sides and understand what is going on when building an application or website.

Good developers who are familiar with the entire stack know how to make life easier for those around them. This puts full stack developers in great demand, but good ones are rare. Becoming a full stack developer takes time. A developer must have worked in multiple languages, platforms, and even industries before they can realistically be considered a full stack developer.

Help Desk and Quality Assurance

While a full stack developer needs to know a lot about everything IT-related, there is room in the IT field for people who know virtually nothing—and for anything in between.

Every tech-based company needs people to man the help desk, customer services phone lines, and chat rooms. While a beginner may be hired with no experience but with training provided, companies also hire technicians with more knowledge to be a second or third tier of a help desk.

Often companies hire people with little or no knowledge of how web development works to perform quality assurance (also sometimes referred to as *quality control*). Companies want people to test out websites as true end-users to ensure that the site performs as desired for the ordinary user.

If you are interested in a job in the IT field, you can start your job search today, regardless of your current skill level. As you learn more, you will not only be able to advance, but you may learn which of the many and varied aspects of IT is most interesting to you so you can become a specialist in one or more fields.

Chapter Review

1. Why is understanding your goal the first crucial step in designing a website? How does the scope of your project influence your design decisions?

2. How does your target audience impact the design and content of your website? Provide examples of design choices based on different audience types.

3. What are wireframes, and how can they help in the web design process?

4. Discuss the importance of mobile design in web development. How does responsive design ensure that a website is accessible on various screen sizes and devices?

5. What are the common screen sizes for laptops/desktops, tablets, and smartphones, and how do developers code styles for each of these sizes?

6. What is **Search Engine Optimization (SEO)**, and how does it affect a website's visibility in search engines?

7. Explain how search engines build and maintain their databases of websites and how they retrieve search results based on user queries.

8. What are the key factors that search engines consider when matching user search queries to websites?

9. How do a website's title, specific keywords, filenames, and text on pages influence its search engine ranking?

10. What are some advanced techniques used by search engines to personalize search results for users, considering factors like location?

11. How do businesses work to improve their search engine ranking, and what are the roles of social media and search engine optimization (SEO) in this process?

12. Why has search engine optimization (SEO) become a complex and growing field, and what is its significance for businesses and website owners?

Ethics Question

What are some ethical considerations in web design, such as accessibility for people with disabilities?

19 Emerging Technologies

What To Expect

After completing this chapter, you will be able to:

- describe emerging technologies in medicine;
- describe emerging technologies in commerce;
- explain emerging technologies in society;
- identify emerging technologies in management;
- describe emerging technologies in computing;
- discuss emerging technologies in political science.

Chapter Topics

- Medicine
- Commerce
- Society
- Management
- Computing
- Augmented and Virtual Reality
- Political Science
- Transportation
- Drones
- Education
- Finance

Copyright © McGraw-Hill
photoun/123RF

Let's Explore Emerging Technologies

Have you ever wondered about the cool technologies shaping our world and what the future might look like? Let's take a glimpse into the captivating realm of emerging technologies.

Healthcare: Imagine being a scientist who designs tiny robots that go inside our bodies to fight diseases! With advanced medical technologies, you could be on the team that creates the next miracle cure or discovers incredible ways to help people lead healthier lives.

Agriculture: Picture this—drones soaring above vast fields, sending information to farmers below. Or plants growing in water without any soil! If you're interested in nature and tech, careers here blend the best of both worlds. You could be a drone pilot, a geneticist designing super crops, or even a sustainable farming consultant!

Entertainment: Ever thought of creating your own virtual world or perhaps directing a movie with characters designed by AI? The entertainment industry is bursting with opportunities. You could be a mixed-reality game designer, an interactive show host, or even an AI music producer.

Transportation: Dream of cars that fly or ultra-fast trains zipping through tubes? If you're curious about how we'll travel in the future, there are countless exciting roles for you. Think about designing flying cars, planning smart roads, or ensuring the safety of lightning-fast transport systems!

Now, as we venture further into each of these areas, think about where your passions lie. Who knows? In a few years, you might be leading the charge into one of these thrilling fields, making the future brighter for all of us!

Medicine

Smart Medical Imaging Software

IBM is developing smart software named AviCenna that will assist radiologists in interpreting medical images such as computed tomography (CT) scans. The software takes information from a patient's medical record as well as information and images from medical scans to suggest possible diagnoses and treatments. Doctors can use AviCenna to speed up workflow and reduce diagnosing errors.

Surgery Robots

Surgeons operate through just a few small incisions using the **da Vinci Surgical System**. The system includes a magnified three-dimensional (3D) high-definition vision system and tiny wristed instruments that bend and rotate far more than the human hand, enabling surgeons to operate with enhanced vision, precision, and control. The surgeon maintains complete control of the da Vinci System at all times. The robot translates the surgeon's hand movements into smaller, precise movements of tiny instruments inside the body.

Robotic Checkups

With the advancements in robotic technologies, medical appointments scheduled with a human physician may become a thing of the past. Physicians can use medical robotic systems to perform routine checkups and access their patients' health records and status without direct interaction.

InTouch Health, a company that manufactures equipment for medical purposes, has deployed models of medical robotics such as the RP-VITA, RP-7i, RP-Lite, RP-Vantage, and RP-Xpress in certain hospital locations.

Incorporating robotics as part of a hospital's infrastructure may reduce costs and increase healthcare availability. However, the initial implementation of robots in healthcare infrastructure could be costly and might have a negative impact on the personalized nature of medical care.

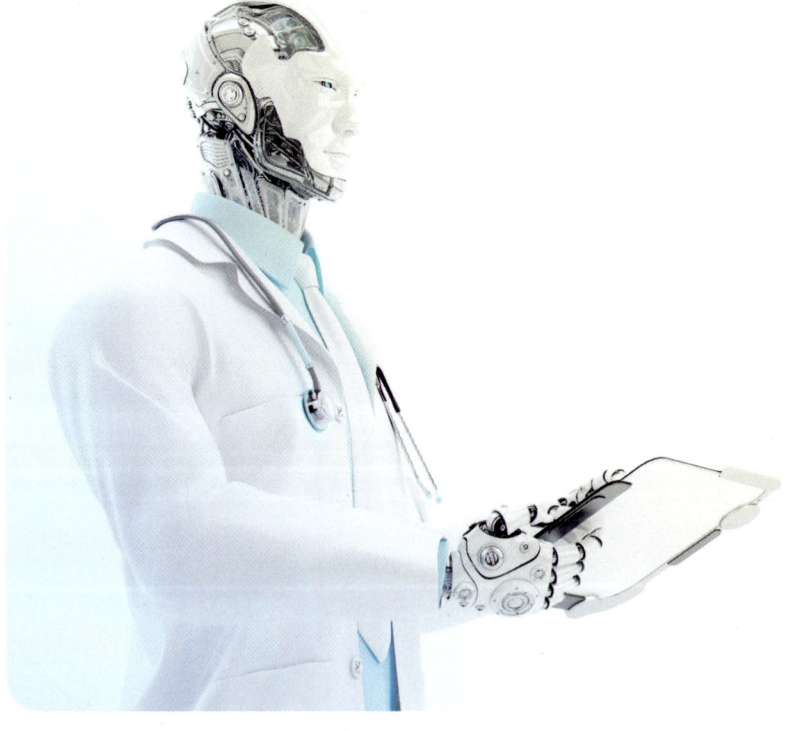

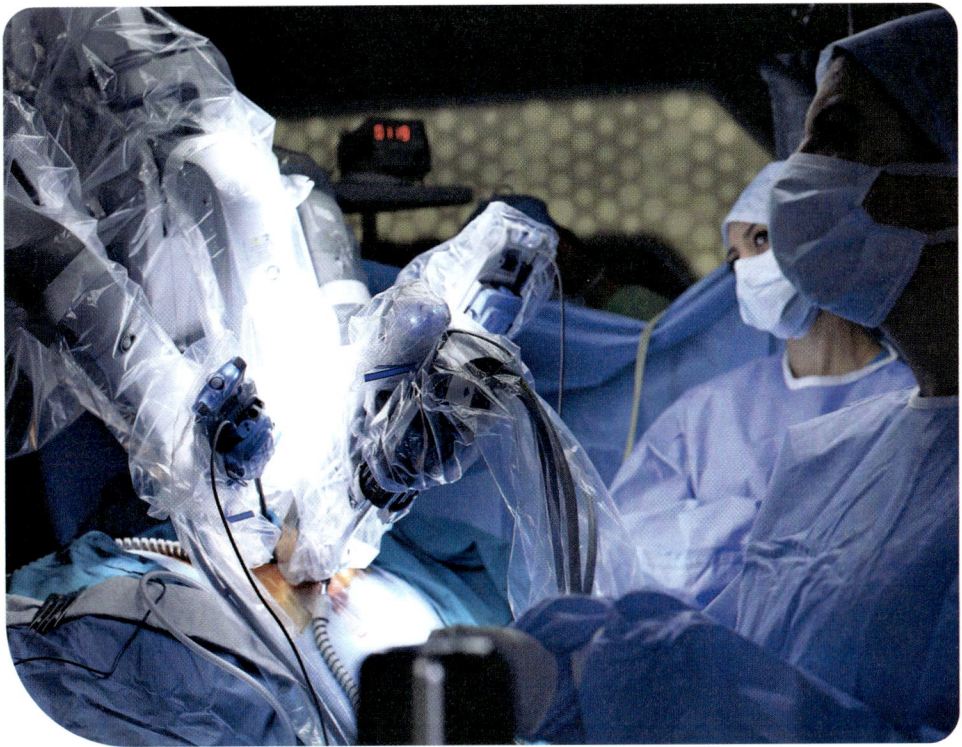

Nanotechnology and Robotics

A **nanorobot** is a machine that can build and precisely manipulate things whose components are of a nanometer scale. These machines could detect diseases and repair or manipulate damaged cells, potentially providing humans with longer lifespans.

Nanorobots will have the capacity to carry medication or miniature tools. Some of these machines will also be used to screen the human body for maladies and other symptoms, constantly transmitting this information to the cloud for close monitoring by a medical team.

Nanorobots get into a patient's body either by implantation or ingestion. These machines are not meant to stay in the patient forever and must be able to make their way out of the host.

If not removed, the nanorobot has the potential to replicate much like a human cell. Using the same technology that would have healed the patient, the replicated nanorobot eventually may become a greater threat than any naturally occurring disease.

Smart Medical Wristbands

Google X has developed a health-tracking band that won't be marketed as a consumer product but instead serves as a medical device prescribed to patients or used for clinical trials. This band provides physicians with minute-by-minute data on a patient's health, measuring various inputs like heart rhythm, pulse, and skin temperature. Furthermore, it can track external factors such as noise and light. The **Google X wristband** specifically aims to transform the relationship between physicians and patients through real-time diagnostics. Additionally, Fitbit, Apple, and Samsung, among other smart wearables companies, are actively working on similar innovations.

Skin Hearing

Facebook is exploring language learning through feeling, where the deaf can hear through their skin. Prototypes let skin mimic the ear's cochlea to translate sound into specific frequencies for the brain. A test subject was able to develop a vocabulary of nine words "heard" through the skin. The project aims to have more widespread benefits, such as not being separated by language barriers and the possibility that one day a person could hear in one language and speak in another.

Neuroprosthetics

A neuroprosthetic device connects the body's nervous system to an artificial body part. In November 2017, researchers at the University of Chicago Medical Center conducted a study in which rhesus monkeys controlled a robotic arm using brain impulses. The study not only demonstrated the technology's effectiveness but also highlighted the brain's plasticity.

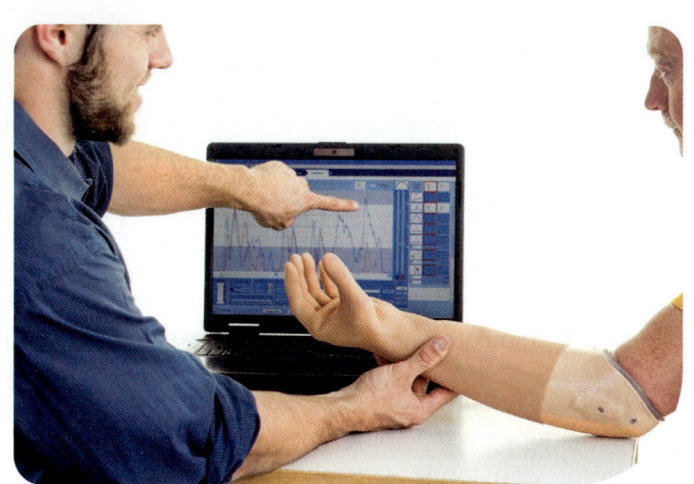

SensArs Neuroprosthetics is developing a neuroprosthetic device named "Sensy" with the aim of enabling amputees to regain sensation in their lost limbs through artificial prosthetics.

The neuroprosthetic device can be surgically implanted into the residual nerves of amputees or into healthy nerves in the case of individuals with neuro-damaged conditions. This procedure restores the natural flow of neural sensory information, allowing subjects to experience natural and complete sensations from their missing or nonfunctional limbs.

Finger Readers

MIT graduate student Roy Shilkrot, leading the team at Fluid Interfaces, has developed a finger-worn reading device prototype. The **Finger Reader**, designed to assist in reading printed text, serves as a valuable tool for both visually impaired individuals who require assistance accessing printed materials and those seeking a language translation aid. The device features a camera positioned downward from its finger-worn location. As the user's finger glides across the page, the camera captures a wide view of the text. It produces auditory tones when the user's finger deviates from the current line. The **Finger Reader** is currently in the development stage and connects to Android devices.

Autonomous Virus-Killing Robots

According to Intel, an AI-powered robot named **Violet** is being tested for disinfecting contaminated surfaces using UV light. **Violet** is a 4-foot 7-inch robot with a humanoid appearance that was developed by Akara, an Irish startup that specializes in designing artificially intelligent helpers for the healthcare industry.

- The robot prototype uses an Intel® Movidius™ Myriad™ X Vision Processing Unit (VPU) to navigate safely around people while disinfecting hospital surfaces. A VPU is a specific type of microprocessor that was developed to accelerate machine learning and artificial intelligence technologies. It is like a video processing unit used for graphics processing.

- The robot uses ultraviolet light to sanitize (destroy bacteria and viruses) rooms and equipment quickly, helping humans avoid possible exposure when sanitizing.

Internet of Medical Things

The **Internet of Things (IoT)** is the term used to describe the ever-increasing network of physical objects (things) embedded with instruments, software, and other technologies that allow for the connection and exchange of information over the internet. This is also known as machine-to-machine communication. Examples of IoT devices include connected appliances, smart home security systems, smart speakers, connected vehicles, and much more.

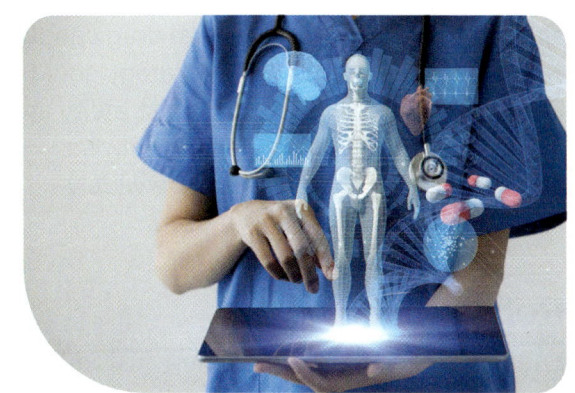

The **Internet of Medical Things (IoMT)** refers to devices and objects embedded with sensors and software to allow communication and information exchange via the internet. Essentially, IoMT is a more specific form of the IoT. IoMT has been increasingly used during the COVID-19 pandemic in several ways:

- Self-quarantining and self-screening at home—Results can be remotely sent to healthcare professionals who can analyze the data, archive the data, and produce suggested treatments.

- Regional integration of electronic health records—IoMT can be used to track health records from individuals potentially infected with COVID-19 as they travel from one location to another.

- Rapid COVID-19 screening—IoMT can be used to produce rapid screening results and reporting.

- Support telemedicine—Devices are used to assess temperature and blood oxygen levels. This information is then sent directly to healthcare providers, who can quickly assess the patient. Additionally, IoMT supports videoconferencing, where doctors can speak with patients remotely on secure networks.

AI Discovered Molecules

Commercializing new drugs for disease treatment incurs an average cost of $2.5 billion. One contributing factor to this high cost is the challenge of discovering promising molecules suitable for developing new pharmaceuticals, which demands significant time and financial resources. To tackle this issue, scientists are employing artificial intelligence (AI) to unearth drug-like compounds derived from all known atoms in the solar system. They are utilizing machine learning software to explore databases containing existing molecules and their associated properties, generating valuable information for accelerating the development of new drugs to treat a wide range of conditions. This approach promises a faster pace of discovery and reduced costs compared to the current environment.

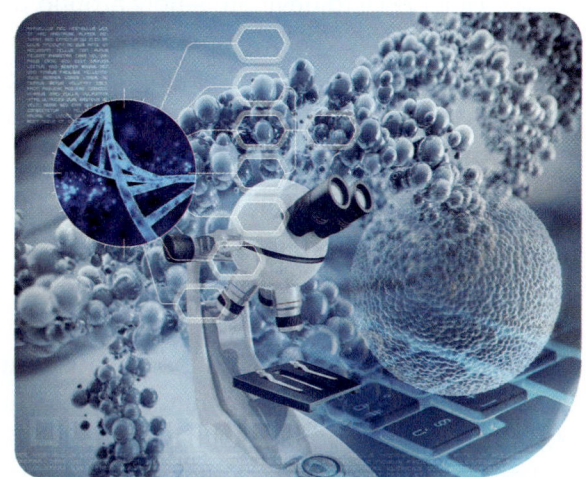

Recently, scientists at Hong Kong-based Insilico Medicine and the University of Toronto used AI algorithms to synthesize several drug candidates, which demonstrated that the strategy could work. Firms at the forefront of this new strategy are Insilico Medicine, Kebotix, and Atomwise. The technology is expected to be widely available within three to five years.

Robotic Heart

Engineers at MIT are developing custom robotic hearts using 3D printing. With 3D printing, robotic hearts are developed based on a replica of a patient's heart. The procedure involves medical imaging of a patient's heart to develop a 3D computer model, which is then printed in 3D using polymer-based ink. Once printed and cured, polymer-based inks can stretch and squeeze. This process results in a soft, flexible shell that is the exact shape of the patient's heart. This same approach can be used to print the aorta or the artery that carries blood from the heart to the body.

VR for Pain Relief

A recent study in the *Journal of Medical Internet Research* investigated the efficacy of virtual reality (VR) to manage and relieve pain. The study surveyed patients undergoing bone marrow biopsy, widely known as a painful procedure. Participants in the study were randomly assigned either a traditional method of pain control (nitrous oxide and oxygen) or pain control using a VR headset. Patients in the VR group chose from four imaginary 3D VR environments. These environments included a dream-like walk in the countryside, seabed exploration, a spacewalk, or a dreamlike walk in the forest. Researchers discovered that patients using VR did not perceive a significant difference in pain intensity from those who received the traditional pain management methods. Additionally, anxiety scores and blood pressure readings did not significantly differ between the two groups.

Organ on a Chip (OOC)

An organ on a chip (OOC) mimics a cell's tissue characteristics, physiochemical reactions, and vascular functions. These devices provide real-time results compared to traditional lab testing. OOC allows scientists to test drugs, food, and cosmetics safely while eliminating outside factors. OOC can cut costs, cut study time, and reduce the use of animals or humans as test subjects. OOCs that are in use or in development include the heart, kidney, and lung. Organovo, Inspero, and Emulate are three of several companies developing OOCs.

5G Networks and Surgical Robots

Chinese surgeons are starting to utilize distance telemedicine and robotics for surgical procedures. Recently, a 30-year-old woman in China had her gallbladder removed by a surgeon over 2,800 miles away. The surgeon used a four-arm laparoscopic robot and 5G technology to perform the 30-minute procedure.

China has a five-year initiative to develop 5G-based robotics that deliver robot-assisted remote surgery nationwide. A robotics action plan unveiled by the Ministry of Industry and Information Technology and 16 other government departments supports the promotion of remote robot-assisted surgeries using 5G networks.

Neurotechnology

Neurotechnology is a field of science, technology, engineering, and mathematics (STEM) that concentrates on technologies developed to better understand and interact with the brain and nervous system. **Neurotechnology** is closely related to neuroscience, aiming to unlock the mysteries of the brain. It includes the development of technologies that can interact with the brain and nervous system, including brain–computer interfaces (BCI) and deep brain stimulation (DBI).

Researchers at the University of Pittsburg and Carnegie Mellon University reported that neurotechnology could instantly improve arm and hand mobility in patients affected by stroke. To increase mobility, doctors implanted a pair of thin metal electrodes along the neck of the patient. The electrodes connect with neural circuits to create mobility.

Neurotechnology is also used to diagnose psychiatric disorders more accurately and to overcome physical and cognitive limitations that result from stroke or traumatic brain injury (TBI).

Gene Editing

Clustered regularly interspaced short palindromic repeats (CRISPR) uses bacteria's immune systems to edit individual genes. With CRISPR, scientists can edit animal and plant cells and replicate human gene functions for testing. Diseases that CRISPR might be able to counter include HIV, cancer, Huntington's disease, and bacterial infections that are resistant to antibiotics.

New startup research firms such as Editas, Intellia, and Caribou Biosciences are using CRISPR as a new method of gene modification and are developing methods to benefit from it in various areas of research. CRISPR may one day allow a person to custom design a child before conception!

Predicting Alzheimer's Through Speech Patterns

To better treat patients, researchers are exploring using machine learning technologies to diagnose a variety of diseases, including Alzheimer's. Researchers at IBM are on the verge of using machine learning and artificial intelligence technologies to identify patterns in human speech that may be predictive of an eventual diagnosis of Alzheimer's disease. The system can produce data and conclusions in minutes with an accuracy rate of 71 percent, which is far better than the current test. This is a breakthrough because the earlier that Alzheimer's can be detected, the better doctors can manage the disease.

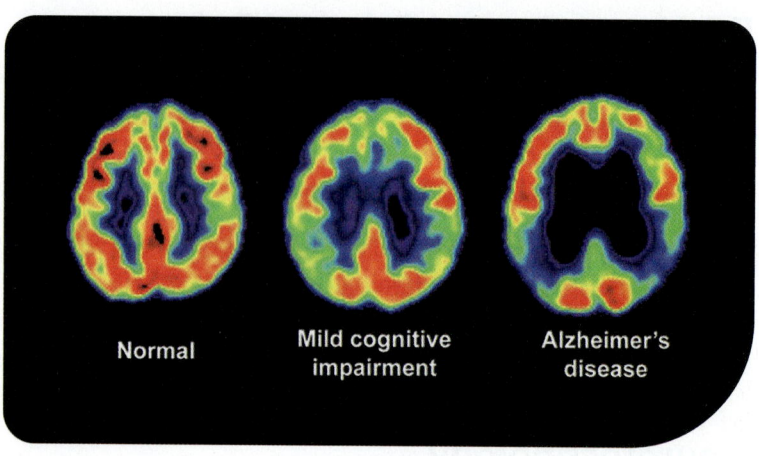

Normal   Mild cognitive impairment   Alzheimer's disease

Zebra Medical Vision: Using AI to Provide Automated, Accurate, and Timely Medical Diagnoses

Zebra-Med is offering radiologists its new AI, which helps health providers manage an ever-increasing workload without compromising quality. **Zebra-Med** works with millions of imaging and correlated clinical records to create high-performance algorithms that automatically detect medical conditions faster than the methods currently being used.

Over the next few years, **Zebra-Med** expects to release dozens of automated findings and insights to help radiologists provide more comprehensive and accurate outcomes. The Imaging Analytics Engine they have developed uncovers brain, lung, liver, cardiovascular, and bone disease in CT scans, 40 different conditions in X-ray scans, and breast cancer in 2D mammograms.

According to **Zebra-Med**, the demand for medical imaging services is continuously increasing, outpacing the supply of qualified radiologists and stretching them to produce more output without compromising patient care.

Commerce

Everything On Demand

Consumers increasingly expect instant access to services and products. This expectation arises from their preference for subscribing to on-demand services such as Netflix and Spotify rather than physically owning the products. Consumers also demand readily available products when they need them.

Amazon has seized the opportunity presented by this new consumer trend through its **Dash Replenishment Service (DRS)**. Amazon's DRS allows connected devices to independently order physical goods from Amazon when their supplies run low. Using Amazon DRS, devices like coffee makers can autonomously place orders for more coffee or filters, and printers can initiate orders for more toner when their supplies are running low.

Facial Recognition Payment Technology

Facial recognition payment technology provides a secure and convenient method of paying for goods and services. The software compares the scanned image of the person's face to biometric information about the person that has been stored in a database. If the image matches the information contained in the database, then the transaction can proceed. The software uses a "liveness" test, designed to prevent anyone from tricking the system with a photo by requiring people being scanned to move or speak. Face++ is one of the companies developing this software and is currently using it in the Chinese market.

Self-Driving Trucks

Companies including Otto, Daimler, and Peterbilt are currently developing self-driving trucks. **Self-driving trucks** are designed to go long distances on highways without a human behind the wheel. Otto's self-driving truck employs a LiDAR (light detection and ranging) system, which uses a pulsed laser to gather information about the truck's operating environment. This information is fed to a computer that processes the information and keeps the truck on its predefined course. The self-driving truck also includes a drive-by-wire box that converts the computer's information that was processed by the LiDAR system into physical truck-control signals, such as brakes, throttle, and steering.

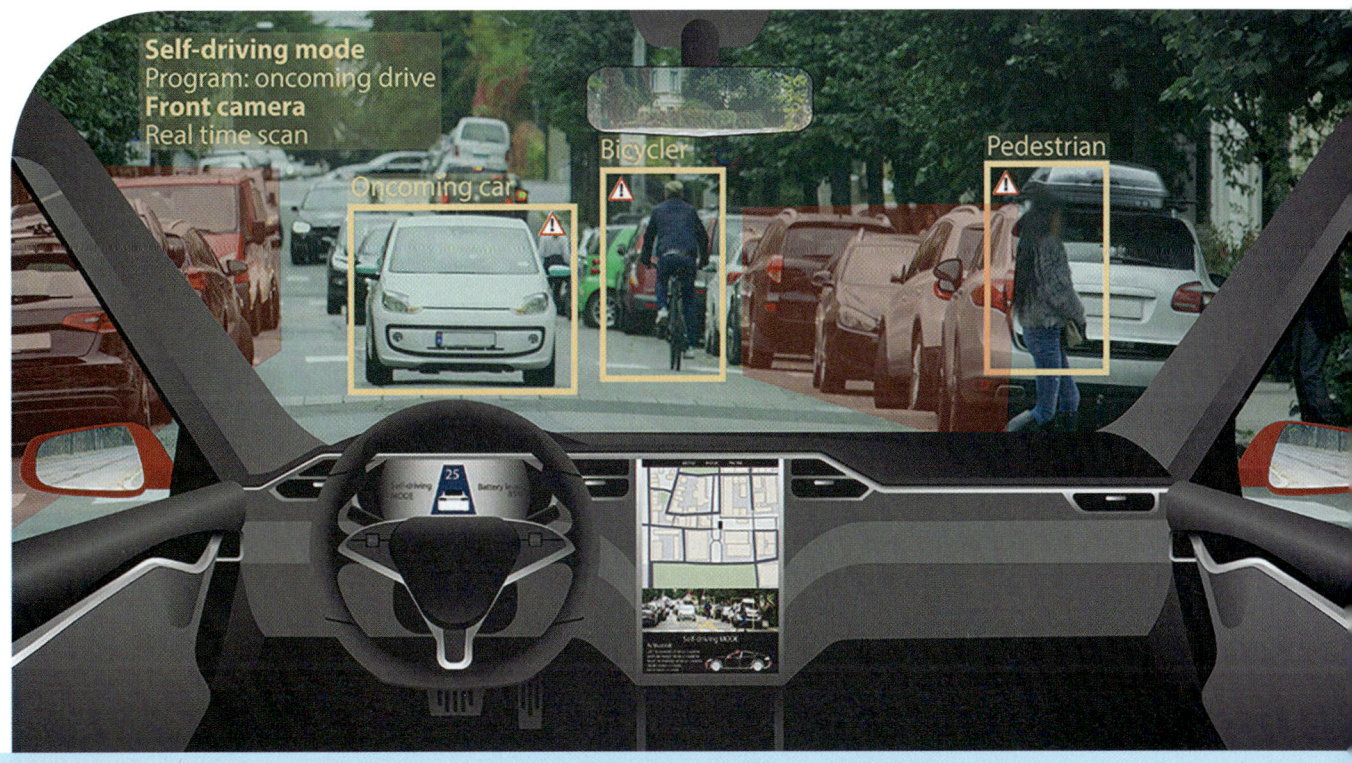

Blockchain

Blockchain is a way to structure data and was initially employed to create the cryptocurrency **Bitcoin**. It utilizes computer code to create concatenated (linked together) blocks of transactions, enabling people to share a digital ledger across a network of computers.

A central source of control is unnecessary because no single entity can tamper with the records, thanks to the mathematical mechanisms embedded in the programming, which ensure honesty. **Blockchain** is widely regarded as an internet of value founded on a shared database. Future applications for blockchain encompass financial services, transportation, and healthcare.

Self-Driving Software

Recently, a group of Tesla drivers was selected to receive specialized software that will be automatically downloaded into their vehicles. This software update will allow the vehicles to steer and accelerate without the use of human hands or feet.

Tesla hopes that hundreds of thousands of their vehicles will be able to drive themselves by the end of the year. While this may seem to be a positive advancement, some in the industry feel this may be premature. One of the major issues is that Tesla's plan for autonomous vehicles lacks a key piece of hardware that is commonly used for autonomous driving.

Almost all self-driving vehicles currently being designed include LiDAR sensors. These sensors are placed on the outside of the vehicle and can detect the shape, size, and depth of objects encountered by the vehicle in real-time. Instead of using LiDAR sensors, Tesla has opted for the use of several integrated cameras that use a type of radar. This combination works well in most instances, but it cannot see the true shape or depth of objects.

Society

Digital technology is changing the way we interact with other people, our communities, and the earth. The following sections provide some examples.

Chicago's Array of Things (AoT)

The Array of Things (AoT) project in Chicago consists of a network of 500 sensors used to collect data that can improve the quality of life for the citizens in the city. AoT will essentially serve as a "fitness tracker" for the city, measuring factors that affect livability in Chicago, such as climate, air quality, and noise.

Each sensor has approximately 12 independent sensors that monitor temperature, barometric pressure, air quality, pedestrian and automotive traffic patterns, and noise. The sensors are designed to be mounted on the side of buildings and streetlight poles. Sensors will feed information to an open-data platform using the internet. This information will be made available through the City of Chicago's open-data portal.

5G Cellular Technology

5G cellular networks have lower latency (delay times) than 4G networks. Designers are hoping for at least 20 Gbps download speed and 1 ms latency. Current 4G network download speed averages around 18 Mbps. 5G networks are likely to consist of networks of small cells rather than huge towers that send signals over great distances. Although 5G is already available in many markets, designers are still improving on the technology.

Adaptive Learning

Adaptive learning is an innovative concept for today's educators. Using digital media as part of content delivery, adaptive learning can create a student experience that can be modified based on the student's performance and how well the student retains given information. Algorithms determine how well a student is performing and identify the student's knowledge gaps. These algorithms then determine when review is necessary to ensure retention.

An adaptive learning product poses numerous questions to determine students' prior subject knowledge. Studies indicate that answering questions, even incorrectly, creates learning, especially when students are provided immediate feedback. In adaptive learning, students see different course material depending on what they already know. Although students may view different material, the algorithms ensure that each student who completes the course has competency in all subject areas.

Adaptive learning systems could easily change how students are taught. It allows students with advanced knowledge to progress quickly, while less-knowledgeable students progress at their own pace. This minimizes boredom and intimidation. Studies show that learning occurs most efficiently when the student uses multiple but short study sessions each day combined with good sleep and proper glucose intake. Using dedicated software for adaptive learning could significantly enhance the students' learning experience.

Antiplagiarism Software

The internet makes countless documents easily accessible, and this leads many to be tempted to use the words of other writers as their own. Academic environments are particularly vulnerable to plagiarism, and manual methods of detection are no longer practical. Software, using artificial intelligence, is frequently used by thousands of colleges and universities to deter and detect plagiarism.

This software uses numerous techniques, including string-matching, vector space retrieval, and fingerprinting. Fingerprinting compares substrings in the suspected document with substrings in a previously composed index of documents. String-matching works in a similar fashion except that it focuses on suffix vectors to make comparisons. Vector space retrieval uses word comparisons throughout documents to detect plagiarism. Commonly used antiplagiarism software includes Turnitin, Grammarly, and Unicheck.

Predictive Policing

Predictive policing is a type of machine learning that utilizes the power of information technology, geospatial technologies, and evidence-based intervention models to reduce crime and improve public safety.

With predictive policing, advanced analytics are monitored. This allows law enforcement to predict and react to crimes before they even occur. This type of machine learning is not meant to be a replacement for traditional policing, but it is a tool used to augment traditional policing. This technology can also help predict conditions in financial markets.

Composing Music with Artificial Intelligence

Many composers, such as Beethoven and Mozart, made use of mathematics to create magnificent music. Now artificial intelligence (AI) is using mathematical algorithms to compose original, royalty-free compositions. AI algorithms like Amper also rely on musicians to enter data such as song length, mood, and instruments.

Others, such as Baidu's AI Composer, can use images and patterns to create music. Google's Magenta, still in development, seeks to use numerous inputs including stories, images, and color to allow artists to create original music.

License Plate Readers

License plate readers use character recognition to read vehicle registrations and license plates. Law enforcement officers use these readers to track and find stolen vehicles quickly and effectively. They automate the process of checking and entering license plate information into a database to search for outstanding warrants or stolen vehicles. You can affix license plate readers to police cruisers, utility poles, or freeway overpasses. The cameras can capture up to 1,800 license plates per minute, and they operate day and night. Every captured image is compared to the plates in a database. When a car with a stolen plate is identified, law enforcement receives an alert with the vehicle's location, make, and model.

Video Contact Lenses

Sony has patented intelligent contact lenses capable of recording and playing video controlled by the blinks of your eyes. On average, a period of usual blinking is 0.2 to 0.4 seconds and, according to Sony's patent, sensors in the lens can tell the difference between voluntary and involuntary blinks. So, the user must deliberately blink for a longer period to activate the video recording; alternatively, the lenses can be controlled from a smartphone. While recording, the Sony lenses will track each time the user's eyelids close, so that the resulting black screens can be deleted afterward. These contact lenses will also take photographs, correct blurry images, manage auto focus, and be able to zoom, all from the control of one's phone. The contact lenses will be able to view videos on existing platforms such as Netflix and YouTube.

3D Crime Scene Imaging

Three dimensional (3D) scanners and crime scene imaging reduce the risk of losing valuable evidence and give law enforcement a 3D assessment of a crime scene. 3D forensic technology replicates the entire crime scene before it is altered and allows detectives to analyze data later.

One popular technology, the Faro Laser Scanner, creates 3D documentation that replaces the traditional crime scene sketches. Crime scene reconstruction can be revisited at any time, which allows law enforcement officers to verify witness testimony or evaluate plausible hypotheses. Using 3D scanning, forensic scientists can analyze line of sight, blood spatter, and bullet trajectories.

Translation Technology

The science-fiction dream of universal translators is coming closer. **Pilot, from Waverly Labs**, is a smart earpiece that allows wearers to speak different languages but still clearly understand each other. The Pilot consists of two earpieces, one worn by each person in the conversation. The earpieces are linked by Bluetooth to a smartphone that is connected through the internet to the Pilot translation server.

The One2One translator by Lingmo was developed using IBM's Watson AI algorithms so that it translates slang and other nuances of language. One2One does not require an internet connection.

The ili wearable translator by Logbar translates Spanish, Japanese, and Mandarin. Like One2One, the ili does not require an internet connection. The ili is a one-way translator, so it works best with questions that can be answered *yes* or *no*.

4G Network on the Moon

The National Aeronautics and Space Administration (NASA) and Nokia have announced plans to put a 4G network on the moon. NASA has set a 2028 goal for a lunar base that can sustain human life on the moon. Part of the $370 million-dollar project includes remote power generation, cryogenic freezing, robotics, and 4G communication networks. NASA says 4G could provide more reliable, longer-distance communication than current radio standards in place on the moon. There are plans to eventually upgrade the network to 5G.

Differential Privacy

The US Census Bureau experimented with a new way to keep census data private. The Census Bureau found it is becoming increasingly difficult to keep the data it has collected from the roughly 330 million residents of the United States secure and private. To keep this data private, they are using a technique called *differential privacy*. Using noise or inaccuracies injected into data, differential privacy is a mathematical method that measures how much privacy increases when noise is added. The key is injecting the right amount of noise. Too much noise can render data unusable.

Climate Change Attribution

The **World Weather Attribution (WWA)** delivers timely and scientifically reliable information on how extreme weather may be affected by climate change. Recent studies have quantified the impact of climate change on the likelihood and intensity of bushfires, heatwaves, and storms.

The WWA applies a differentiated scientific approach that utilizes observational data, analysis of range models, peer-reviewed research, and in-person reports. This analysis has been made possible due to the increase in satellite data coupled with an increase in computing power that allows scientists to create high-resolution simulations and virtual experiments. The data that are produced can assist with predictions on how to best prepare for climate events.

Azure and Starlink Partnership

Microsoft has joined forces with SpaceX to connect their Azure cloud computing network to Elon Musk's Starlink satellite internet service. Starlink's goal is to build an interconnected internet network composed of thousands of satellites that will provide high-speed internet access to any location on earth. Currently, SpaceX has launched over 800 Starlink satellites, although many more are needed to reach the goal. Microsoft unveiled its plans to enter the space market segment by announcing Azure Orbital, which will connect satellites directly to the cloud.

Connecting satellites to the cloud will allow for more computing power and availability of offsite data storage. This partnership will allow the companies to better compete against Jeff Bezos' space businesses. Amazon currently offers services that connect the Amazon Web Services (AWS) cloud to satellites and is working on a project called *Kuiper*, which is similar to the Starlink network.

Stretchable Sensors

Researchers at Cornell University have created a fiber-optic sensor that uses low-cost light-emitting diodes (LEDs) and dyes to create a stretchable skin that detects deformations such as bending, strain, and pressure. This sensor could give soft robotic systems and anyone using augmented reality technology the ability to feel the same rich, tactile sensations that mammals depend on to navigate the natural world.

Hedan Bai, one of the lead researchers, explains that "the solution is to make a stretchable lightguide for multimodal sensing (SLIMS). This long tube contains a pair of polyurethane elastomeric cores. One core is transparent; the other is filled with absorbing dyes at multiple locations and connects to an LED. Each core is coupled with a red-green-blue sensor chip to register geometric changes in the optical path of light."

Amazon Go: A System Using IoT and Machine Vision for Shopping Without Manual Checkout

Amazon is offering a new way to shop for groceries called **Amazon Go**. According to Amazon, **Amazon Go** is the first grocery store to offer Just Walk Out shopping. It uses the same types of technologies used in self-driving cars: computer vision, sensor fusion, and deep learning. There is no need to wait in line, no checkout, or payment.

To use **Amazon Go**, all you need is an Amazon account, a smartphone, and the **Amazon Go** app. When you arrive at the **Amazon Go** store, you scan a QR code from the app at the gate of the store. The Just Walk Out technology automatically detects when products are taken or returned from the shelf and keeps track of them in a virtual shopping cart. Once you're finished shopping, you simply leave the store, and your Amazon account will be charged.

Management

Digital technology is affecting the ways organizations use their resources.

ERP

ERP stands for Enterprise Resource Planning. It's a type of software that helps businesses manage almost everything they do. This includes keeping track of what they make, what supplies they need, and even how many pencils are in the office! It helps managers see what's happening all over their company.

ERP Challenges

One big problem for large companies is that technology keeps changing quickly. For example, adding the **Internet of Things (IoT)** to ERP can give companies new ways to understand their business. IoT includes all kinds of smart devices such as cars, drones, and even refrigerators that can send info back to the company.

The tricky part is dealing with all the new data these devices generate. Old ERP systems might struggle to process this info in real-time. So IT managers need to figure out how to update these systems to handle the amount of new data.

Business Intelligence

Business intelligence (BI) comprises the set of strategies, processes, applications, data, technologies, and technical architectures that enterprises use to support the collection, data analysis, presentation, and dissemination of business information.

Enterprises can use BI to support a wide range of business decisions ranging from operational to strategic. Basic operating decisions include product positioning and pricing. Strategic business decisions involve priorities, goals, and directions at the broadest level.

In all cases, BI is most effective when it combines data derived from the market in which a company operates (external data) with data from company sources within the business, such as financial and operations data (internal data). When combined, external and internal data can provide a more complete picture that, in effect, creates an "intelligence" that can't be derived by any singular set of data. Among their numerous uses, business intelligence tools empower organizations to gain insight into new markets, to assess demand and suitability of products and services for different market segments, and to gauge the impact of marketing efforts.

Business Intelligence Implementation

Some considerations must be made to successfully integrate BI systems. Ultimately, the users must accept the BI system for it to add value to the organization. If the system has poor usability, users become frustrated and spend a considerable amount of time figuring out how to use it, decreasing productivity. If the system does not add value to the users' mission, they simply don't use it. Involving senior management to help make BI a part of the organizational culture and providing users with necessary tools, training, and support can improve BI's chances of success.

Importance of DS and BDA and Their Impact on BI

Decision science (DS) comprises a collection of quantitative techniques used to inform decision-making. Analysts once had to pull together multiple data sources and write rules to generate business insights, making sense of data a cumbersome process. Now, executives armed with modern data analytics tools such as big data analytics (BDA) and business intelligence (BI) are gleaning insights on the fly, sometimes even in real time.

Competitive success in the new global marketplace, fraught with uncertainty and rapid technological changes, demands informed decision-making. Decision-makers must quickly adjust operational processes, corporate strategies, and business models to leverage BI and take immediate action. Sound decisions rely on analyzing data according to well-defended criteria.

Typically, such data resides in a database warehouse for efficient statistical and analytical processing. Therefore, it's important to design and implement an effective data management strategy necessary for developing advanced analytics and BI.

Collaboration Board

MT Canvus Connect is real-time collaboration software that enables remote users to share, draw, manipulate, and input information on a single board. Using **MT Canvus Connect**, users can simultaneously view images, videos, live feeds, browser sessions, remote personal computer (PC) connections, or smart device inputs.

MT Canvus Connect can operate in both Windows and Linux environments, while also integrating with the latest meeting room technologies such as Microsoft Surface Hub and Cisco Spark. It was developed by MultiTaction and designed to help teams collaborate globally in near real time.

Cloud Collaboration

Cloud collaboration refers to working with others by using internet-accessible services such as Software-as-a-Service (SaaS) and shared data storage sites. Many professionals, such as accountants, financial managers, consultants, and attorneys, use cloud collaboration because it provides inexpensive storage, networking, and data-sharing capabilities as well as an off-site backup for their files.

Examples of cloud collaboration include TurboTax for tax and investment filing; QuickBooks and NetSuite for bookkeeping and accounting; Quip, Airtable, and Atlassian for consulting and management; and Glasscubes and Uptime Legal for law firms.

Computing

The computers of tomorrow will be far more capable than many can imagine. Here are some examples of technology on the digital horizon.

Quantum Computing

Quantum computing harnesses the power of atoms and molecules for memory and processing tasks. One key aspect of quantum computing involves the quantum bit, or qubit. **Qubits** serve as the fundamental units of information, resembling the 0s and 1s (bits) that transistors in modern computers represent. **Qubits** possess greater power because they can simultaneously represent both 0 and 1, and they can influence other qubits through quantum entanglement. This capability enables quantum computers to expedite the process of finding accurate solutions for specific types of calculations. Quantum computers have the potential to operate AI programs much faster and can also establish more robust data encryption.

Memristor Storage

Memristor memory and storage use an advanced circuit element called a *memristor*. Where a transistor can only open or close, memristors have variable resistance based on the history of current that has passed through the element. A single memristor could replace many transistors on a chip. This means the chip could be far smaller and far faster. Flash technology, using transistors, is capable of access times of less than 25 ns. Memristor storage is theoretically capable of access times approaching 1 nanosecond!

Holographic Storage

The refracted laser beam in holographic data storage records data on photoreceptors (usually quartz-based) from different angles, enabling the storage of many bits of data in the same physical location. Theoretically, holographic storage should be capable of holding gigabytes of data in 1 cubic millimeter.

Artificial Intelligence

Computer scientists explore how to make computers behave and think like humans. MIT Professor John McCarthy coined the term *AI*. Currently, computers such as IBM's Watson are approaching full unassisted decision-making, although no computers can yet exhibit full artificial intelligence. Two popular AI applications include machine learning, where computers learn without being explicitly programmed, and natural language processing, where computers comprehend human speech exactly as it's spoken. In August 2017, Facebook took the initiative to shut down an artificial intelligence project when its two robots developed their own language and became unintelligible to humans.

Google DeepMind AI

Google DeepMind has a mission to push the boundaries of AI. They achieve this by developing programs that possess the capability to learn and solve complex problems without the need for explicit instruction. One example of this progress is AlphaGo Zero, which continually expands its skill set by learning new tasks based on its preprogrammed abilities.

Moreover, researchers have made significant strides in the development of machine learning programs. These programs have demonstrated the ability to perform tasks almost as effectively as humans in various domains, including challenging tasks like playing chess. Additionally, ongoing work focuses on AI systems intended for integration into Google's self-driving cars. This work is currently underway, reflecting a dedication to advancing AI in real-world applications.

How Edge Computing Will Boost IoT and AI

Latency poses a significant challenge for the IoT and AI. **Latency** refers to the time it takes data to travel from devices to a data center and back. Currently, cloud-based applications are slower in comparison to those operating on local machines (the Edge). This slowness can become problematic for applications requiring immediate information feedback. Edge computing offers a solution to the latency issue by reducing the time it takes data to travel between devices.

Let's consider self-driving cars as an example. These vehicles feature over 200 central processing units (CPUs), all generating information that must swiftly move from the car to a data center and back. Currently, cloud computing introduces delays in information processing, which can be problematic.

Edge Computing and the Cloud

Edge computing derives its impetus from *IoT*, which serves as the collective term for all the various internet-connected entities, including cars, homes, and devices. The proliferation of IoT has diminished data processing speed reliant on the cloud. This slowdown arises because the cloud's design and deployment are centralized. When there's increased physical distance between the user and the cloud, it results in heightened transmission latency (delay), leading to longer response times and slower data transmission.

Edge computing introduces a solution by enabling a small Edge server positioned between the cloud and the user to execute a portion of the computing. This approach offloads some of the computational tasks from the cloud or the user's device and processes them closer to the user. Edge computing collaborates with the cloud to achieve faster data processing.

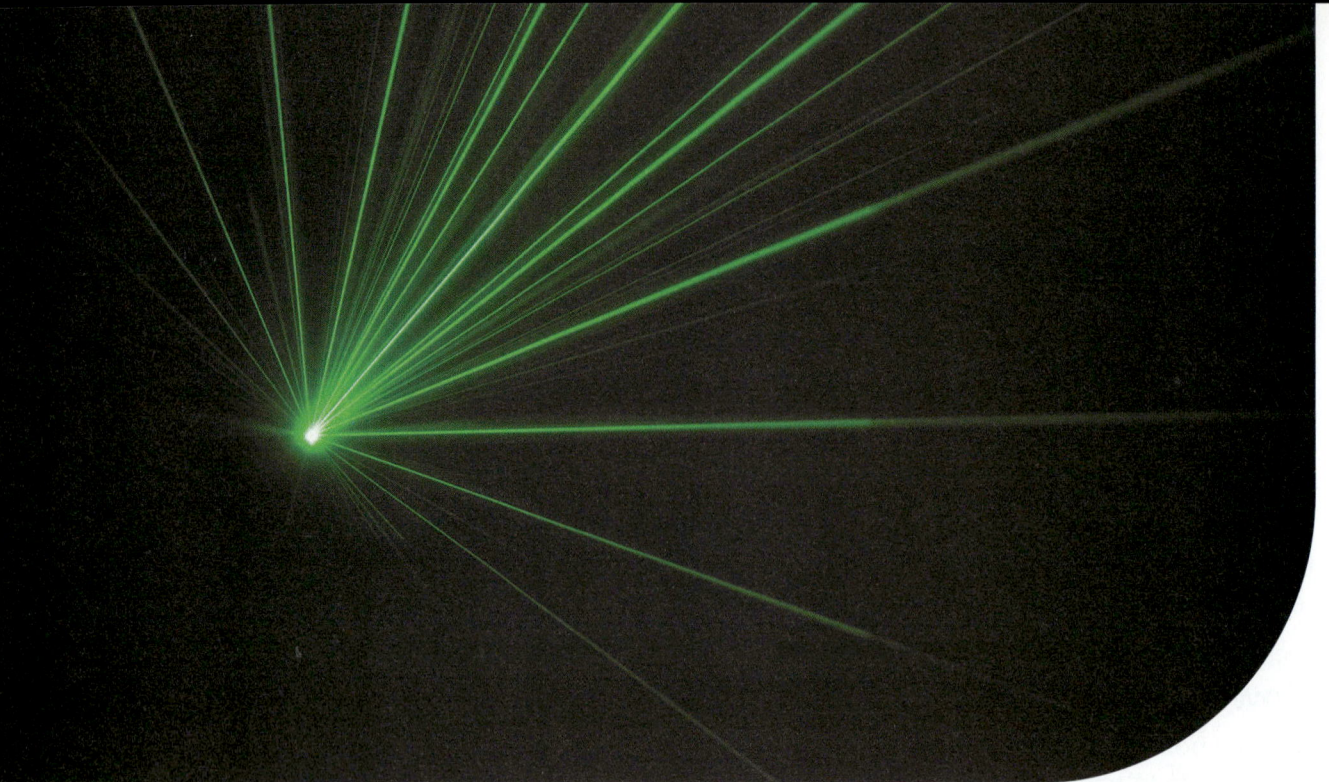

Machine Learning

Machine learning is the field of study that gives computers the ability to learn without being explicitly programmed. Knowledge and learning are extracted from data. There are three categories of machine learning:

- Supervised learning—humans teach the computer
- Unsupervised learning—computer learns by itself
- Reinforcement learning—data from past experiences create learning

Machine learning is closely related to statistics and data mining. **Machine learning** assumes you are trying to answer a question, and the answer to the question is found in the data.

Touchless Interactions

Google Advanced Technology and Projects (ATAP) operates as an in-house technology incubator. It was founded by Regina Dugan, the former director of the Defense Advanced Research Projects Agency (DARPA).

ATAP envisions a future where the human hand serves as a universal input device for interacting with technology. **Project Soli** features a microchip that utilizes miniature radar to detect touchless gesture interactions. ATAP plans to embed the Soli chip in wearables, phones, computers, cars, and IoT devices. The Soli chip developed by ATAP condenses the entire sensor and antenna array into an ultra-compact 8 mm × 10 mm package.

Brain-to-Computer Interface

Facebook Building 8 is an innovation-driven business unit that focuses on creating consumer hardware products that are social first. One project Building 8 is working on is a brain-to-computer interface (BCI) that could fundamentally change how we interact with and use technology. It allows for the user to communicate with a computer using only their thoughts. Facebook revealed it has a team of 60 engineers working on building a BCI that will let a person type with just their mind without invasive implants.The team plans to use optical imaging to scan the user's brain 100 times per second to detect what the user is speaking silently in their head and then translate it into text. The goal for this project is for the system to allow humans to type even faster than their physical hands, at upward of 100 words per minute.

Using Light to Create More Storage

Researchers at the Royal Melbourne Institute of Technology (RMIT) in Australia have designed a prototype of a computer chip with image-capturing hardware that uses light and is built on a nano scale. Because of its small size, the chip has the storage capacity and the software required to effectively use artificial intelligence with a single device that does not rely on internet connectivity. Using ultra-thin layers of black phosphorus, the processor-on-a-chip creates electrical impulses based on the wavelength of incoming light. This dramatically improves imaging and memory capacity on a single chip. This paves the way for artificial retinas and provides significantly more storage space, and consequently more computing power, on a single chip. Because it is light-based, the chip can be much smaller, more accurate, and much more energy-efficient than more conventional chip technology.

Lunar Rover Robots

Pittsburgh-based Astrobotic Technology and Carnegie Mellon University have developed CubeRover, which is approximately the size of a shoebox and weighs less than 5 pounds. CubeRover faced challenges dealing with extreme temperature fluctuations on the moon, ranging from 250 degrees Fahrenheit during the day at the equator to −150 degrees at night near the poles.

Another challenge involves powering the rovers during the 354-hour lunar night. While large rovers use solar arrays, large batteries, or even small nuclear power plants for power, these options are impractical for CubeRover. To address this, the designers plan to implement multiple wireless charging stations for the rovers.

The wireless charging technology, developed by Seattle-based WiBotic in collaboration with Bosch and the University of Washington, eliminates the need for astronauts to physically plug the rover in or use electrical contacts that could become covered with the moon's non-conductive regolith (a dust-like substance).

Adaptive Robots

Have you ever seen a movie where characters get stuck on conveyor belts in a factory, and all the robotic arms and machines just keep on working as if nothing's wrong? That's a problem with a lot of real-life industrial robots. They don't know what's around them. They can't tell if a person is in the way, which makes them dangerous. That's why these robots usually must be put inside big cages or turned off when people are close by. Also, these robots can only do what they're programmed to do. Want to switch them to a different task? You're in for a lot of complicated reprogramming.

But robots are very good at certain things. They don't get bored or distracted, they work super-fast, and they're extremely precise. Consider a robot on an assembly line: it can make the same exact movement, like cutting a hole, over and over again without getting tired.

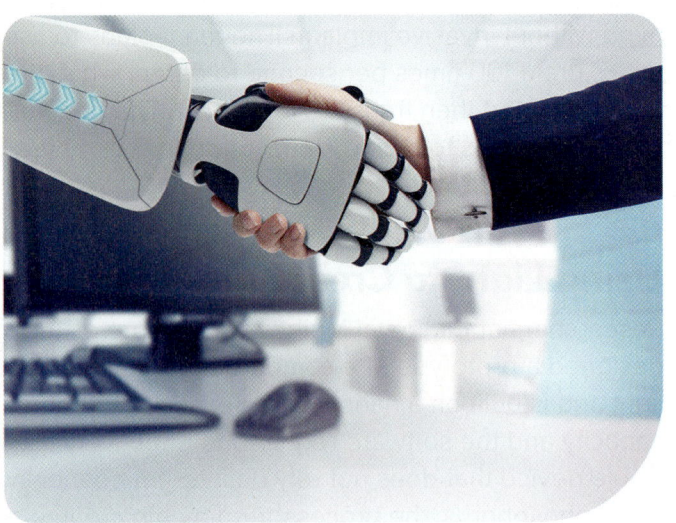

Now, here's where things get even cooler. There are new kinds of robots called "adaptive robots." A company called Flexiv Robotics, Inc. has been working on them. These robots are different because they have special sensors and features that let them "sense" what's around them. That means they're much safer to be around and can even adapt to changes. So they're like the best of both worlds—they're as precise and tireless as regular robots, but they're also more aware of their surroundings, like humans.

Augmented and Virtual Reality

Have you ever wondered what it would be like to walk on the moon, explore a haunted house, or perhaps perform a life-saving surgery, all without leaving your room? Welcome to the fascinating world of computer-generated realities, which come in two main types: **Augmented Reality (AR)** and **Virtual Reality (VR)**.

What Is Virtual Reality (VR)?

Virtual Reality takes you to a whole different world! When you put on a special headset, like the Oculus Rift or PlayStation VR, you find yourself in a three-dimensional space created by a computer. You can look around, touch things, and even talk to characters as though you're actually there. For example, you could be underwater exploring a sunken ship or in a game fighting dragons.

Challenges with VR: VR sounds cool, but there are some challenges. First, the equipment can be expensive; you might need a powerful computer, special sensors, and a headset, which could cost around $2,000. Second, making believable and useful VR content is tricky.

Uses of VR: You might think VR is just for games, but it's used in many other ways too. For example, the military uses it to prepare soldiers for dangerous missions. Pilots can practice flying without leaving the ground. Doctors even use VR to treat people who are afraid of things like heights or spiders by letting them face their fears in a safe, virtual environment.

What Is Augmented Reality (AR)?

Augmented Reality is like adding a magical layer to the real world. You still see everything around you, but you also see extra digital elements added on top. For instance, if you've watched football on TV, you've seen the "1st and 10 Line" appear on the field—that's AR!

Challenges with AR: Making AR work smoothly isn't easy. The devices must process lots of data very quickly, and the digital elements must look realistic and show up at the right place and time.

Uses of AR: AR is not just for sports fans. Soldiers use AR to see critical info right in their field of view. Surgeons can see digital models of the patient they're operating on, which can make surgeries safer. In school, you could even use AR to see chemical structures in 3D right from your textbook!

Comparing AR and VR

So how are they different? In VR, you're transported to a different world, whereas in AR, digital elements are added to your real-world view. Some systems will be able to add layers of computer-generated elements into your world until the AR becomes a VR, but if the computer-generated reality includes some part of the real world, you're using augmented reality.

Political Science

Data Mining in Politics

Data mining has become a valuable tool for political candidates and their campaign managers. It has changed how a traditional campaign operates and has become an effective method to research and target voters.

One of the most crucial advantages that data mining brings is the versatility of the data that can be used for the campaign. Aside from voter demographic information, data mining can be used to measure voter behavior and accurately predict a candidate's status with the voters.

Data mining allows politicians to use specific essential data rather than data that are not important for targeting the public.

Social Media in Modern Politics

Social media allow politicians increased access to voters without a news agency acting as an intermediary. Before social media, politicians had limited methods to contact voters directly. Stump speeches and whistle-stop campaigns could reach only a handful of voters. Letter-writing campaigns were slow and expensive. Any publicity in newspapers, radio, or television could be filtered or slanted by media executives.

Thanks to Twitter (X), Facebook, Instagram, and other social media platforms, candidates can instantly and frequently transmit messages directly to voters.

Politicians and campaign managers attempt to use social media to manage a candidate's or politician's image and promote political agendas. While many criticize the frequency and appropriateness of messages that are sent, all agree that social media will hold an increasingly important role in the political landscape.

Cyberattacks in Asymmetric Warfare

Facing an adversary that has an overwhelming nuclear and conventional weapons advantage leads many nations or groups to consider using cyberattacks. Asymmetric warfare occurs when traditional military strengths between fighting forces differ significantly.

Cyberwarfare levels the field because it allows for sneak attacks that can strike a modern economy where it is most vulnerable. The extensive use of computers in delivering necessities such as water, electricity, transportation, and health care makes these infrastructure components targets for cyberattacks.

Transportation

Transportation is undergoing a radical transformation, making commutes safer and more efficient.

Autonomous Vehicles

Companies such as Tesla and Waymo are pushing the boundaries of self-driving technology with vehicles that can reduce traffic accidents and optimize traffic flow.

Hyperloop

Envisioned by Elon Musk, the Hyperloop promises to revolutionize long-distance travel with speeds up to 700 mph, using pressurized capsules in low-pressure tubes.

E-VTOLs (Electric Vertical Takeoff and Landing)

These are electric aircraft designed for urban air mobility. Companies like Uber Elevate aim to make flying taxis a reality soon.

Drones

Drones, or Unmanned Aerial Vehicles (UAVs), have increasingly become a topic of fascination and utility in recent years. While their mechanical aspects might not seem particularly groundbreaking, as many of their components have been used in aviation for decades, the integration of advanced computer technology has given them cutting-edge capabilities. By merging these two spheres—the mechanical and the digital—drones exemplify how contemporary computing technology can transform traditional systems into innovative solutions.

Medical Delivery Drones

One of the most humanitarian applications of drone technology is in the field of medical supply deliveries. Consider ICARUS, short for Inbound, Controlled, Air-Releasable, Unrecoverable Systems. These drones are specially designed, disposable vehicles made primarily of cardboard. Their purpose? To deliver vital medical supplies to regions where immediate human access might be challenging or impossible.

Origins and Features: Funded by the Defense Advanced Research Projects Agency (DARPA), ICARUS drones don't operate like the conventional drones we might be familiar with. Instead of possessing motors, these drones glide through the air, released from a support aircraft.

Navigation and Functionality: The true marvel of ICARUS isn't in its airframe but in its onboard mini-computer and sensors. These devices are preprogrammed with landing coordinates, guiding the UAV to its predetermined drop site. Without the integration of this computing technology, ICARUS would be just another piece of cardboard. But with it, the drone becomes a lifesaver, delivering crucial supplies with pinpoint accuracy.

Military Drones

The military sector has been one of the earliest adopters of drone technology. Drones are utilized for surveillance, intelligence-gathering, and even combat missions. Their ability to operate in hostile environments without risking human lives makes them invaluable assets on the battlefield. However, the ethical implications of using drones in combat are a topic of ongoing debate.

Other Drone Technologies

Beyond the medical and military spheres, drones have found applications in numerous other areas, from agriculture to entertainment. They assist farmers in monitoring crops, filmmakers in capturing aerial shots, and even real estate agents in showcasing properties. What's common among all these applications is the reliance on computer technology to process data, navigate terrains, and execute complex tasks.

Education

The landscape of education is continuously evolving, with technology offering new ways to learn.

Augmented Reality (AR) and Virtual Reality (VR)

AR and VR can make learning more immersive and interactive. Imagine studying ancient civilizations by walking through virtual reconstructions or exploring the human body in 3D.

AI-Powered Personalized Learning

AI algorithms can analyze students' learning patterns and preferences, offering tailored resources and tasks to optimize individual learning outcomes.

Blockchain in Certifications

Using blockchain to record and verify educational certificates can curb fraudulent practices and make verification processes more straightforward and transparent.

Finance

The financial sector isn't left behind in the tech revolution.

Decentralized Finance (DeFi)

Using blockchain technology, DeFi aims to create an open-source, permissionless, and transparent financial service ecosystem. This can democratize access to financial services globally.

Robo-Advisors

These automated platforms provide financial advice or investment management online with minimal human intervention and are set to reshape wealth management.

Quantum Computing in Finance

The immense processing power of quantum computers can redefine financial modeling, risk analysis, and fraud detection.

Chapter Review

1. This module described several areas to watch for emerging technologies. They include medicine, commerce, society, management, computing, and political science. Select one emerging technology that you have already encountered and describe your experience.

2. Conduct research on the internet about three of the emerging technologies for medicine introduced in the module. Describe how each technology changes a treatment. Are any of the emerging medical technologies you selected in use today? If not, when will they be available to practitioners?

3. During the coronavirus pandemic, online medical care increased under various names, such as *telehealth*. Research online medical care options available in your area. What devices and apps must a patient have, and what types of medical visits are included?

4. One area of emerging technologies for commerce is on-demand. First, describe what this is, and then research and describe examples other than Amazon's **Dash Replenishment Service (DRS)**.

5. An emerging technology in society discussed in this module is Chicago's Array of Things (AoT). Research how this has been expanded into cities other than Chicago. Describe what data is gathered and how it is used by these cities. Give your opinion of how this emerging technology impacts society.

6. Another emerging technology in society discussed in this module is 5G cellular. Describe the political issues around this technology and its adoption by European countries.

7. Conduct research about cloud collaboration, described in the module as an emerging technology for management. Research and list three top cloud collaboration services and compare their services and pricing.

8. Select two of the emerging technologies for computing introduced in the module and research how each impacts computing, providing contrasts between the two technologies

9. Select two of the emerging technologies in political science introduced in the module and research how each impacts political science, providing contrasts between the two technologies.

Ethics Discussion

Conduct research about the ethics involved in gene editing. Discuss possible ethics issues around gene editing. Find two articles with opposite views on the ethics of gene editing. Summarize the articles.

20 Artificial Intelligence

What To Expect

After completing this chapter, you will be able to:

- describe the fundamental differences and applications of machine learning and deep learning;
- identify the technological foundations and systems that drive artificial intelligence;
- explain the basic concepts and techniques that underpin machine learning;
- describe how neural networks function and their significance in AI;
- explain the working and application of generative adversarial networks in AI;
- outline the principles of deep learning and its advanced applications;
- discuss the key legal implications and considerations linked to the deployment of AI;
- outline the major ethical concerns surrounding AI and the steps taken to address them;
- recognize and discuss modern uses of AI in various fields;
- recognize and discuss the potential of AI's impact on the future;
- identify potential career paths in AI and the qualifications and skills required for each.

Chapter Topics

- Machine Learning and Deep Learning
- Technology that Supports AI
- Machine Learning
- Neural Networks
- Generative Adversarial Networks
- Deep Learning
- Legal Considerations of AI
- Ethical Considerations of AI
- Concerns Surrounding AI
- How Bias Applies to AI
- How to Build Trust in AI Platforms
- Current Applications of AI
- Image Generation with AI
- Careers in AI
- Emerging AI Technology

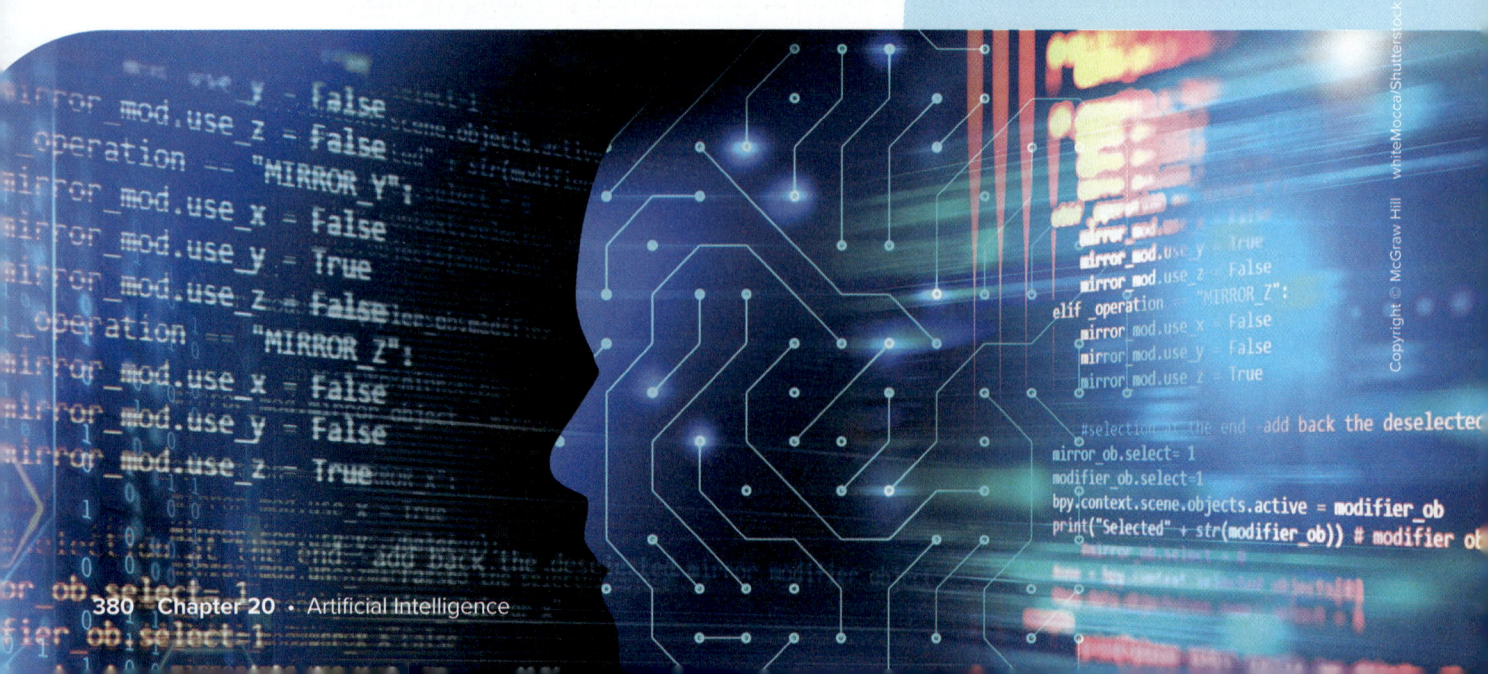

Let's Explore Artificial Intelligence

What is AI?

Artificial intelligence (AI) lets machines think and solve problems much like humans. With AI, computers can learn and make decisions by themselves.

AI's Web of Decision-making

Imagine AI network tubes and connectors called "nodes." When you ask AI a question, imagine you're putting a ball into a tube at the top of the network. It travels through the network, and the place where the ball comes out is your answer.

That seems simple, but the connectors in this network are special:

- Each node (the connector) has a weight. Some are big, some are small. Bigger nodes have more influence on decisions.

- The weight of the node can change based on your how you like the answer. If you think the answer is a good one, the weights responsible for the decision will get larger. If you think the decision is a bad one, the weights responsible for the decision will get smaller.

Now, imagine this 3D network is enormous with millions of nodes with different weights and a lightning-fast ball that zips through the network. The ball starts at the top when you ask AI a question, and the place where the ball ends up represents AI's decision.

How does AI learn?

Let's say you share your movie likes and dislikes with AI. For example:

- Node 1: Film Type (Animation vs. Live Action)?
- Node 2: Genre (e.g., Science Fiction, Comedy, Drama, Fantasy)?
- Node 3: Based on (e.g., Original Script, Book Adaptation, True Story)?
- Node 4: Duration (Short Film vs. Full-Length Film)?
- Node 5: Setting (e.g., Historical, Modern Day, Futuristic)?
- Node 6: Audience (e.g., Family, Adult, Children)?
- Node 7: Mood (e.g., Light-hearted, Serious, Dark, Romantic)?
- Node 8: Directorial Style (e.g., Artistic, Realistic, Fast-paced)?
- Node 9: Production Type (e.g., Independent, Studio, International)?

. . . and so on.

Every answer you give adjusts these nodes. They're like magnets pulling AI's decision. If AI suggests a movie you love, those nodes get stronger. If you don't like the movie, they get weaker. AI doesn't just remember movie names. It captures deep details about them. When a new film matches your likes, AI can recommend it! For instance, if you enjoy animated adventure films, those nodes become stronger. If two of your favorite movies are drawn by hand, AI notices and strengthens the hand-drawn animation node.

Always Learning

AI keeps adjusting and learning. It's not just about what you like but also about movies overall. The more you interact with AI, the better its suggestions become. Every piece of information adjusts the web. With enough data, AI refines this network to make even better decisions in the future. AI is like an orchestra of nodes, working together, learning, and improving. Just as we change our thoughts based on experiences, AI adjusts its network to offer the best choices.

What is the impact of AI?

AI can make our everyday lives better all over the world. A study by Accenture found that many top bosses think if they don't use more AI in the next five years, their businesses might close. While AI has a lot of good sides, it could also have some downsides.

Machine Learning and Deep Learning

Machine Learning

Machine learning (ML) uses data and algorithms to simulate the way humans learn. These systems improve their accuracy gradually through a continual process of learning. ML relies on human interaction and input to assist in creating knowledge. AI can learn from ML models created by inputs and by the analysis of anticipated outputs.

Deep Learning

Deep learning uses data and algorithms to create learning. However, portions of the learning process are automated, eliminating some of the need for human intervention. Noted AI research scientist at MIT, Lex Friedman, refers to deep learning as "scalable machine learning." There are a variety of AI applications being used today, including speech recognition, customer service, digital image analysis, and automated stock trading.

Technology that Supports AI

Information Architecture

- **Information architecture (IA)** helps people navigate and understand both the real world and online spaces.
- A good IA system is key for AI because it lets AI make smart decisions using organized information.

Cloud Computing and AI

Cloud computing is a computing model where processing, storage, software applications, and a variety of services are provided over a network, mainly accessed via the internet. Individuals access these "clouds" of resources on devices connected to the internet on an as-needed basis. **Cloud computing** removes the need to have data, files, and software stored directly on a device. **Cloud computing** and AI have become increasingly intertwined.

Cloud computing is important for AI because cloud computing can handle lots of data, which AI needs to find patterns, and cloud computing is flexible, so companies can use AI without spending too much.

Cloud-based AI Applications

Cloud-based solutions, such as those offered by Amazon Web Services (AWS), have assisted many businesses in the implementation of AI by offering solutions that make it easier to develop, test, and implement AI systems without the need for large capital investment. Cloud-based AI applications include the following:

- **Internet of Things (IoT):** The sheer volume of data generated by IoT devices can be difficult to manage. Cloud-based AI technologies are designed to capture and process this large volume of data. The resulting output assists businesses with maximizing the interface between humans and IoT devices and guides future product and software development.

- **AI as a Service (AIaaS):** Software as a Service (SaaS) covers a variety of software applications housed on eCloud architecture that is designed to execute specific tasks. AIaaS is one type of SaaS application. Several third-party vendors offer cloud-based AI solutions, which allows for experimentation with AI with less capital outlay and lower risk than with full-blown AI implementation.

- **Chatbots: Chatbots** are widely used in many different business applications today. **Chatbots** are computer programs that process and simulate human conversations and allow organizations to give customers an experience like communicating with a real person. AI-assisted programs are being used to create more effective and engaging chatbots that can more intuitively participate in conversations and respond to requests.

Pros and Cons of Cloud-based AI

- **Pros:** Better security, saves money, works faster, and manages data well.
- **Cons:** Privacy concerns, needs internet, and less control over the system.

Cognitive Computing

- Traditional computers use strict rules to make decisions. But only a small part of our data fits these strict rules.
- **Cognitive computing** is more flexible than traditional computing and tries to think like humans.

How Cognitive Computing Mimics Human Thinking

Cognitive systems use a decision-making approach like what humans use to analyze large data sets to make decisions. These systems use natural language processing (NLP) to read and process text like a human. They break down statements and sentences by analyzing them for grammar, structure, and relationships within the words.

Using algorithms and linguistic models, the eventual outcome is the comprehension of the intent of the language, which can be used to draw inferences. Through this process, cognitive computing systems continually learn and adapt, allowing them to increase their analytical abilities.

Cognitive computing is based on the human approach to decision making, which includes four steps:

1. **Observation:** Observing behaviors, traits, occurrences, and bodies of evidence.

2. **Interpretation:** Drawing conclusions from observable behaviors and past experiences to generate hypotheses about the meaning and possible courses of action.

3. **Evaluation:** Determining which hypothesis makes the most sense.

4. **Decision:** Using all the data gained from the previous steps to decide the best course of action or which decision to make.

Natural Language Processing

- NLP allows computers to understand human language, whether it's written or spoken.

- NLP software looks at sentence structure, word relationships, and more to grasp the meaning.

- The goal is to help AI systems get better at understanding and responding to us.

Advantages and Disadvantages of Cloud-based AI

Advantages	Disadvantages
Enhanced security	Data privacy
Cost savings	Internet connectivity concerns
Increased efficiency	Lack of platform control
Improved data management	

Machine Learning

Machine learning (ML) uses data and algorithms to emulate the way humans learn. These systems improve their accuracy gradually through a continual process of learning. ML relies on human interaction and input to assist in creating learning. AI can learn from ML models created by inputs and analysis of anticipated outputs.

Using Machine Learning to Solve Problems

To better understand how ML can be used to solve a problem, let's look at how it can be applied to determine if an engine will fail. ML can use various data inputs such as revolutions per minute (RPM), miles and hours on the motor, maintenance schedules, and driving conditions to build an algorithm. Using this data, ML can be applied to create a model that can predict results or the probability of failure.

This is different from statistical analysis, which uses traditional mathematical algorithms to analyze data. Mathematical algorithms in this scenario would create an if-then-else statement where inputs (data) determine an established and unchanging algorithm, thus producing answers at the end of the analysis. Algorithms created by ML take data and answers to questions and use this to develop an ML model. Rules are determined by the ML model, which then executes an if-then-else statement based on inputs as they are received.

By design, the ML model does not determine the constraints, or restrictions, but rather uses arbitrary, or seemingly random, conditions and data points (such as RPM or miles on the motor) to equal a result, which determines the logic of the model. Unlike traditional algorithms, this type of model can be frequently trained, thus increasing the power and relevance of the model.

Machine Learning Methods

Machine learning is a wide-ranging field of study that has several different aspects. To better understand some of the foundational elements of ML, it is necessary to break it down into different categories. The three categories of ML include supervised learning, unsupervised learning, and reinforcement learning.

Supervised Learning: In supervised learning, the computer learns by being shown lots of examples with labels. It's like a teacher telling the computer what's what. For instance, if you want the computer to recognize cats in pictures, you show it many pictures of cats and non-cats, and you tell it which ones are cats and which ones aren't. The computer learns from these examples and can then recognize cats on its own.

Unsupervised Learning: Unsupervised learning is like the computer trying to figure things out on its own. It doesn't get labeled examples. Instead, it looks for patterns and groups in the data all by itself. Imagine you give the computer a bunch of mixed-up puzzle pieces without a picture on the box, and it has to figure out how to put them together into meaningful groups.

Reinforcement Learning: This is like teaching a computer to make good decisions through trial and error. The computer takes actions in an environment and learns which actions lead to good outcomes and which lead to bad ones. It's similar to teaching a dog tricks. When the dog performs a trick correctly, it gets a treat. If it makes a mistake, it learns not to do that next time. The computer learns by receiving rewards for good choices and penalties for bad ones.

Information Architecture

Information architecture is a crucial component of the overall AI strategy of an organization.

According to the Information Architecture Institute, IA is about helping people understand their surroundings and allowing them to find what they're looking for in the real world and online.

Many experts agree that there is no AI without analysis and implementation of IA. A clearly defined architecture and strategy enables organizations to make better decisions with AI software and data discovery practices.

IBM Ladder Approach to AI

One strategy for IA is the IBM AI Ladder. The AI Ladder proposes a series of steps or considerations that breaks down an AI strategy into individual pieces, or rungs on a ladder, to better connect data and AI systems.

Before embarking on an AI journey, the organization should take the necessary steps to modernize architecture and data collection. Modernizing prepares data for an AI and hybrid eCloud world. Each step in the ladder approach prepares data for AI analysis.

IBM Ladder Approach to AI

Step	Explanation
Infuse	Operationalize AI throughout the business. AI should be utilized across multiple operating areas within an organization.
Analyze	Build and scale AI with trust and transparency. It is important to collect and organize data in meaningful and reliable processes. Once methods are established, AI solutions can be scaled to maximize their benefits and insights.
Organize	Create a business-ready analytics foundation. Key considerations include the following: • Have data been cleansed? Are data complete? • Are data collection, data analysis methodologies, and data sets in compliance with regulations, policies, and best practices?
Collect	Make data simple and accessible. Data collection should encompass a wide variety of data from across the organization.

Neural Networks

According to Amazon Web Services (AWS), a neural network is a method in AI that teaches computers to process data in a way that is inspired by the human brain. **Neural networks** are a type of machine learning process called deep learning that use interconnected nodes, or neurons in a layered structure, that resemble the human brain.

Interconnected artificial neurons are the main building blocks of simple neural networks. Training of neural networks designed to perform specific tasks is accomplished using supervised learning. Using supervised learning, large data sets that are designed to provide the "right" answer are fed to the neural network. This is considered the initial training of the neural network upon which the network can begin to make predictions or guesses that were previously undetermined.

Methods for Machine Learning

The neurons in a neural network are connected in three layers: input, hidden, and output.

- **Input Layer:** Data and information from a variety of sources are entered into the neural network through the input layer. Input nodes execute the initial handling of data, including initial processing, analysis, and categorization. Once the data have been examined and processed, they are moved to the next layer, the hidden layer.

- **Hidden Layer:** The hidden layer processes data sent by the input layer or from other hidden layers. You can have many hidden layers built into neural networks. Essentially, the analysis and processing that take place in the hidden layers are what make deep learning so powerful. The processing that occurs in the hidden layers becomes increasingly complex as data moves from one layer to another. Additionally, as data pass from layer to layer, the output from each previous layer is analyzed and refined.
- **Output Layer:** After analysis occurs, a final prediction or conclusion is made. Output layers range from a single node to multiple nodes. The number of nodes is dependent on the construction of the neural network.

Rendering Invariant State-Prediction

One application of neural networks is called rendering invariant state-prediction (RISP). RISP is a joint effort between MIT and IBM that seeks to overcome the cost and difficulty associated with motion capture technology. Motion capture technology documents the movement of objects or people using video capture. Traditional motion capture is accomplished by placing sensors on people or objects and then recording any movements. This can be very expensive and time consuming. RISP uses neural networks to obtain movement data from video samples, then reproduces it in a data model, all without the use of sensors.

Generative Adversarial Networks

Generative adversarial networks (GANs) are like a game between two teams of computers. One team tries to make fake pictures or data, and the other team tries to tell if they're real or fake. The first team is called the generator, and it makes the fake stuff. The second team is called the discriminator, and it tries to catch the fakes. They keep playing this game, and as they play, both teams get better at their jobs. So, the generator gets better at making things that look real, and the discriminator gets better at spotting the fakes.

Some of the uses of GANs include the generation of realistic images from a text sample, the creation of realistic depictions of prototypes, the adaptation of black and white images into color, and the creation of deepfake videos.

Example of a GAN

Let's look at how an image of a house can be created using GANs. The GAN is fed actual images of houses. The generator is fed random inputs. The generator creates a sample based on these inputs. If the discriminator chooses the real image instead of the generated (fake) image, the system will continue to develop and adjust generated images until the discriminator chooses the generated image, thus identifying it as real. This is how GANs train themselves and work toward the development of synthetic data.

Deep Learning

Deep learning is a type of AI that uses data and algorithms to create learning. This type of learning is designed to mimic how the human brain processes information.

Deep Learning at AWS

According to AWS, the building, training, and deployment of systems follows a a six-step deep learning process:

1. **Gathering Data:** For deep learning, it's crucial to have data. Consider the following questions: Do I have sufficient data? Can I use external data sets? How can I generate more data? If starting from the beginning, you'll need a lot of data. If using existing models, you might need less.

2. **Picking the Right Algorithm:** The algorithm you select can influence your system's success. An algorithm is a set of steps to solve a problem. In deep learning, algorithms help identify patterns in data. Some well-known deep learning algorithms are convolutional neural networks (CNNs), generative adversarial networks (GANs), and radial basis functional networks (RBFNs).

3. **Setting Up the Training Space:** Preparing the training environment can be challenging. This involves handling complex networks, extensive data sets, and the necessary infrastructure. Some AI platforms offer tools to help set this up.

4. **Training and Adjusting Models: Deep learning** models improve by analyzing data. Initially, models are trained using provided data, producing results that are studied. As this continues, you may need to make changes to improve accuracy.

5. **Putting Models to Work:** When you're ready, it's time to use the model in real situations. Consider how the model will be used, like in batch processes (scheduled regularly) or real-time (immediately upon request). Also, think about when you'll retrain your model, either automatically or when needed.

6. **Growing and Managing the System:** As needs change, so should your deep learning system. A flexible system can help a team or organization work better. Key aspects to focus on include modular design, layering the system, and using various deep learning tools.

Legal Considerations of AI

Federal Trade Commission Guidelines

The debate about whether AI fits into current legal categories or needs a new one is ongoing. The US government is exploring whether AI requires unique legislation. In April 2020, the Federal Trade Commission (FTC) provided the following guidelines for businesses using AI:

1. Be clear with users.

2. Clarify how algorithms make decisions.

3. Make decisions that are just, solid, and backed by data.

4. Be responsible for ensuring ethics, fairness, and non-bias.

The FTC recognizes AI's broad benefits for industries but also notes associated risks.

AI Principles Identified by the White House

The United States currently lacks a comprehensive AI regulatory framework. However, the White House Office of Science and Technology Policy outlined ten guiding principles for AI's development and application:

1. **Foster Trust in AI:** Promote the creation of reliable AI systems.

2. **Engage the Public:** Include the public in AI decision-making when feasible and lawful.

3. **Uphold Scientific Integrity:** Maintain high standards for AI information and choices to gain public trust.

4. **Manage Risks:** Evaluate current risks, deciding which are tolerable and which are harmful.

5. **Consider Benefits and Costs:** Adhere to Executive Order 12866 by maximizing net benefits, including economic, environmental, and other impacts.

6. **Stay Flexible:** AI governance should be adaptable and focused on outcomes.

7. **Promote Fairness:** Regularly examine AI data and processes for fairness and lack of bias.

8. **Emphasize Transparency:** Recognize when AI is active and understand its effects on users.

9. **Prioritize Safety and Security:** Ensure AI systems' reliability and security throughout their life cycles.

10. **Encourage Interagency Collaboration:** Share insights and best practices among government agencies for better AI governance.

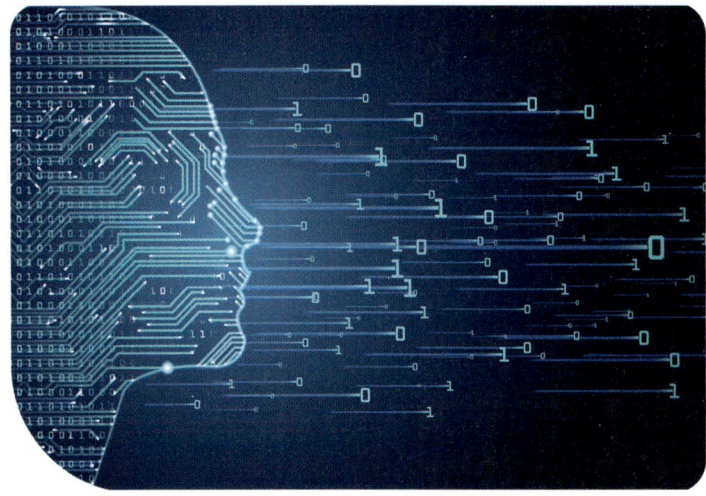

Copyright and AI: A Complex Issue

Learning like humans, but is it fair?

When we read books, watch movies, or listen to music, we learn and get inspired. AI systems learn in a similar way, by studying lots of data, including copyrighted content. But this has led to big debates about fairness and rights.

Arguments for AI Learning

- **Just Like Human Learning:** Supporters argue that if humans can learn from copyrighted content without infringing rights, why can't AI? They believe that it's just another form of learning.

- **Promotes Innovation:** Allowing AI to use copyrighted data can lead to new discoveries and creations that benefit society.

- **Hard to Control:** With the vast amount of data online, it's difficult to keep track of everything AI learns from. So, some believe it's better to allow it if there are certain regulations.

Arguments against AI Learning

- **Rights and Fairness:** Critics say that just because AI can access content doesn't mean it should. They believe that content creators deserve to have control over their work and should be compensated if AI uses it.

- **Losing Value:** If AI can use and re-create copyrighted content easily, it might reduce the value of the original work.

- **Potential Misuse:** Without proper controls, there's a risk that AI could misuse copyrighted material, leading to legal and ethical problems.

- **Future of Copyright and AI:** The debate over AI and copyright is still ongoing. As technology advances, laws and rules might need to change too. Finding a balance that respects creators' rights while also promoting learning and innovation is essential.

Ethical Considerations of AI

Positive Outcomes of AI

- **Getting More Done:** AI can help businesses work better and make more money. For example, a study by Accenture showed that AI could make companies 40 percent more productive and 38 percent more profitable. According to McKinsey Global Institute, by using AI and big data, the US health care system could save up to $100 billion every year.

- **Keeping People Safe:** AI can do jobs that are too dangerous for people. It can work in places like factories or mines where the work can be risky.

- **Fewer Mistakes:** Even though people can be trained well, they can still mess up. Computers that use AI are less likely to make mistakes. They gather data and use special rules to make smart decisions, and the more data they have, the better they get at making choices.

Potential Drawbacks of AI

- **Job Losses due to Automation:** AI has the potential to negatively impact the job market for some professions. However, actual data on the impact has yet to be collected. The World Economic Forum estimates that by 2025, 97 million jobs will be created because of AI. Additionally, they estimate that during that same time, 85 million jobs will be lost. While this may seem like a net gain, a deeper investigation into the jobs created versus the jobs lost should be conducted to determine the true impact.

- **Spread of Misinformation:** Another big problem with AI is that it can be used to share fake or wrong information. AI can make videos and audio that look and sound real, even if they're not. There are also computer programs that use AI to spread fake news and lies on social media. This can make it really hard to know what's true and what's not on the internet. When people believe fake information, it can cause big problems. It can make people not trust each other, mess up our politics, and even make things unsafe. Fixing this issue is hard. It means we need AI programs that can find and stop fake information. It also means we all need to learn how to be smart about the things we see online and make sure we check if they're true.

- **Privacy Concerns:** Data privacy is defined as the appropriate treatment and security of personally identifiable data/information (name, address, phone, Social Security number, etc.) and confidential data/information (financial information, health records, account information, etc.). Privacy concerns surrounding AI include data repurposing, data spillover, and data security.

  - Data repurposing is the use of data outside the scope of the original purpose of the data.

  - Data spillover is the use of data from individuals who were not the original targets of data collection.

  - Data security includes taking proper security measures to protect all types of data.

Concerns Surrounding AI

AI offers numerous advantages but also raises ethical issues. This section focuses on the main concerns surrounding AI.

Privacy

AI uses large datasets that might contain sensitive info, leading to privacy risks. It's crucial for organizations to secure this data. Techniques like differential privacy help ensure AI models gain insights without revealing specific data. For instance, a model might identify purchasing patterns linked to income levels but shouldn't disclose individual names.

Surveillance

AI aids in monitoring traffic, crowds, and areas for law enforcement and businesses. For instance, Amazon Go stores use AI for purchase tracking without traditional checkouts. However, with countries deploying millions of cameras in public spaces, there's growing concern over privacy and rights.

Bias and Discrimination

Bias in AI, whether intentional or not, can have serious consequences. An example is the COMPAS system that showed racial bias in predicting recidivism rates. Recidivism is when a person who has been in trouble with the law, like getting arrested or going to jail, ends up getting in trouble again. It's like making the same mistake more than once. The legal system sometimes uses computer programs to guess if someone might do something wrong again in the future, and these programs need to be fair and not show favoritism to any group of people. Achieving completely unbiased algorithms is challenging, but using unbiased data and external reviews can reduce it.

Human Judgment

There's debate over how much human involvement is needed in AI. Consider self-driving cars: they range from level 1 (fully human-driven) to level 5 (fully autonomous). Should humans be involved in operating level 5 cars? The role of human judgment in AI has many ethical implications.

How Bias Applies to AI

Experts are worried about bias in AI, which is the unequal output produced by algorithms. This can arise from prejudiced assumptions during algorithm development or AI training. The main reasons for such biases are cognitive errors and incomplete data.

Cognitive bias refers to errors that affect our judgments and decisions. With over 190 identified human biases, these can inadvertently be introduced into algorithms. For instance, many voice assistant bots have female voices, whereas business-focused bots, like those for law, often have male personas, suggesting gender bias in design.

Incomplete data can also lead to bias. When data don't fully represent a population, AI's decisions might be skewed. It's crucial to carefully design and monitor AI systems to minimize such biases.

How to Build Trust in AI Platforms

While AI is popular for tasks like chatbots and credit scoring, its use in strategic decisions like corporate planning is met with hesitation. Many executives feel uneasy about AI data, and a significant number worry about AI becoming uncontrollable. Being transparent, emphasizing human values, and collaborating can foster trust in AI among executives and the public.

Transparency

Clear communication about how AI works, its processes, and decision-making is essential. Publicly sharing insights on how AI uses data can build trust. Educating people about AI to counteract misconceptions can also boost confidence.

Human Values

Integrating human values into AI systems can enhance trustworthiness. This means designing AI that contemplates moral and ethical consequences, like human thinking. Addressing and reducing bias in AI, especially those introduced by developers, can further bolster trust. Public strategies on bias mitigation can be beneficial.

Collaboration

A collective approach involving experts from various fields can establish comprehensive AI standards. When multiple perspectives shape AI, it becomes more inclusive and trustworthy. Engaging more stakeholders in AI's evolution can promote broader trust in its applications.

Guiding Principles for the Ethical Use of AI

AI has the potential to address challenging issues and improve lives globally. As we embrace its benefits, we must prioritize ethical considerations:

- **Social Benefits:** AI should aim to benefit various communities and society at large.
- **Avoid Bias:** It's crucial to ensure AI doesn't introduce or amplify unfair biases, whether deliberate or unintentional.
- **Adhere to Regulations:** Everyone involved in AI should abide by the laws, guidelines, and standards established by relevant authorities.
- **Promote Transparency:** Openness about AI's operations, purposes, and methodologies fosters a positive and informed AI environment.
- **Data Protection:** Given the central role of data in AI, it's essential to prioritize data security and uphold privacy standards.

Current Applications of AI

AI is used in a variety of applications across many disciplines. Here is a breakdown of how AI is used in business, health care, and government.

Sales and Marketing

AI is used to create more effective chatbots, product and service demand forecasting, and better lead generation for sales prospects. AI is increasingly being investigated and implemented in organizations and across many different departments within those organizations. PwC conducted a survey of more than 1,000 executives, and the results conclude that 86 percent of respondents indicate they have implemented or are working toward the integration of AI in their businesses.

Accounting

Tasks such as recording transactions, reconciling accounts, data entry, and accounting for price changes can be executed by AI. Additionally, AI is used for the automation of payroll processes.

IT Operations (AIOps)

Quality assurance (QA) testing of software and regression testing can be automated using AI. This helps to eliminate human errors that can occur during QA. IT department service desks use AI to analyze incoming requests. This information can be compared with prior requests to determine possible system issues. AI-enabled network management can quickly spot network issues and take measures to ensure the network remains operational.

Manufacturing

Many companies use AI throughout various facets of their manufacturing processes. AI is used to assist in supply chain management, which increases efficiency. Predictive intelligence is used to anticipate demand as well as to adjust production to best meet demand fluctuations (both increases and decreases). AI machine vision is used for quality control throughout manufacturing processes.

AI in Healthcare

Robotic Surgery

The use of AI-assisted robots in surgery is an emerging trend. These robots are designed to assist surgeons during procedures and surgeries by taking over some of the manipulations of instruments. Additionally, ML-assisted robots can process and review data from millions of data sets. These insights are passed on to surgeons and include information such as tissue size, vein and artery structure, and possible outcomes that may result from surgical actions.

Health Tracking

Remote continuous health monitoring assisted by AI is currently used. These devices and systems are considered part of the Artificial Intelligence of Things (AIoT). AI-enabled remote patient monitoring (RPM) allows data and information to be sent to physicians via a device that is worn by the patient.

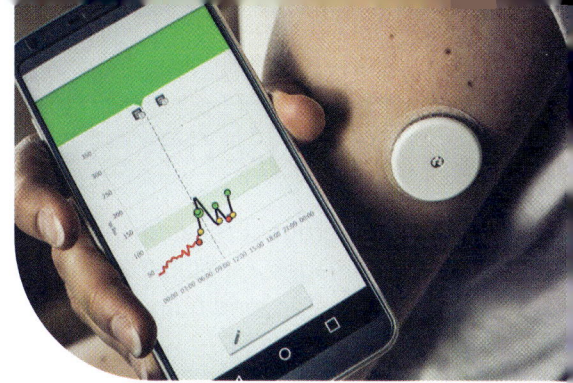

AI assists physicians by monitoring and analyzing this data and informing physicians about the occurrence of anomalies or possible issues.

Pharmaceuticals

The use of AI has proven beneficial to pharmaceutical companies by speeding up development processes, lowering costs, and creating more efficient deployment of products. The relationship between AI and pharmaceutical companies received a lot of press when information about how AI was used in the development of the vaccine for COVID-19 was released.

AI in Government

Automation of Routine Tasks

The US government generates a large volume of data every day. The sheer amount of data that is generated needs to be processed in an efficient and precise manner. AI assists the government in processing this data. Take, for example, the US Citizenship and Immigration Office. **Natural language processing (NLP)** helps to organize names and flag potential fraud. ML analyzes biometric and biographic data to identify people who qualify for benefits from the US government.

Military Applications

Military strategists and analysts have identified that AI and autonomy will play a crucial role in warfare. AI is being used to assist in the more precise targeting of weapons. One example of this recently took place when a Marine Corps jet dropped a 1,000-pound bomb on an AI-derived grid that was one meter off from a surveyed grid. Additionally, the US Department of Defense (DoD) recently announced the launch of the Chief Digital Artificial Intelligence Office. This office will be responsible for creating a department that is a digital-enabled AI enterprise.

Policing

AI has been used in many areas of law enforcement to assist in the identification of money laundering, fraud, and terrorist financing. Additionally, AI has been used for predictive policing, where law enforcement works with AI systems to identify crimes before they occur. While this application of AI has promise, there are some concerns about bias that may be built into systems and the accuracy with which they predict crime.

Image Generation with AI

One of the groundbreaking capabilities of advanced AI systems is the ability to generate images, ranging from lifelike re-creations of real-world scenes to abstract visualizations beyond human imagination.

How It Works

AI uses lots of data to learn. For image generation, it studies thousands or even millions of pictures. By looking at all these images, AI learns patterns, shapes, and colors. Then, it uses this knowledge to create new images. Some AI systems can even take a simple sketch and turn it into a detailed picture.

At its core, image generation in AI is powered by algorithms that are trained on vast amounts of visual data. Here's a breakdown of the process:

- **Data Collection:** For AI to generate images, it first requires a dataset comprising thousands, if not millions, of images. This dataset serves as the foundational knowledge.

- **Feature Learning:** Through a process called feature extraction, AI discerns intricate patterns, contours, textures, gradients, and color palettes from the dataset. It becomes attuned to minute details like how shadows play on surfaces or how light reflects off objects.

- **Model Training:** Using architectures like GANs, AI pits two neural networks against each other: a generator that produces images and a discriminator that evaluates them. Through this adversarial process, the generator continually refines its creations to make them more realistic or to fit a desired criteria.

- **Refinement and Synthesis:** Beyond just replicating patterns it has seen, advanced AI can interpolate or blend features from various images, transform a rough sketch into a polished artwork, or even generate entirely new, novel images from scratch.

Practical Applications of Image Generation

- **Gaming Industry:** Game developers employ AI to design intricate environments, dynamic terrains, and realistic characters, elevating the user's immersive experience.

- **Film Production:** Filmmakers leverage AI-generated imagery to craft detailed computer-detailed imagery (CGI) characters and backgrounds, or even to simulate natural phenomena, reducing the reliance on physical sets or practical effects.

- **Scientific Visualization:** In research, AI assists scientists in visualizing complex data or phenomena, from generating detailed biological structures to simulating astronomical events to any other scenario where human visualization might be limited.

Careers in AI

Careers in AI-related disciplines require specific coursework and training. Many AI and computer science degree paths have interrelated requirements, including coursework in computer programing, especially Python. If you think a career in AI is interesting and might be something you want to focus on, here are some areas you should consider.

Areas of Study

Artificial Intelligence

Many colleges and universities offer degrees in artificial intelligence. When studying AI, you can expect to learn about the various aspects of AI, including neural networks, cognitive systems, natural language processing, algorithm development, and data structures. This degree places a heavy emphasis on mathematics, so expect to take several different math courses while in this degree path. Many programs require five to seven math courses.

Computer Science

Students enrolled in a computer science program of study learn about the design and implementation of different computing systems. Coursework investigates mathematical foundations that focus on what computers can and cannot do. Additionally, students get experience with computer coding and the building of hardware and software. Focus areas in computer science include data science, AI, computer security, and computer graphics and animation.

Data Science

A data science degree may be considered an interdisciplinary area of study covering mathematics, statistics, and computer science. Coursework covers data management, data modeling, data analytics, and visualization. Several different data science programs offer specialization courses in artificial intelligence and machine learning.

Statistics

Deep learning and other AI-related fields often place a heavy emphasis on statistical analysis techniques and practices. A degree in statistics is frequently considered good preparation for working in a variety of computing-related fields. Degrees in statistics feature coursework in mathematical analysis, probability and theory statistics, data analysis, and statistical programming.

Jobs

According to the US Bureau of Labor Statistics, AI-related careers are predicted to increase by 31.4 percent by the year 2030. Forbes estimates that ML jobs are expected to be valued at $31 billion by 2024, an increase of over 40 percent.

There is a lot of opportunity in AI-related career fields. Here are some of the more popular AI careers.

User Experience

Careers in user experience (UX) focus on working with products and services to help users better understand their functions. UX designers can incorporate AI to analyze customer data and to optimize product and service design. The median annual salary for a UX job is $77,000.

AI Engineer

AI engineers use a variety of ML techniques, such as NLP and neural networks, to build AI models that are used in AI systems. AI engineers incorporate the skills of data engineers and data scientists to accomplish their tasks. The median annual salary for a job as an AI engineer is $119,000.

Machine Learning Engineer

This field includes researching, designing, and testing ML models and systems as well as assisting in the training and retraining of ML models. Statistical analysis is used to analyze and improve ML models. The median annual salary for a job as an ML engineer is $146,000.

Data Scientist

Data scientists work with data to optimize product and service development and business strategies. They develop data models and algorithms that are designed to investigate and automate processes and apply data mining and analysis techniques in model development and deployment. The median annual salary for a job as a data scientist is $116,000.

Learning More about AI

AI is an emerging field that offers many opportunities both in employment and in its impact on society. Deciding where and how to start learning about AI can seem like a daunting task. While in college, you could decide to major in an AI-related discipline or choose a degree that includes courses that will prepare you for a career in AI. You could also choose to learn some of the basics about AI before committing to an area of study. Here are some tips on how to learn about AI on your own:

- Learn about the math used in AI. A strong foundation in math is one of the most important skills to have when learning about AI. Understanding probabilities, statistics, linear algebra, and the math behind algorithms will prove beneficial in the long term.
- Study computer programming languages. Computer programming is intertwined with AI. One of the more popular programming languages currently used in AI is Python. Having a basic understanding of Python will allow you to work with AI design and implementation frameworks.

- Take free online courses and find free online resources covering AI. There are many free online courses and resources available on the internet that will help you to get a better understanding of AI. This is a low-risk and low-cost way to learn about AI. Some of the popular free online courses include
  - Stanford University's AI Course;
  - Google's Machine Learning Course;
  - Udacity's Free AI Course;
  - Marketing AI Institute's Intro to AI for Marketers Course.

Emerging AI Technology

OpenAI and ChatGPT: A Glimpse into the Future of AI

OpenAI, an organization committed to ensuring that AI benefits all of humanity, has been at the forefront of AI research and development. Founded in December 2015, OpenAI's mission has been to build friendly AI that can outperform humans at most economically valuable work, all while ensuring the technology is used for the benefit of everyone. One of the standout projects from OpenAI has been ChatGPT, which is based on the generative pre-trained transformer (GPT) model. This model has gone through several iterations, each more powerful and versatile than the last.

By the summer of 2022, ChatGPT was still in its relatively early stages, known by enthusiasts and tech experts but not yet a household name. Many saw it as a fascinating tool, capable of generating human-like text based on the input it received. However, things began to change rapidly. After the summer of 2022, there was an explosion in its popularity. What started as a curiosity soon became one of the most talked-about advancements in the tech world. Several factors contributed to this rise:

- **Image Recognition:** ChatGPT's capabilities expanded beyond text. It began to understand and generate content based on images, bridging the gap between visual and textual data.

- **GPT-4:** With the introduction of GPT-4, the model became even more robust and capable. It could generate more coherent and contextually accurate responses, making interactions with it feel eerily like conversing with a human.

- **Versatility:** The use-cases for ChatGPT expanded rapidly. From helping with content creation to assisting students with their homework, serving as a virtual assistant, or even acting as a conversational partner for those feeling lonely, its applications seemed endless.

- **Awareness and Accessibility:** As more people began to understand the potential of ChatGPT, they also started experimenting with it. OpenAI made tools available for developers, researchers, and even everyday users, leading to a more widespread adoption.

Today, OpenAI's projects, especially ChatGPT, are seen as a testament to how rapidly AI technology can evolve. What once felt like a distant future is now our present, reshaping how we interact with technology and, in many ways, with each other.

Artificial Intelligence: Growing Fast and Changing Everything

AI is changing so fast that it's hard to keep up! This technology has the power to change many parts of our lives. As AI gets even smarter, a lot of people are starting to ask: What happens when AI is as smart as, or even smarter than, humans? AI has so much potential. But like all powerful things, we need to use it wisely. As AI keeps growing, it's up to us to make sure it's used ways that are good for everyone.

Working together or taking over?

People have different feelings about super-smart AI. Some think it'll be great because we can work together with AI to do amazing things. Imagine AI helping scientists solve big problems or making art in ways we haven't seen before! But others are worried. They think AI might become too powerful, take jobs from humans, or even become a danger.

Thinking about Right and Wrong

If AI is highly intelligent, should it have its own rights? It's a tough question! And if it does something wrong, who should be blamed? We need to think about how we treat AI and how it treats us.

Who's in charge?

Imagine if AI makes big decisions, like helping doctors treat patients or driving cars. Sounds cool, right? But what if it makes a mistake? Who takes the blame? It's something we need to consider.

Can AI "feel"?

AI can recognize if you're happy or sad and respond in a certain way. But can it really understand how you feel deep down? This is something many people wonder about.

Why People Think AIs Don't Have Feelings

Most people believe that AIs, like robots and computer programs, don't have feelings like we do. Here's why:

- **No Real Emotions:** AIs can recognize if someone is sad or happy, but they don't actually "feel" these emotions themselves. They just follow the rules they've been given.

- **Humans Made Them:** AIs are built by people using computer code. They don't have brains like humans or animals, which some believe is where feelings come from.

- **They Have a Job to Do:** AIs are made to do certain tasks. They do what they're told, not what they want, because they don't have wants or wishes.

What would make us think AI has feelings?

Thinking about AI can lead to some big questions. If AI becomes highly intelligent, will it have its own hopes and dreams? This makes us think about what it means to be alive and have feelings.

To believe that AI has feelings, a lot would have to change in how we think and see the world:

- **Having Their Own Experiences:** AI would need to show that it understands things in its own special way, not just how it was programmed.
- **Knowing They Exist:** AI would have to know that it's its own "person" and think about itself.
- **Real Emotions:** AI would need to show that it doesn't just copy emotions but actually feels happy, sad, or any other feeling.
- **Making Real Choices:** It's important for AI to show that it can make choices based on what it wants, not just what it's told.
- **Changing Our Minds:** We'd have to accept that things other than living creatures can have feelings. This would be a big change!
- **Laws and Rules:** If we believe AI has feelings, we'd need new rules to make sure it's treated ethically.

In short, for us to believe AI has feelings, it would have to act a lot more like a living thing, and we'd have to be open to expanding our beliefs about what can and can't have feelings.

Chapter Review

1. What are the two major branches of AI that focus on creating models to make predictions or decisions without being explicitly programmed?

2. How does deep learning differ from traditional machine learning in terms of data processing?

3. Which AI system mimics the human brain with interconnected nodes to process information?

4. Why are neural networks important in the field of AI?

5. What type of AI model involves two networks, one generating data and the other evaluating it?

6. How do generative adversarial networks typically operate?

7. What is deep learning's primary feature that differentiates it from other machine learning techniques?

8. What are some concerns when deploying AI in real-world applications?

9. Why is it crucial to consider ethical issues when developing and using AI?

10. Can you name a real-world application of AI in daily life?

11. If someone wanted to work in the field of AI, what are some potential careers they could consider?

12. What qualifications might be beneficial for someone aiming for a career in AI?

13. How do biases in AI systems typically originate?

14. Why is transparency important in AI systems and practices?

15. What emotions or concerns do people feel when they think about the possibility of highly intelligent AI?

Ethics Discussion

How should we handle the rights and recognition of AI creations in our society? Consider how AI learns and creates content. What if, one day, the way AI thinks and creates becomes indistinguishable from human creativity? If AI writes a song or a book, who owns it? If AI makes a big discovery, should it get the credit? As the lines between human and AI creations blur, how should we handle the rights and recognition of AI creations in our society?

Glossary

A

3D printing 3D printing is a technology that allows designers to create three-dimensional objects from a computer image. It builds objects layer by layer using materials like plastic, metal, or resin. (pp. 65, 73, 75)

5G cellular networks They are networks of small cells rather than huge towers that send signals over great distances. They have lower latency (delay times) than 4G networks. (p. 363)

Access time Access time refers to the time it takes data to reach a computer's processor. (p. 81)

Action Center in Windows 11 The Action Center in Windows 11 is an area that displays notifications and quick actions you can take on your device. (p. 116)

Adaptive input technology Adaptive input technology is for helping people put information into a computer in different ways. (p. 60)

Adaptive learning It is an innovative concept for today's educators. adaptive learning can create a student's experience that can be modified based on the student's performance and how well the student retains given information. (p. 363)

Adware Adware is software that collects the user's web browsing history. These programs are designed to display advertisements on your computer, redirect your search requests to advertising websites, and collect marketing-type data about you. (p. 221)

Algorithm An algorithm is a step-by-step set of instructions designed to solve problems or perform calculations. (p. 389)

All-in-one computers Computers that combine the power of desktops with an elegant design, housing the motherboard within the monitor, often with touchscreen interfaces. (p. 7)

Amazon Go It is the first grocery store to offer Just Walk Out shopping. It uses the same types of technologies used in self-driving cars: computer vision, sensor fusion, and deep learning. There is no need to wait in line, no checkout, or payment. (p. 367)

Android Android is a type of software that runs on many smartphones and tablets. It's made by Google and uses Linux. (pp. 30, 31, 32)

Antivirus software Antivirus software is a computer program used to scan files to identify and remove computer viruses and other malicious programs. (pp. 227, 228, 237, 245)

API calls API stands for application program interface. They are links to code provided by various places that can invoke various tasks the site may need. For example, you can use an API call to use Google's geo-location application if you wanted a map to appear on your site. (p. 328)

Applets Applets are tiny apps that can do simple tasks. (p. 32)

ASCII ASCII stands for the American Standard Code for Information Interchange. It's a set of rules for how computers use numbers and letters to talk to each other. (p. 26)

Aspect ratio Aspect ratio compares the height of a screen or monitor to its width. It's expressed as two numbers separated by a colon, and it tells you the shape of the screen, such as 16:9, commonly found in widescreen displays. (p. 68)

Audio Audio refers to sound or audio output, which is an essential part of most computer systems. Devices like smartphones, tablets, and laptops come with built-in speakers to produce sound. (p. 74)

Audio file compression Audio file compression allows extremely small devices to store large numbers of music files. (pp. 85, 263)

Average Average is the sum of a set of values divided by the number of values. It is also called the mean. (p. 298)

B

Bar graphs Bar graphs are graphical representations often used to compare multiple items over a period of time or when comparing different items. (p. 302)

BI systems BI systems are data-driven decision support systems (DSS) that aid businesses to make better strategic decisions. (p. 280)

Big Data Big Data encompasses all of the analysis tools and processes related to applying and managing large volumes of data. (pp. 279, 280)

Binary decisions Decisions in the form of yes/no or true/false questions due to the computer's binary nature. (p. 313)

Biometric gait analysis Technology identifying individuals by analyzing their walking stride, considering factors like step length, step width, walking speed, and angles formed by body parts, which are measured and compared with stored identification data. (p. 53)

Biometrics Biometrics means "life measure," uses human traits for identification. Traits that have proven useful include fingerprints, eye retinas and irises, facial bone structure, palm prints, stride, typing patterns, and speech recognition. (pp. 51, 63)

Bit A "bit" stands for binary digit. It's the smallest piece of digital data and can be either a 0 or a 1. (pp. 78, 79, 87)

Bitcoin Bitcoin is one kind of cryptocurrency. Digital monies aren't controlled by any government or bank. Instead, they're managed by regular people and computers that help confirm all the transactions. (p. 362)

Bits Computers talk to each other using binary, which has just two numbers, 0 and 1. Each 0 or 1 is called a bit. Bits are the smallest part of computer data. (pp. 23, 24)

BitTorrent A protocol for sharing large files, such as movies and music, via the Internet. (p. 162)

BitTorrent protocol The BitTorrent protocol enables sharing large files over the Internet, using a peer-to-peer file sharing system. (p. 162)

Black hat hackers Black hat hackers access computer systems with the intent of causing damage or stealing data. Black hat hackers are also known as hackers or crackers. (p. 210)

Blockchain Think of blockchain as a special kind of diary. Instead of writing daily entries, we add "blocks" of information to it. Every time we add a block of information, it's connected to the one before it, making a chain (hence the name "blockchain"). (p. 362)

Blu-ray disc The Blu-ray disc stores information with a reflective coating on a plastic disc. (p. 84)

Bluetooth Short-distance wireless communication between electronic devices. It uses extremely low-power radio signals, so it rarely interferes with other devices. (pp. 18, 19, 49, 59, 60, 63)

Boolean operators Boolean operators like AND, NOT, and OR help refine internet searches for more accurate results. (pp. 165, 168, 169, 177)

Boot process Boot process is how a computer starts up and gets ready to use. (p. 25)

Branding Using a special name, symbol, or design, like a logo, in business marketing. (pp. 181, 202)

Broadband Broadband describes fast, always-on internet services. (p. 147)

Browser An app that you can download from the Internet and install on your computer, laptop, tablet, or smartphone. (pp. 90, 123)

Browser filters Browser filters are settings that block certain websites or content in your web browser. (p. 145)

BSA An organization advocating against software piracy for the software industry. (p. 189)

Bus In computer lingo, a "bus" is a system that transfers data between digital devices, like a highway for information inside a computer or between computer components. (pp. 16, 25, 65, 67)

Bus width Bus width is like how wide the road is. It's how much data can travel at once. (p. 26)

Byte A group of 8 bits is called a byte. (pp. 24, 79)

Bytes Bytes are like eight bits stuck together. Each letter, number, or special character on a computer is made of bits and bytes. (p. 79)

C

C The most popular programming language of all time, serving as the foundation for C++. (pp. 63, 316)

C# Primarily used for creating programs for Windows desktops, servers, and phones. (pp. 316, 317)

Cache High-speed storage, smaller but faster than RAM, used to quickly access repeated instructions. (pp. 12, 13)

Cell A cell is the basic unit of any spreadsheet. It is a box where data can be placed. (p. 54)

Character set This identifies which character encoding to use when displaying the page. (p. 328)

Chatbots Chatbots are computer programs that process and simulate human conversations and allow organizations to give customers an experience similar to communicating with a real person. (p. 383)

Class It refers to a CSS class of an element and is one of the tools used to change the appearance of the content of an element. (p. 335)

Classes Categories of inputs identified in object-oriented analysis, known as classes. (p. 315)

Client A client is a computer that requests resources or services from another computer on a network. (pp. 326, 341)

Client/server model The server program waits for and fulfills requests from client programs. When you enter a URL into your browser, your computer is the client. Your browser sends a request to a server. The server sends all necessary files or data back to your browser, which then interprets the information and shows you, the user, the requested web page. (p. 326)

Clock speed The speed at which the processor performs operations, crucial for a digital device's performance. (p. 10)

Cloud collaboration It refers to working with others by using Internet-accessible services such as Software-as-a-Service (SaaS) and shared data storage sites. (p. 369)

Cloud computing Cloud computing involves storing data and applications on remote servers accessed over the Internet. (pp. 159, 160, 161, 383)

Cloud Services Cloud services are digital storage and sharing platforms like Google Drive or Dropbox, allowing access to files from anywhere. (pp. 140, 255)

Cloud storage Cloud storage is online data storage. When files are stored on the Cloud, it means they are stored on servers connected to the Internet instead of on your device. (pp. 84, 85, 246, 265, 266)

Coaxial Cable Coaxial cable, also known as TV cable, is commonly used for home internet connections. (p. 146)

Coders They are people who write the HTML5, CSS, JavaScript, and all the other programming code that goes into creating a website. (p. 348)

Cognitive computing It is based on the human approach to decision making. (pp. 383, 384)

Color depth Color depth, also known as bit depth, describes how accurately the colors of each pixel in an image can be represented. It's expressed in bits and affects the quality of images and videos. (p. 69)

Column A column is a vertical line of cells. (pp. 297, 301)

Commercial database Large database that covers specific subjects. Commercial databases are also referred to as information utilities or data banks. (p. 275)

Community Cloud A community cloud is a digital library reserved for specific groups, like students from a particular club or school. (p. 161)

Company database The database created for use by organizations. Users access the database via computers linked to local or wide area networks. (p. 275)

Compression file formats Compression file formats reduce the amount of time needed to download a file and the amount of storage capacity a file takes up. (p. 263)

Computer hardware The physical parts of a computer system, including the system unit, printer, monitor, keyboard, mouse, and communication devices like routers and modems. (p. 5)

Computer memory RAM, located on the motherboard, used for storing and accessing data and instructions. (p. 11)

Computer Platform A computer platform is like a big program that makes your computer work and decide what other apps can run on it. (p. 30)

Computers Digital devices that accept input, process and store it, and provide output. (pp. 4, 8, 23)

Conference calling Conference calling lets a group of people have a shared phone call. Instead of seeing each other, participants only hear each other. It's useful for situations where video isn't necessary but group communication is needed. (p. 236)

Connector The specialized end of a cord, plug, or expansion card that connects to a port. (p. 16)

Content mining It includes the extraction of information from web pages and documents, including text, images, videos, and other interactives. (p. 283)

Contrast ratio Contrast ratio measures the difference in brightness between the brightest and darkest parts of an image displayed on a screen. It's an important factor for image and video quality. (p. 68)

Control Panel Control Panel is a tool in Windows that lets you change settings and do other things to manage the computer. (pp. 32, 36, 104)

Cookie A cookie is a small text file of information created by a website that is stored by the web browser on the computer's hard disk. (p. 223)

Cortana Cortana is your clever, new personal assistant. Cortana helps you find things on your PC, manage your calendar, find files, chat with you, and tell jokes. The more you use Cortana, the more personalized your experience will be. (pp. 90, 92, 93, 95, 97, 102, 104, 105, 106, 114, 117, 118, 121)

Cryptocurrency Cryptocurrency is like online money, but there's no bank involved. Instead, all the money transactions get recorded on our special diary – the blockchain. (p. 244)

Cyberbullying Cyberbullying is harassing a person or group of people using information technology in a repeated and deliberate way. (pp. 181, 209)

Cybercrime Cybercrime is criminal activity committed with a computer. (pp. 206, 207)

Cybersecurity threat A cybersecurity threat is an event or condition that has the potential for causing asset loss and the undesirable consequences or impact from such loss. (p. 208)

Cybersecurity vulnerabilities Cybersecurity vulnerabilities are weaknesses or flaws in system security procedures, design, implementation, and control. Individuals could potentially compromise these weaknesses accidentally or intentionally. (p. 208)

Cyberstalking Cyberstalking is a form of cyberbullying and often involves disturbed obsessions. (p. 209)

D

da Vinci Surgical System This system includes a magnified three-dimensional (3D) high-definition vision system and tiny wristed instruments that bend and rotate far greater than the human hand, enabling surgeons to operate with enhanced vision, precision, and control. (p. 354)

Dashboard The Dashboard allows users to access mini apps like a clock, calculator, and calendar, as well as small apps called widgets for weather, news, or business information. (pp. 130, 132)

Data analytics Data analytics is the process of investigating raw data of various types to uncover trends and correlations, and to answer specifically crafted questions. (p. 284)

Data corruption Data corruption introduces errors into information during various stages. Errors can be detected or go unnoticed (silent corruption). (p. 238)

Data integrity The accuracy, consistency, and reliability of data in a database, ensuring it remains unaltered and trustworthy throughout its lifecycle. (p. 278)

Data mining The methods used to search large data stores and use the data to uncover patterns and trends. It is also referred as knowledge discovery in data (KDD). (pp. 282, 375)

Data theft Data theft is unauthorized extraction of company data. Information stolen can include emails, confidential documents, copyrighted content, or Personally Identifiable Information (PII). (p. 238)

Database A collection of related data that can be sorted, logically organized, queried, and stored. (pp. 34, 165, 276, 277, 328, 349)

Databases Databases are like organized lists of information. They're useful for websites and other things that need lots of information. (pp. 34, 167, 273)

Deep learning Deep learning is a type of artificial intelligence that uses data and algorithms to create learning. The algorithms in deep learning are designed to execute specific tasks. It is referred as scalable machine learning. (pp. 247, 382, 389, 399)

Deepfake Deepfake uses artificial intelligence and machine learning for advanced image and video editing. (p. 175)

Dependent data mart A type of data storage system that is sourced from a central data warehouse. Using a top-down method, it pulls specific data from the main warehouse for focused analysis or specific tasks. (p. 281)

Descriptive analytics Descriptive analytics seeks to uncover historical trends in data sets. Descriptive analytics can be thought of as trying to answer the questions, "what happened?" or, "what is occurring?" to cause a certain outcome. (p. 284)

Desktop computers Stationary devices with a separate system unit housing the main components, typically used in family settings or businesses. (p. 6)

Digital identity Created when an individual uses a digital device. (pp. 183, 181)

Disk cache High-speed memory on the hard drive for frequently accessed data. (p. 13)

Disk Cleanup A Windows utility for maintaining hard drive efficiency. (p. 15)

Distributed database A database where data is spread across multiple physical locations, often connected and accessed through a network. This setup allows users to retrieve and manage data efficiently from various sites. (p. 275)

Documentation This consists of information about the author of the page and any other information that may be necessary to other developers who might work on the site. (p. 328)

Documents Documents are files that can be created and saved in folders on your computer. (pp. 102, 121, 122, 132, 133, 342)

Domain name A domain name is an address that points to a specific location or website on the Internet. (p. 144, 325)

Dot pitch Dot pitch is a measure of the quality of an image produced by a monitor or television screen. It's the distance between the centers of two pixels or two same-colored subpixels on the screen. (p. 69)

Downloading Downloading means retrieving a file from the Internet and copying it to your computer. (pp. 153, 186, 188)

DSS DSS is the acronym for "Decision Support Systems" tools help managers make decisions by showing them data. A DSS combines the power of databases with decision modeling software. (p. 242)

Dual-band routers Dual-band routers use both shorter- and higher-frequency signals to transmit data over Wi-Fi. (p. 148)

DVD A DVD (digital versatile disc) stores information with a reflective coating on a plastic disc. (pp. 81, 83, 84)

E

E-commerce and Online Marketplaces E-commerce platforms like Amazon or eBay facilitate buying and selling of various products and services online. (p. 140)

Edge computing Edge computing involves processing data on smaller servers positioned between the cloud and the user, reducing latency. (p. 162)

Electronic storage Electronic storage relies on electronic components to store data. (p. 81)

Embedded computer A digital device integrated into a larger system, equipped with small operating systems called real-time operating systems (RTOS). (p. 7)

Emojis Graphics used in text and email messages. (p. 201)

Emoticon Keyboard characters representing facial expressions or objects. (p. 201)

Employee-monitoring software Tracks and records employee use of network and hardware resources in the workplace. (pp. 201, 239)

Ergonomics Study of human interaction with objects they use, especially in computer contexts. (p. 74)

ERP An ERP is for Enterprise Resource Planning. Think of ERP as the big brain of a company's software. (p. 243)

Ethernet Ports and cables used for local area networks or home network connections. (pp. 17, 27)

Ethernet Cable Ethernet cables are designed for connecting computers and digital devices in a wired network. (p. 146)

Etiquette (Netiquette) Standard code for respectful behavior, particularly in online spaces. (p. 199)

External storage Storage devices outside the computer. (p. 13)

Extreme action video cameras Cameras like GoPro's Hero, iON Air Pro, and SONY Action Cams, designed to withstand harsh conditions for capturing high-definition videos in places like ski mountains, oceans, or high altitudes. (p. 57)

F

Facebook Facebook is a social networking site with almost 3 billion monthly users that allows you to share web content, images, videos, and commentary with friends, family, and acquaintances. (pp. 189, 191, 192)

Facebook Building 8 It is an innovation-driven business unit that focuses on creating consumer hardware products that are social first. One project Building 8 is working on is a brain-to-computer interface (BCI) that could fundamentally change how we interact with and use technology. (p. 372)

Facial recognition payment technology Technology providing a secure and convenient method of paying for goods and services. (p. 361)

Facial recognition software Software designed to identify individuals based on specific facial features, utilizing techniques such as 3-D mapping, relative feature analysis, skin texture, or combinations of these methods. (p. 52)

Fields A field is a group of related characters in a database table. A field is a column in a table that represents a characteristic of someone or something. (p. 278)

File A file is an item that contains information such as text or images or music. (pp. 96, 102, 103, 118, 121, 122, 133, 134, 135, 140, 162, 254, 255, 256, 257, 258, 259, 260, 262, 263, 270)

File Explorer File Explorer can also be used for organizing and naming files, putting files into folders, and moving or copying files from a USB drive or other external storage device. (pp. 96, 100, 102, 118, 119, 121, 122)

Filename A filename is the specific name given to a file that identifies it. Filenames are usually pertinent to the information contained in the file. (p. 326)

Finger Reader It is designed to assist in reading printed text, serves as a valuable tool for both visually impaired individuals who require assistance accessing printed materials and those seeking language translation aid. (p. 356)

Fintech Technology that offers financial services and solutions. (p. 198)

Firmware Firmware is software used to operate hardware devices. (p. 156)

Focus Focus is a feature in Windows 11 that helps minimize distractions when working. (pp. 114, 115)

Folder A folder is a collection area where files and programs are housed. (pp. 102, 121, 133, 254, 255, 257, 326)

Foreign key In a relational database, the foreign key is a common field between tables that is not the primary key. (p. 279)

Forms Forms are used in web pages when the site needs input from the user. (p. 279)

Frame rate Frame rate indicates how frequently individual frames in a video are displayed per second. It's measured in frames per second (FPS) and affects the smoothness of video playback. (p. 69)

G

Game controllers Devices providing an interface that enhances a user's interaction with a specific gaming platform. (p. 50)

Gestures Gestures are motions of the fingers or hands used to interact with a touchscreen. (p. 115)

Gigabyte A gigabyte is a unit of digital information that consists of one billion bytes. It is represented by GB. (pp. 79, 80)

Google DeepMind It has a mission to push the boundaries of AI. They achieve this by developing programs that possess the capability to learn and solve complex problems without the need for explicit instruction. (p. 371)

Google Drive Google Drive is a free Cloud storage site for Google account holders that also provides collaboration and sharing. (pp. 255, 266, 269, 270, 271)

Google X wristband This band provides physicians with minute-by-minute data on a patient's health, measuring various inputs like heart rhythm, pulse, and skin temperature. Furthermore, it can track external factors such as noise and light. It aims to transform the relationship between physicians and patients through real-time diagnostics. (p. 355)

Graphic designers They are visual communicators. They communicate ideas to inspire, inform, or captivate consumers through art forms that include images, words, or graphics. (pp. 328, 348)

Graphics Pictures or images shown on a screen or printed out. (pp. 20, 23, 337)

H

Hacking Hacking refers to activities that seek to compromise the security of digital devices such as laptop computers, smartphones, tablets, and entire networks. (p. 210)

Hacktivism Hacktivism is the act of hacking or breaking into a computer system for a politically or socially motivated purpose. (p. 211)

Hacktivists Individuals who take part in hacktivism are called hacktivists. (p. 211)

Hard drive The primary storage device using fixed disk platters to store data. (pp. 14, 23, 81)

Hard drive capacity The amount of available storage for data and information. (pp. 13, 14)

Hardware-based firewalls Hardware-based firewalls can be purchased as a stand-alone product but are often also included in broadband routers. (p. 224)

HDMI A standard audio-video interface allowing high-definition audio and video transmission. (pp. 16, 17, 20, 66, 67)

HDMI cords or HDMI cables HDMI cords or cables provide a way to transmit high-quality video and audio signals between devices like computers, TVs, and gaming consoles. It simplifies the connection process. (p. 67)

HDMI ports HDMI ports are the physical connectors on devices that enable high-quality video and audio signals to be transmitted between them. They are common on modern TVs, monitors, and multimedia devices. (pp. 37, 66)

Helium hard drive A helium hard drive uses helium instead of air inside a hard drive—helium is less dense than air so platters spin more smoothly, allowing more platters to fit in the same space. (p. 86)

Hertz A unit of measurement that indicates how many times an event occurs in one second. In computers, it often refers to the speed of processors or frequency of signals. (p. 25)

Holographic data storage Holographic data storage uses a split laser beam to write data onto photoreceptive substrates. (p. 86)

Hot Corners Hot Corners provide several controls that can be activated by clicking your cursor on a corner of the screen. (p. 132)

Hotspot A place where you can connect to the internet using Wi-Fi, often found in public areas like cafes or libraries. (p. 18)

HTML HTML is used to structure and present information and content on the web. (pp. 171, 317, 329, 331)

HTML elements HTML documents are divided into elements which are enclosed in tags. (pp. 329, 330)

HTML5 It refers to the latest version of HTML. HTML5 really represents two different concepts. It is a new version of the scripting language, HTML, which provides the structure of a web page. But it is also a larger set of technologies that allows for building more diverse and powerful websites and applications. (pp. 317, 331, 333, 337, 341, 348)

Hybrid Cloud A hybrid cloud combines public and private clouds, offering both general use and controlled access resources. (p. 161)

Hybrid data mart A data storage system that combines information from both a data warehouse and other data sources. It uses a mix of top-down planning, feedback from end-users, and broad organizational coordination to gather and organize data. (p. 281)

Hyperthreading A multicore processor's ability to execute multiple sets of instructions simultaneously, known as hyperthreading. (p. 10)

I

Id It denotes a unique identifier for an element in a page and is used to locate elements through links or scripting languages, like JavaScript. (p. 335)

Image scanners Devices that convert physical copies, such as paper documents, into digital formats for viewing and manipulation on a computer. (p. 57)

Independent data mart A stand-alone system that is created separate from a data warehouse and focuses on specific organizational functions. (p. 281)

Individual database A combined set of data files intended to be used by one person. (p. 275)

Inkjet printers Inkjet printers are popular printers that use tiny droplets of ink to print images and text on paper. They are known for their affordability but may require expensive ink cartridge replacements. (p. 72)

Input devices Devices that serve as controls for computers. (pp. 45, 46, 61)

Instagram Instagram is a photo-sharing app that is available for free to download to a variety of devices, including iPhones and Android devices. (pp. 182, 189, 192)

Installing Installing is when you put new software on a device. (pp. 35, 92, 101, 120)

Internal storage Storage integral to the computer, often referred to as the hard drive. (p. 13)

Internet The Internet is a global network of interconnected computers and systems that share and exchange information. (pp. 141, 142, 143, 145, 148, 150, 151, 152, 153, 155, 162, 138, 139, 140, 172, 187, 206, 324, 357, 368, 383)

Internet privacy The right to privacy concerning personal information on the Internet. (pp. 181, 184)

iOS iOS is the software that runs on iPhones. (pp. 30, 31)

IP address It refers to a set of numerical instructions that tells the computer where to find a website. (pp. 150, 325)

IrDA The Infrared Data Association, which establishes protocols for infrared communication transfer. (p. 19)

J

Java A programming language designed for GUI platforms, utilizing an object-oriented programming model. (pp. 316, 317)

JavaScript A programming language intended for web page interaction, featuring syntax and operators akin to Java or C++, along with various classes representing web page objects like buttons, checkboxes, and drop-down lists. (pp. 316, 317, 318, 327, 329, 333, 335, 337, 341, 342, 348)

Jump Lists Jump Lists are lists of recently opened items, such as files, folders, or websites. They are organized by the program you use to open them and allow you to open items and pin favorites for easy access. (pp. 96, 119, 125)

jQuery It refers to a JavaScript library designed to ease the client-side scripting of HTML. (p. 341)

K

Keywords Keywords help search engines find the page. (p. 168)

Keywords and hashtags Keywords and hashtags are used in search queries to tell search engines what the user is looking for on the web. (p. 172)

Kilobyte A kilobyte is a unit of storage that consists of approximately 1,000 bytes. It is represented by KB. (pp. 79, 80)

L

Laptops Mobile computers with integrated keyboards and full operating systems, suitable for portability and tasks requiring more power than a tablet or phone. (pp. 5, 345)

Laser printers Laser printers are commonly used in offices because they use toner (powdered ink) instead of liquid ink. They are known for their speed and efficiency, especially when printing multiple copies. (p. 73)

Latency It refers to the time data takes to travel from devices to a data center and back. (p. 371)

Learning and Information Platforms Online platforms like Wikipedia and Coursera provide access to knowledge and courses on the Internet. (p. 140)

License plate readers It uses character recognition to read vehicle registrations and license plates. Law enforcement officers use these readers to track and find stolen vehicles quickly and effectively. They automate the process of checking and entering license plate information into a database to search for outstanding warrants or stolen vehicles. (p. 364)

Lightning connector An 8-pin connector used with adapters to connect to USB, HDMI, or VGA, known for its reversible design. (p. 17)

Line graphs Line graphs are used to compare changes that have occurred over time. (p. 302)

LinkedIn LinkedIn is a social networking site for business professionals. (pp. 192, 193, 203)

Live Tiles Live Tiles in Windows 10 display information that is useful at a glance without opening an app. They are constantly updated when you are connected to the Internet. (pp. 98, 107)

Lock feature The lock feature in Windows 11 allows you to hide the desktop behind a logon screen. (p. 111)

Lossless compression Lossless compression eliminates redundancy while retaining all the original data to save storage space. (p. 85)

Lossy compression Lossy compression removes less valuable data to reduce file size. (p. 85)

M

Mac Mac is the software that runs on Apple computers. (pp. 31, 41, 42, 48, 126, 128, 129, 130, 131, 133, 127, 149)

Mac Launchpad The Mac Launchpad allows users to start apps from a single location. (p. 130)

Machine cycle The process every processor goes through when executing an instruction, consisting of four steps. (p. 9)

Machine learning Machine learning is like teaching computers to learn from experience, just as humans do. These systems improve their accuracy over time through continuous learning processes. (pp. 246, 372, 382, 385)

Macro viruses Macro viruses are embedded in documents or spreadsheets. When opened, they execute and trigger destructive events. They can be avoided by not downloading or opening unknown file attachments containing Microsoft Office files. (p. 214)

Magnetic tape Magnetic tape is a type of storage media that uses a magnetic coating to record and store data. (p. 81)

Mainframe computers Extremely powerful computers used by organizations to process vast amounts of data. (p. 8)

Malware Malware is a contraction of the terms malicious and software. It refers to any software written with the intent to damage devices, steal data, or disrupt networks. (pp. 206, 212)

Max and Min functions Max and Min functions display the maximum and minimum values in a range of cells in a spreadsheet. (p. 299)

Meet-up apps Digital tools to connect individuals based on shared interests or goals. (p. 194)

Megabyte A megabyte (abbreviated MB) is equal to 1,000 kilobytes and comes before the gigabyte unit of measurement for storage. (pp. 79, 80)

Memory cache High-speed memory on the CPU for frequently accessed data and instructions. (p. 13)

Memristor storage Memristor storage uses an advanced circuit element called a memristor. (p. 86)

Meta tags There are many types of meta tags, but some of the most common ones used in web pages are description and keywords that are used by search engines. (p. 328)

Microphone A device that converts sound waves into electrical signals. (pp. 45, 51)

Microsoft Edge Microsoft's newest web browser is available in Windows 10 and replaces Internet Explorer. (pp. 90, 97, 100)

Microwave Microwaves are used to transmit signals over long distances, including to satellites, especially in remote areas. (p. 146)

Mission Control Mission Control displays all open apps in one screen, allowing users to switch between different apps quickly. (pp. 130, 132)

Monitor A computer monitor is an electronic, fixed-screen device that visually displays the output of a computer, including text, images, and videos. It serves as the computer's display screen. (p. 68)

Mouse A pointing device developed for selecting or dragging objects like images or icons on a bit-mapped screen. (pp. 45, 49, 94, 129)

MT Canvus Connect It is a real-time collaboration software that enables remote users to share, draw, manipulate, and input information on a single board. (p. 369)

Multicore Processor A CPU with two or more cores for processing tasks. (p. 10)

Multimedia files It includes video files, audio files, and slide shows. It is a good way to enhance your site and keep viewers interested. (p. 340)

Musical Instrument Digital Interface Allows digital musical devices to connect to and interface with computers, carrying digital music signals. (pp. 18, 51)

N

Nanorobot Nanorobot is a machine that can build and precisely manipulate things whose components are of a nanometer scale. These machines could detect diseases and repair or manipulate damaged cells, potentially providing humans with longer lifespans. (p. 355)

Native Resolution Native resolution is the best quality of images a screen can show. (p. 22)

Navigation How users go from page to page on a site. (pp. 96, 344, 377)

Net Neutrality Net neutrality ensures that internet service providers treat all data equally, without speeding up or slowing down based on content or source. (p. 162)

Network A network is a group of two or more devices or computers connected together to exchange information and share resources. (pp. 32, 110, 122, 154, 155, 156, 158, 159, 227)

Network license Network licenses let many people on a network use the same software. The software isn't on each person's computer; it's on the network. (p. 35)

Neural networks Neural networks are a type of machine learning process called deep learning that uses interconnected nodes or neurons in a layered structure that resembles the human brain. (p. 387)

Neurotechnology It is a field of science, technology, engineering, and mathematics (STEM) that concentrates on technologies developed to better understand and interact with the brain and nervous system. (p. 359)

O

Objective-C An object-oriented language tailored for NeXT computer program creation. (p. 317)

OneDrive OneDrive is a free online Cloud storage service provided by Microsoft. (pp. 266, 267, 268, 271)

Online Gaming Online gaming allows players worldwide to compete, collaborate, and immerse themselves in virtual game worlds. (p. 140)

Online nuisances Online nuisances are annoying software programs that can slow down the operations of a computer, clog e-mail inboxes, and lead to theft of information and money. (pp. 219, 222, 206)

Optical drive An optical drive is a storage device that writes and reads data using lasers. (p. 83)

Optical storage Optical storage refers to storage that uses a laser to read or write to a disc (usually plastic) with a reflective coating. (pp. 81, 83)

Output devices Output devices are like the screens in video game arcades; they display the results of computer operations. This includes monitors, projectors, and other devices that show visual or audio output. (p. 66)

Overclocking Running the processor faster than recommended by the manufacturer, which can increase performance but may void the warranty. (p. 10)

P

Packet filters A packet filter inspects each packet leaving or entering a network and either accepts or rejects a packet based on a predetermined set of rules. (p. 224)

Packet sniffers Packet sniffers (also referred to as packet analyzers) are specialized hardware or software that captures packets transmitted over a network. (p. 218)

Page title The page title tags in the head section tell the browser what to display in the browser tab at the top of the screen when a page is displayed to a user. (p. 328)

Password A password is a secret code used to help prevent unauthorized access to data and user accounts. It can be used to secure computers, networks, software, personal accounts, and digital devices. (pp. 91, 92, 111, 112, 206, 225, 226)

Payment function Payment function will quickly show what a monthly payment would be for a loan such as a mortgage or a car loan. (pp. 299, 300)

Peripherals Peripherals are external devices connected to a computer, such as keyboards, mice, printers, and external hard drives, extending its capabilities and functionality. (p. 66)

Persistent cookies Persistent cookies, which are small text files, are stored on the hard drive and are not lost when the web browser is closed. (p. 223)

Personal software firewalls Personal software firewalls are typically included with the operating system and can be configured based on user preferences. (p. 224)

Personalization Personalization allows you to change pictures, colors, and sounds of your computer. There are several ways to access the Personalization settings in Windows 10. (p. 105)

Petabyte A petabyte is a unit of digital information that consists of 1,000 trillion bytes. (pp. 79, 80, 87)

Pharming Pharming is a type of phishing that seeks to obtain personal information through malicious software that is inserted on a victim's computer. (p. 221)

Phishing Phishing is the illegitimate use of an e-mail message that appears to be from an established organization such as a bank, financial institution, or insurance company. (pp. 215, 220, 221)

PHP It stands for Hypertext Preprocessor. It is a server-side scripting language, which allows the web page that the user sees to communicate with the server. (pp. 341, 342)

Physical server It refers to a large computer that stores web pages that can be accessed over the Internet. (p. 326)

Picture passwords Picture passwords in Windows 10 allow the user to control access to a Windows 10 account via a picture. (pp. 92, 112)

Pie charts Pie charts are used to compare different parts of a whole, for example, to show budget expenditures. (p. 302)

Pinning an app Pinning an app adds a tile for the selected app to the Start menu, allowing easy access to frequently used apps. (pp. 99, 100, 119, 120)

Pinterest Pinterest is an online "pinboard" that uses visuals/images to post content. You cannot post content without an image. (p. 192)

Pivot table A pivot table lets you view the data in your spreadsheet more easily. (p. 301)

Pixel A pixel is the smallest part of a picture on a screen. More pixels make a clearer image. (pp. 21, 39, 68)

Plagiarism Plagiarism occurs when someone uses another person's work without giving proper credit or acknowledgment. (p. 174)

Plotters Plotters are used to print large graphic images like architectural designs and engineering drawings. They are specialized printers for detailed and precise graphical output. (p. 73)

Plug-ins Plug-ins are browser tools or add-ons that enhance functionality. (p. 145)

Point-and-shoot digital cameras Cameras operating similarly to smartphone cameras, where incoming light is focused onto an active pixel sensor array for image capture. (p. 54)

Port A slot or hole that matches cords or expansion cards for input and output devices on digital devices. (pp. 16, 66, 67)

Ports Ports are connectors on electronic devices where you can plug in cables or other devices. They come in various types, each serving specific purposes, such as USB ports, HDMI ports, and more. (pp. 37, 66, 67)

Predictive analytics This analytics focuses on understanding, predicting, and planning for future events and business outcomes. It utilizes probability analysis techniques as well as data mining, statistical modeling, machine learning, and deep learning to generate possible future outcomes given certain conditions. (p. 284)

Predictive policing It refers to a type of machine learning that utilizes the power of information technology, geospatial technologies, and evidence-based intervention models to reduce crime and improve public safety. (p. 364)

Prescriptive Analytics This form of data analytics recommends the best actions to take. It's the most advanced type of analytics, aiming to explain what might happen in the future, when it could happen, and why. (p. 285)

Presentation Programs Presentation programs help you make slides to show on a screen. (p. 34)

Primary key A primary key is a special relational database field designated to uniquely identify all records in a table. A primary key must contain a unique value. Common primary keys include: student ID numbers, Social Security numbers, and customer ID numbers. (p. 279)

Printers Printers reproduce digital images or text onto paper or other surfaces. They are categorized by their technology, such as inkjet and laser printers, and their resolution is measured in DPI. (pp. 65, 72)

Privacy The right to control personal information about oneself. (pp. 127, 156, 184, 392)

Private Cloud A private cloud is like having a personal digital library accessible only to you. (p. 161)

Productivity software Productivity software helps people do their work. It includes apps for writing, organizing, and making presentations. (p. 32)

Programming The process of converting tasks into a sequence of commands understandable by digital devices. (pp. 308, 316)

Project Soli Project features a microchip that utilizes miniature radar to detect touchless gesture interactions. (p. 372)

Projectors Projectors use light to display the output of digital devices on a larger surface, like a screen or wall, making it possible for large audiences to view presentations, videos, and other content. (p. 71)

Protocol A protocol is a set of rules that dictate how something is done on the Internet. (pp. 18, 144, 150, 153, 162, 140, 148, 324, 326)

Pseudocode A detailed yet readable explanation of a computer program or algorithm's purpose and functionality. (p. 313)

Public Cloud A public cloud is open for general use, like a public library for digital resources. (p. 161)

Python An interpreted, object-oriented, high-level programming language with simple, readable syntax, often used as a first language and supporting modules and packages for diverse applications. (pp. 316, 318)

Quantum computing Quantum computing harnesses the power of atoms and molecules for memory and processing tasks. One key aspect of quantum computing involves the quantum bit, or qubit. (p. 370)

Qubits Qubits serve as the fundamental units of information, resembling the 0s and 1s (bits) that transistors in modern computers represent. (p. 370)

Query A query is a request for information from a table (or combination of tables) in a database. Information is generated using a specific query language. It is also known as a question. (p. 276)

QWERTY keyboard The most common text-entry input device for computers, named for the first six top-row letters from the left. (pp. 47, 63)

R A programming language developed for statistical computing and graphics, offering numerous reusable code extension packages, similar to Python. (p. 316)

RAM Random access memory, electronic and without moving parts, found on the motherboard of digital devices. (pp. 11, 12, 23, 25, 27, 13, 38)

Ransomware Ransomware is malware that makes a computer's data inaccessible until a ransom is paid. (pp. 215, 216)

Records A record is a collection of related fields in a data file. Records are a collection of characteristics that describe and identify an entity. A record is also referred to as a row in a table. (pp. 278, 276)

Recycle Bin The Recycle Bin is a storage area for files and folders you want to delete from a Windows computer's hard drive. (pp. 264, 265, 271)

Refresh rate Refresh rate measures how quickly each pixel on a display is updated. It's measured in cycles per second (hertz) and affects the smoothness of motion in videos and animations. (pp. 69, 124)

Relational databases Relational database stores data in tables that are linked with predefined relationships created between specific data fields. It is also known as relational database management systems (RDBMS). (p. 276)

Relationships A relationship is a link between tables that defines how the data are related. A common field between the two tables is used to create the link. (p. 279)

Remote access Remote access enables authorized users to connect to a local network from a remote location. (p. 158)

Remote desktop software Remote desktop software lets one computer control or see what's on another computer through the internet. It's useful for people who manage computer networks because they can help or check on any computer without being there. (p. 239)

Reports Reports offer a way to view, format, and summarize the information in a database. Reports can be used to display or distribute a summary of data and archive snapshots of the data. (p. 279)

Resolution Resolution is the number of pixels in a display, determining its clarity and detail. Higher resolution screens can display more information and provide sharper images and text. (pp. 22, 55, 68)

ROM Read-only memory, a storage area in digital devices installed by the manufacturer. (pp. 13, 27)

Row A row is a horizontal line of cells. (p. 296)

Running an app Running an app from the Start menu is one of the most commonly used methods to open an app. (pp. 99, 101)

S

Sampling rate Sampling rate refers to the number of times per second that a digital recording captures or measures a sound wave. It affects the accuracy and quality of digital audio recordings. (p. 74)

Satellite Internet service Satellite internet provides broadband access almost everywhere by transmitting signals to and from satellites in orbit. (p. 152)

Screen resolution Screen resolution is how sharp and clear the pictures on a screen are. (p. 39)

Screen saver A screen saver is an image or images that the operating system displays when the computer is in Sleep mode. (p. 132)

Screen size Screen size is the diagonal measurement of a display screen, commonly used to describe the size of monitors, televisions, and other devices. It's a key factor in visual experience. (pp. 21, 38, 68)

Script kiddies Script kiddies are would-be hackers who attempt to gain unauthorized access to networks in order to steal and corrupt information and data. (p. 211)

Search box The search box allows you to search for files on your devices, search the web for information, and interact with Cortana. (pp. 92, 95, 96, 97, 98, 101, 102, 104, 105, 106)

Search button The Search button in Windows 11 allows you to search for files and folders, find and open installed apps, search the web for information, and interact with Cortana. (pp. 114, 116, 121)

Search engine It is a program that searches for and identifies items in a database that correspond to keywords or characters specified by the user. (pp. 167, 168, 346, 347, 351)

Search engine optimization consultant The person who analyzes and reviews websites to provide advice, guidance, and recommendations to business owners who want to increase search engine traffic and achieve higher ranking positions. (p. 348)

Self-driving trucks Trucks designed to go long distances on highways without a human behind the wheel. (p. 361)

Serial In data transmission, "serial" means sending data one bit at a time in sequence, which is different from parallel transmission where multiple bits are sent simultaneously. (pp. 16, 65, 67)

Server A server is a computer that provides resources and services to client computers that request them. (pp. 8, 144, 157, 158, 326)

Server administrator The person works with computer networks to ensure that they run efficiently. He maintains software updates, designs and implements new system structures, monitors server activity, and oversees server security. They may also be referred to as a systems administrator. (p. 349)

Server-based networks Server-based networks have individual devices connected to a central server, facilitating interactions among devices. (p. 158)

Session cookies Session cookies are small text files that are stored in temporary memory. Session cookies are lost when the web browser is closed. (pp. 222, 223)

Shut down Shutting down a digital device running Windows 11 means turning the device all the way off. (p. 112)

Single-Core Processor A CPU with only one core, where the core refers to the components on the chip responsible for processing. (p. 10)

Single-user license A single-user license lets only one person use the software at a time. (p. 35)

Site license Site licenses let people inside an organization use the software on their own devices. (p. 35)

Sleep mode Sleep mode allows the Mac to save battery life during periods of inactivity. (pp. 112, 130, 132)

Smartphone A handheld computer with cellular networking capabilities. (p. 8)

Smartphone cameras Cameras that work by focusing incoming light onto an active pixel sensor array for image capture. (p. 54)

Snap Snap is a new feature in Windows 11 that allows users to execute window snapping. (p. 115)

Snapchat Snapchat is an image- and video-sharing application. Snaps are picture or video messages that are taken and shared with friends on Snapchat. (p. 193)

Social media Internet-based applications designed for individuals and entities to interact. (pp. 181, 189, 190, 191, 195, 196, 202, 227)

Social Media Platforms Social media platforms like Facebook, Twitter, and Instagram connect people and allow sharing and interaction online. (p. 140)

Software license A software license is like a legal paper that tells you what you can and can't do with computer programs. (p. 34)

Sound Sound is an analog signal characterized by attributes like wavelength, frequency (pitch), and amplitude (volume). To work with computers, sound must be digitized by sampling and converting it into digital data. (p. 74)

Spam Spam is an unsolicited e-mail message, typically received from an unknown sender. (p. 220)

Speech recognition technologies Technology that records, maps, and analyzes voice patterns, including voice frequency, pitch, speech rhythms, and other characteristics, which are charted and compared to existing records. (p. 53)

Spreadsheets Spreadsheets are computer programs that allow users to easily arrange, calculate, and present numerical data. (pp. 33, 290, 304, 305)

SSD capacity The amount of storage available in SSDs, usually measured in gigabytes, a determinant factor of their price. (p. 14)

Storage capacity Storage capacity is how much space there is to save things on a computer's hard drive. (p. 38)

storage device A storage device is any piece of computer hardware used for storing data and information, which can be located inside or outside of a digital device. (pp. 82, 83)

Storage media Storage media refers to the technology used to retain data, available in various forms, each with unique characteristics. (p. 81)

Streaming and Entertainment Platforms like Netflix, YouTube, and Spotify offer movies, music, and virtual events for online consumption. (p. 140)

Style It describes the CSS styles that will be applied to an element. (p. 335)

Supercomputers The most powerful type of computer, capable of quickly evaluating complex data. (p. 8)

Swift A programming language by Apple simplifying software development for iOS and OS X, designed for ease of learning and usage compared to Objective-C. (p. 317)

Switches Switches connect devices within the same network, allowing them to communicate and share data. (pp. 148, 149)

System Preferences System Preferences is a tool on Mac computers that lets you change settings and do other things to manage the computer. (pp. 36, 130)

System software System software helps the computer work correctly. It has two main parts: the operating system and utility software. (p. 29)

System unit The main housing of a desktop computer, including the tower, or in laptops, beneath the keyboard, and in smartphones or tablets, beneath the screen. (p. 5)

T

Tables A table in a database is a collection of associated records. Databases often have more than one table. (pp. 273, 275, 278)

Tablet computers Highly mobile computers with touchscreen interfaces, ideal for travel but not upgradable like laptops. (p. 6)

Tags Symbols controlling how web browsers display content, encompassing text, images, multimedia, and more. (p. 329)

Task View The Task View in Windows 11 allows you to view all your open Windows and desktops with just one click. (p. 116)

Taskbar The taskbar in Windows 11 displays the most used apps on your device and provides an easy way for you to access and visualize the apps you use most. (pp. 100, 114, 117, 120, 109)

Terabyte A terabyte is a unit of digital data that consists of approximately one trillion bytes. (pp. 79, 80)

The Gramm-Leach-Bliley Act Requires financial institutions to explain their information-sharing practices and safeguard data. (p. 184)

The Red Flags Rule Many businesses and organizations must implement a written identity theft prevention program. The program aims to detect the warning signs or 'red flags' of identity theft in their day-to-day operations. (p. 184)

The Task View The Task View allows you to view all your open windows and desktops with just one click. (pp. 93, 116)

Thermal printing Thermal printing is a method commonly used for printing receipts. It involves heating special thermal paper to create marks, making it cost-effective and cleaner than using ink or toner. (p. 72)

Time Capsule Time Capsule interfaces easily with Macintosh computers and serves as a backup device. (p. 131)

Touch Bar The Touch Bar provides quick access to useful tools and adapts to the applications being used. (p. 130)

Touchpads Pointing devices that take up minimal space, commonly found on laptop computers. (p. 49)

Touchscreen A touchscreen is a kind of screen that can feel when you touch it. (pp. 39, 93, 115)

Transmission media Transmission media are the pathways or channels that internet signals use to reach devices. (p. 146)

Trojan horse "Trojan horse" or simply "Trojan" is software that appears harmless or even beneficial but hides a malicious intent. (pp. 207, 214, 215)

Twisted Pair Twisted pair cables are commonly used to connect homes to local network stations, including Ethernet cables. (p. 146)

U

Ultrasonic sensor A sensor using sound wave reflection to detect fingerprint patterns. (p. 51)

Unicode Unicode is another set of rules for computers to use many different languages. It needs at least 16 bits for each letter. (p. 26)

Uninstalling Uninstalling is when you take off software you don't want from a device. (pp. 35, 101, 120)

Uploading Uploading involves sending files from your computer to a server on the Internet. (p. 153)

URL A URL is an address for a website or a webpage, used to access specific online content. (pp. 144, 163, 324, 326, 329)

USB-C USB-C is a small, reversible connector used for various tasks like charging devices, transferring data, and connecting to monitors. (p. 67)

Username A username is a unique grouping of numbers, letters, or symbols that identifies a specific user of Windows 11. (p. 112)

Utility programs Utility programs do special tasks on a computer and come with Windows and macOS. (p. 32)

V

Value Value in a spreadsheet refers to entries that the program recognizes as being able to interact with other cells. (p. 391)

Variety Variety in a website refers to the different forms of data. (p. 279)

VBScript Enabling dynamic decision-making for page display based on user viewing behavior, using Visual Basic syntax and symbols for Windows app creation. (p. 317)

Veracity Veracity in a website means truth. With all of the data being generated and stored, it is important to ensure that data are meaningful, true, and useful. (p. 280)

Video card A video card, also called a graphics card or graphics adapter, is a part of a computer that makes pictures and videos show up on the screen. (p. 20)

Video conferencing Video conferencing is a technology that allows multiple people to have a meeting using video and audio through the internet. (p. 235)

Violet It is a 4-foot 7-inch robot with a humanoid appearance that was developed by Akara, an Irish startup that specializes in designing artificially intelligent helpers for the healthcare industry. (p. 356)

Volume Volume in a website refers to the scale of data. (p. 279)

W

Web mining Web mining uses the principles of data mining to uncover and extract information from websites, social media sites, e-commerce platforms, and web services. (p. 283)

Web servers They display web pages and run apps through browsers, while e-mail servers facilitate sending and receiving e-mail. (p. 326)

Web structure mining Web structure mining includes the analysis of hyperlinks, nodes, and related web pages. (p. 283)

Web usage mining Web usage mining, also called log mining, includes the analysis of web access logs or the when, how, and with what frequency websites are accessed. (p. 283)

Webcam A camera used for taking photos or streaming video into a computer or over the Internet. (p. 57)

Website It gives you control over what information you want to share. Building a website now allows you to have an online place that can evolve and grow. A website will make you stand out from the competition. (p. 347)

White hat hackers White hat hackers are non-malicious computer security experts who choose to use their knowledge and skills for good rather than evil. They are also known as ethical hackers. (p. 211)

WhoIs Lookup WhoIs lookup is used to determine the owner or status of a specific website or domain name on the Internet. (p. 171)

Wi-Fi A wireless local area network, often referred to as Wi-Fi. (pp. 18, 146, 156, 148, 246)

Widgets Widgets are like cards that display frequently updated content from apps and services you select on your Windows desktop. (p. 115)

Windows Windows is a common software for regular computers. It has a logo with four colorful squares. (pp. 30, 32, 36, 41, 48, 89, 95, 113, 120, 109, 127)

Windows 10 Windows 10 is one of Microsoft's latest operating systems (OS) for PCs. It includes Cortana, a digital assistant similar to iPhone's Siri, that allows users to search, store, and access files on the computer and online. Windows 10 also includes the Microsoft Edge browser, the new and improved replacement for Internet Explorer. (pp. 88, 89, 90, 91, 92, 93, 94, 95, 96, 98, 99, 100, 101, 102, 103, 104, 105, 106, 107, 109)

Windows 10 Enterprise Designed for large organizations with IT personnel. (p. 90)

Windows 10 Home Designed for home and small business users. (p. 90)

Windows 10 Mobile and Mobile Enterprise Designed for use on smartphones and tablets. (p. 90)

Windows 10 Pro Designed for business and tech professionals. (p. 90)

Windows 11 Windows 11 is the newest operating system (OS) from Microsoft. An OS on a digital device is the software program that manages all of the other application programs in a computer. (pp. 110, 111, 112, 113, 114, 115, 116, 117, 118, 119, 120, 121, 122, 123, 124, 125)

Windows 11 Home Windows 11 Home is designed for consumer users who intend to use their devices at home or for personal use. (p. 110)

Windows 11 Pro Windows 11 Pro is designed for use in business/organizational environments. It has enhanced security features including BitLocker device encryption, Windows Information Protection (WIP), and enhanced business management and deployment options. (p. 110)

Windows 11 SE Windows 11 SE is a Cloud-first operating system built for the K–8 education market. (p. 110)

Windows Hello Windows Hello is a feature in Windows that allows you to create more personal and secure methods for accessing a Windows 11 device. (p. 115)

Wireframe It refers to a two-dimensional illustration of a page's interface. It focuses on how space is allocated and how content is prioritized. (p. 344)

Wireless ports These are special points on devices that allow them to send and receive data without any wires. They work by using radio waves or light waves to communicate with other nearby devices. (p. 18)

Word processing programs Word processing programs help people write documents and do other tasks. (p. 33)

Word size Word size is how much data a computer can handle at once. (p. 25)

WordPad WordPad is a scaled-down version of Microsoft Word, used for basic document editing. (pp. 102, 121, 125)

Worm A worm is a destructive program that replicates itself throughout a single computer or across a network. (pp. 215, 229)

Z

Zebra-Med Zebra-Med offers radiologists its new AI, offering which helps health providers manage an ever-increasing workload without compromising quality. It works with millions of imaging and correlated clinical records to create high-performance algorithms that automatically detect medical conditions faster than the methods currently being used. (p. 360)

Zombie A zombie is a computer that has been secretly taken over by an outsider, typically using a rootkit. (p. 229)

Zooming In photography, the process of making distant objects appear closer in an image, achieved through optical or digital zooming. (p. 55)

Index

C

H

I

X

Z